Mechanisms and Machine Science

Volume 52

Series editor

Marco Ceccarelli
LARM: Laboratory of Robotics and Mechatronics
DICeM: University of Cassino and South Latium
Via Di Biasio 43, 03043 Cassino (Fr), Italy
e-mail: ceccarelli@unicas.it

More information about this series at http://www.springer.com/series/8779

Mehmet Ismet Can Dede
Mehmet İtik · Erwin-Christian Lovasz
Gökhan Kiper
Editors

Mechanisms, Transmissions and Applications

Proceedings of the Fourth MeTrApp Conference 2017

 Springer

Editors
Mehmet Ismet Can Dede
Department of Mechanical Engineering
Izmir Institute of Technology
Izmir
Turkey

Mehmet İtik
Department of Mechanical Engineering
Karadeniz Technical University
Trabzon
Turkey

Erwin-Christian Lovasz
Faculty of Mechanical Engineering
Polytechnic University of Timisoara
Timisoara
Romania

Gökhan Kiper
Department of Mechanical Engineering
Izmir Institute of Technology
Izmir
Turkey

ISSN 2211-0984 ISSN 2211-0992 (electronic)
Mechanisms and Machine Science
ISBN 978-3-319-60701-6 ISBN 978-3-319-60702-3 (eBook)
DOI 10.1007/978-3-319-60702-3

Library of Congress Control Number: 2017943050

© Springer International Publishing AG 2018
This work is subject to copyright. All rights are reserved by the Publisher, whether the whole or part of the material is concerned, specifically the rights of translation, reprinting, reuse of illustrations, recitation, broadcasting, reproduction on microfilms or in any other physical way, and transmission or information storage and retrieval, electronic adaptation, computer software, or by similar or dissimilar methodology now known or hereafter developed.
The use of general descriptive names, registered names, trademarks, service marks, etc. in this publication does not imply, even in the absence of a specific statement, that such names are exempt from the relevant protective laws and regulations and therefore free for general use.
The publisher, the authors and the editors are safe to assume that the advice and information in this book are believed to be true and accurate at the date of publication. Neither the publisher nor the authors or the editors give a warranty, express or implied, with respect to the material contained herein or for any errors or omissions that may have been made. The publisher remains neutral with regard to jurisdictional claims in published maps and institutional affiliations.

Printed on acid-free paper

This Springer imprint is published by Springer Nature
The registered company is Springer International Publishing AG
The registered company address is: Gewerbestrasse 11, 6330 Cham, Switzerland

Preface

The first MeTrApp (Mechanisms, Transmissions and Applications) was organized in Timisoara, Romania, in 2011, as a workshop. Following the first MeTrApp, the next ones are organized every two years. The second one in the series of this conference was organized in Bilbao, Spain, in 2013, and the third one was organized in Aachen, Germany, in 2015. MeTrApp 2017 is the fourth one of the conference series organized in Trabzon, Turkey, between the dates July 3–5, 2017.

This book gathers the technical papers that are submitted and presented in the fourth Conference on Mechanisms, Transmissions and Applications co-organized by Karadeniz Technical University and Izmir Institute of Technology, Turkey.

MeTrApp conference series is organized under the patronage of International Federation for Promotion of Mechanism and Machine Science (IFToMM), and it is joint event of two technical committees of IFToMM namely "Linkages and Mechanical Controls" and "Gearing and Transmissions." MeTrApp 2017 is the first conference of IFToMM conference series that is organized in Turkey.

MeTrApp 2017 aims to bring together the researchers, scientists, industry experts, and students from all over the world to provide them a platform to share their state-of-the-art work in the fields of mechanisms, transmissions, and their applications. Hence, promoting the exchange of ideas and collaboration is one of the expected outcomes of this conference. Another outcome of the conference is this edited proceedings book within the Springer Mechanism and Machine Science series, which enlarges the sharing platform to all researchers, scientists, industry experts, and students that are interested in the fields of mechanisms and transmissions. The content of this book is presented in six parts for the reader to easily navigate between the proceedings that fall into certain subject areas as follows: Mechanism Design, Cam Mechanisms, Parallel Manipulators, Control Applications, Mechanical Transmissions, and Dynamics of Machinery.

A total of 30 submissions and two invited talk papers were received from authors of 10 different countries for MeTrApp 2017. The invited speakers were Prof. Dr. Yavuz Yaman, Middle East Technical University, and Dr. Murat Gültekin, Aselsan Inc. After an initial plagiarism check and rigorous two-stage double-blind review process by at least two reviewers for each paper, 29 papers were accepted to appear in this book and to be presented in MeTrApp 2017. We sincerely thank all the reviewers for their contribution to the review process timely with their strong expertise and scientific background.

We also thank all the authors for their strong will to contribute to this conference during the tough period that Turkey is going through. We believe that this displays the true spirit of the founding fathers of IFToMM—"promoting science without borders." We thank all who helped in organizing this book and MeTrApp 2017. We thank Karadeniz Technical University, MAKTED (IFToMM Turkey), and Izmir Institute of Technology for all their support and sponsorship. Finally, we thank all the staff at Springer for all the help they provided in editing this book and their quick response times with minimal oscillations.

Our wish is that the excellent works that are presented in MeTrApp 2017 will be cherished by a larger number of readers with the help of this book, and participation to the next MeTrApp will be much higher.

April 2017

Mehmet Ismet Can Dede
Mehmet İtik
Gökhan Kiper
Erwin-Christian Lovasz

Organization

Conference Chairman

Mehmet İtik

Conference Co-chairmen

Mehmet Ismet Can Dede	Izmir Institute of Technology, Turkey
Gökhan Kiper	İzmir İnstitute of Technology, Turkey
Erwin-Christian Lovasz	Politehnica University of Timisoara, Romania
Daizhong Su	Nottingham Trent University, UK

Program Committee

Burkhard Corves	RWTH Aachen University, Germany
Marco Ceccarelli	University of Cassino and South Latium, Italy
Martin Bilek	Technical University of Liberec, Czech Republic
Domenico Mundo	Università della Calabria, Italy
Takaaki Oiwa	Shizuoka University, Japan
Victor Petuya	University of the Basque Country UPV/EHU, Spain
Doina Pisla	Technical University of Cluj-Napoca, Romania
Hidetsugu Terada	University of Yamanashi, Japan
Giuseppe Quaglia	Politecnico di Torino, Italy
Yan Chen	Tianjin University, China
İbrahim Uzmay	Erciyes University, Turkey
Eres Söylemez	Middle East Technical University, Turkey

M. Kemal Özgören   Middle East Technical University, Turkey
Miroslav Václavík   VÚTS, Czech Republic
Rasim Alizade   Azerbaijan Technical University, Azerbaijan

Local Organizing Committee at Karadeniz Technical University

Ertan Baydar   Karadeniz Technical University, Turkey
Murat Eray Korkmaz   Karadeniz Technical University, Turkey
Mustafa Yavuz Coşkun   Karadeniz Technical University, Turkey
Caner Sancak   Karadeniz Technical University, Turkey

Under the Patronage of IFToMM

Marco Ceccarelli   President
Teresa Zielinska   Secretary-General
Erwin-Christian Lovasz   Chair of the Technical Committee for Linkages and Mechanical Controls
Daizhong Su   Chair of the Technical Committee for Gearings and Transmissions

Contents

Mechanism Design

Morphing Wings and Control Surfaces: A New Approach in Aircraft Design 3
Yavuz Yaman

Designing Human Powered Balers for Straw Bale Construction in Developing Countries: The Case of Haiti 11
Carlo Ferraresi, Walter Franco, and Giuseppe Quaglia

Guiding Linkages with Remote Centre of Rotation for Thermal Cutting Processes 21
Carsten Teichgräber, Jörg Müglitz, and Maik Berger

Structural Synthesis of 2R1T Type Mechanisms for Minimally Invasive Surgery Applications 31
Abdullah Yaşır and Gökhan Kiper

3R1H Pseudo-Rigid-Body Model for Compliant Mechanisms with Inflection Beams 39
Yue-Qing Yu and Shun-Kun Zhu

A Spatial Four-Bar Linkage RSPS for Ball-Bar to Test R-pair 48
Delun Wang, Zhi Wang, Xiaopeng Li, Huimin Dong, and Shudong Yu

Leg Mechanisms Motion Characteristics 56
Adriana Comanescu, Elisabeta Banica, and Dinu Comanescu

Design and Dimensional Optimization of a Novel Walking Mechanism with Firefly Algorithm 67
Özgün Selvi and Samet Yavuz

Cam Mechanisms

Improving the Kinematics of Motion Curves for Cam Mechanisms Using NURBS 79
Thi Thanh Nga Nguyen, Stefan Kurtenbach, Mathias Hüsing, and Burkhard Corves

Assessment of the Rolling Contact Fatigue 89
Monika Hejnová

The Stress Distribution in the Contact Region of a Cam Mechanism General Kinematic Pair 99
Jiri Ondrášek

Parallel Manipulators

5 DoF Haptic Exoskeleton for Space Telerobotics – Shoulder Module.. 111
Dan Margineanu, Erwin-Christian Lovasz, Corina Mihaela Gruescu, Valentin Ciupe, and Santra Tatar

Kinematic Design of a Tripod Parallel Mechanism for Robotic Legs 121
Matteo Russo and Marco Ceccarelli

Parallel Manipulators: Practical Applications and Kinematic Design Criteria. Towards the Modular Reconfigurable Robots 131
Alfonso Hernández, Mónica Urízar, Erik Macho, and Victor Petuya

Two-Degree-of-Freedom Special Parallel Manipulator for Laser Interferometry-Based Tracker...................... 141
Baris Celik, Takaaki Oiwa, Kenji Terabayashi, and Junichi Asama

Control Applications

Design of Artificial Neural Network Predictor for Trajectory Planning of an Experimental 6 DOF Robot Manipulator 153
Şahin Yıldırım and Burak Ulu

Fault-Tolerance Experiments with a Kinematically Redundant Holonomic Mobile Robot 161
Osman Nuri Şahin, Onur Çelik, and Mehmet İsmet Can Dede

Dual-Loop Motion Control for Geometric Errors and Joint Clearances Compensation of a Planar 2-PRP+1-PPR Manipulator..... 171
Santhakumar Mohan, Jayant Kumar Mohanta, Mathias Huesing, and Burkhard Corves

The Effects of Admittance Term on Back-Drivability 181
Ogulcan Işıtman, Orhan Ayit, and Mehmet İsmet Can Dede

Design and Proposed Model Reference Trajectory Control of a Snake like Robot 191
Şahin Yildirim and Kirakoya Abdoulaye Ben-Aziz

Mechanical Transmissions

Free Vibration and Sensitivity Analysis of RV Reducer 203
Yuhu Yang and Chuan Chen

Proof of Existence of a Gear Variator as Wheelwork with Constant Engagement of Toothed Wheels 212
Konstantin S. Ivanov

Parameterized Substructure Model of PGT for Finite Element Contact Analysis 224
Huimin Dong, Chu Zhang, Jili Zhang, and Delun Wang

On the Impact of Transmission Error on the Dynamic Behavior of Geared-Linkages 232
Domenico Mundo, Shadi Shweiki, and Piervincenzo Catera

Effect of the Coil Shape on Magnetic Field of an Electromagnet for Contactless Power Transmission to Microrobots 240
Abdulkareem Alasli, Nail Akçura, and Levent Çetin

Dynamics of Machinery

The Edge of Chaos in Kinematics and Dynamics of Mechanism 251
Zhaohui Liu, Jin Xie, and Yong Chen

Dynamics of Orthogonal Mechanism of Vibrating Table in View of Friction 261
Zharilkassin Iskakov, Kuatbay Bissembayev, and Nutpulla Jamalov

Design of Neural Network Predictor for Vibration Analysis of a Drill Column Machine During Drilling Plastic Work-Pieces 269
Şahin Yıldırım and Emir Esim

Railway Vehicle Model Developed by ASELSAN 279
Mustafa Nicem Tanyeri and H. Murat Gültekin

Author Index 287

Mechanism Design

Morphing Wings and Control Surfaces: A New Approach in Aircraft Design

Yavuz Yaman[✉]

Middle East Technical University, Ankara, Turkey
yyaman@metu.edu.tr

Abstract. This paper details the fully morphing wings and control surfaces. The idea of morphing is particularly attractive in aircraft technologies. Various approaches like span increase, camber increase and/or decrease, twist, and sweep are finding extensive applications in aeronautical structures.

Keywords: Morphing wings · Unconventional control surfaces · Aerodynamic efficiency

1 Introduction

It is a general belief that, in unmanned aerial vehicles, the wings having unconventional wings and control surfaces are heavier as compared to the conventional wings due to the weight and the complexity of the additional actuation mechanisms involved. However, the aerodynamic efficiencies attained at different phases of the flight usually pays off its dividends and still make the utilization of unconventional designs attractive.

The conventional wings of the aerial vehicles are designed in such a way that the aerodynamic performance is intended to be maximized for the flight regime at which the aerial vehicle spends most of its mission time. Therefore, in off-design conditions, the aerial vehicles require additional mechanisms such as flaps to eliminate the deficiencies inherent to the conventionally designed wings. But it is known that the existing gaps between the wing and the flaps cause to decrease aerodynamic efficiency and increase aerodynamic noise. Additionally, the fuel consumption of the aerial vehicles with conventional wings increase due to performance deficiencies and this ultimately results in greater emissions of harmful CO_2 and NO_x gases to environment.

This paper analyzes the fully morphing wings and control surfaces. Various approaches like span increase, camber increase and/or decrease, twist, and sweep will be detailed and the current trends will be outlined [4].

2 Morphing Wing Studies

Figure 1 indicates the natural excellence and also highlights some examples of man-made attempts to mimic the nature [1, 6]. As given in Fig. 1a the eagle uses its wings and even single feathers differently for each phase of its flight regime.

The pioneering studies in the field of morphing started with NextGen Aeronautics and Lockheed Martin companies and examples are given in Figs. 2 and 3. Figure 2

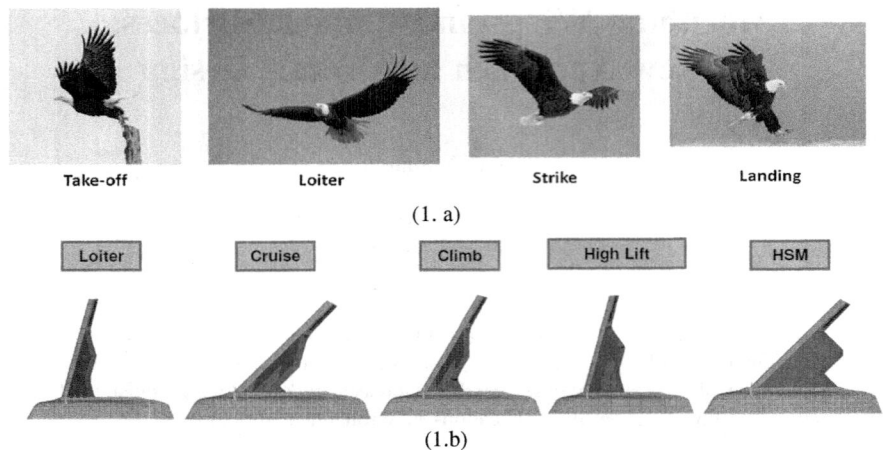

Fig. 1. (a) Excellent flight performance in nature [1], (b) Studies to mimic those performances [6].

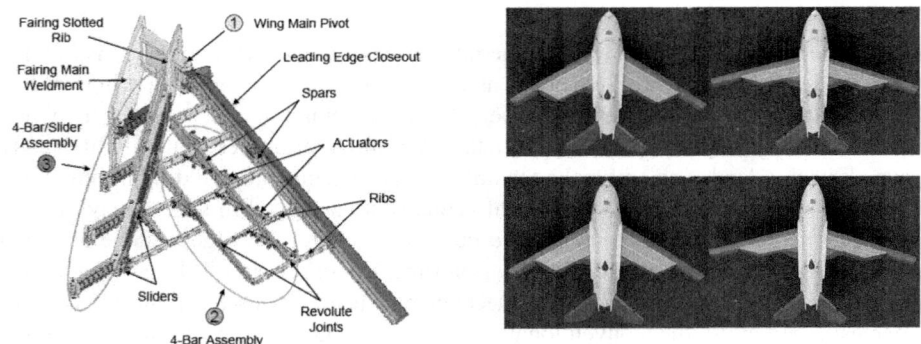

Fig. 2. NextGen Aeronautics concepts [5].

also gives the intended internal mechanism of NextGen wing. The synthesized mechanisms inside the wing allowed significant changes in the planform which will be aerodynamically effective for different flight conditions.

The Lockheed-Martin Company folding wing concept is illustrated in Fig. 3.

Fig. 3. Lockheed Martin folding wing concepts [11].

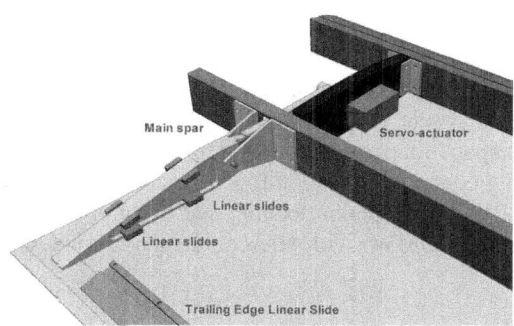

Fig. 4. Politecnico di Milano sliding rib concept [9].

The sliding rib studies originated in the Politecnico di Milano is given in Fig. 4. These studies were effective in the camber change of the trailing edges.

Some EU FP7 projects also dealt with morphing aerospace structures. SADE project designed a mechanism and applied it to the leading edge of an aircraft. NOVEMOR project on the other hand studied various blended wing configurations. Figures 5 and 6 give examples from those EU FP7 projects.

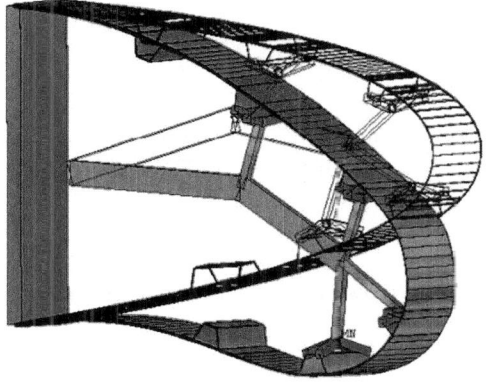

Fig. 5. SADE (Smart High Lift Devices for Next Generation Wings) droop nose concepts [10].

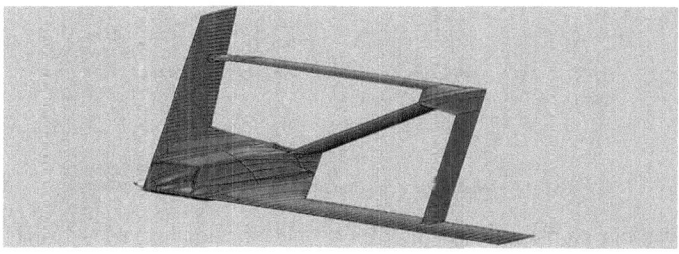

Fig. 6. NOVEMOR (Novel Air Vehicle Configurations: From Fluttering Wings to Morphing Flight) concepts [8].

One of the current studies in this field is European FP7 project CHANGE (*Combined morpHing Assessment software usiNG flight Envelope data and mission based morphing prototype wing development*) [15]. The objective of CHANGE is to develop, build and flight-test a novel morphing Unmanned Aerial Vehicle (UAV), with the aim of performance improvement over a range of flight missions. The CHANGE wing combines telescopic span, different leading and trailing edge systems in order to change the camber. The wing concept is given in Fig. 7. It contains several regions such as an inner fixed wing (IFW) with different leading and trailing edge morphing devices and an outer morphing wing (OMW) with telescopic span and fixed cross section.

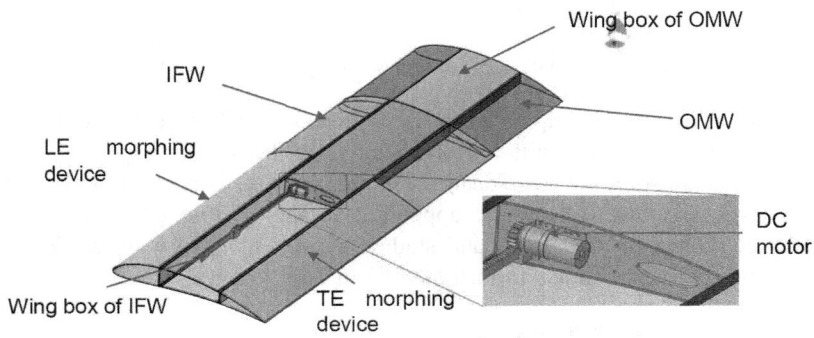

Fig. 7. Schematic of the modular morphing wing used in CHANGE [17].

The trailing edge design and analysis of CHANGE wing had been undertaken by METU Department of Aerospace Engineering. The METU design had to fulfil camber changes [2, 3, 7, 12–14, 16].

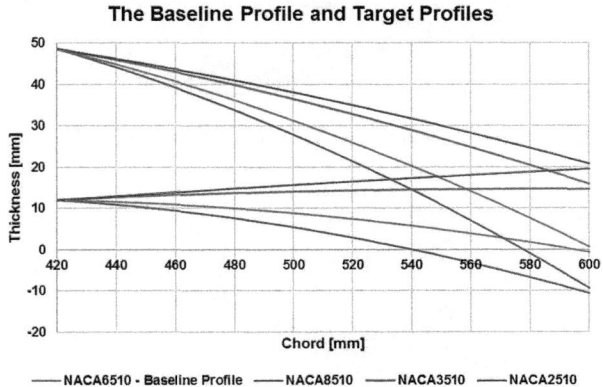

Fig. 8. The baseline profile and target profiles of CHANGE trailing edge of METU design [16].

In order to achieve these target profiles given in Fig. 8, a trailing edge had been indigenously designed at METU, Department of Aerospace Engineering Figure 9 represents the trailing edge designed. Figure 9(b) shows the servo actuators. Three actuators (First, third and fifth from the left) are applied for the lower part of the control surface and two actuators (Second and fourth from the left) are acting for the upper part. All actuators are always applied in order to have tension in the fully compliant part. The differential effect results in camber or decamber.

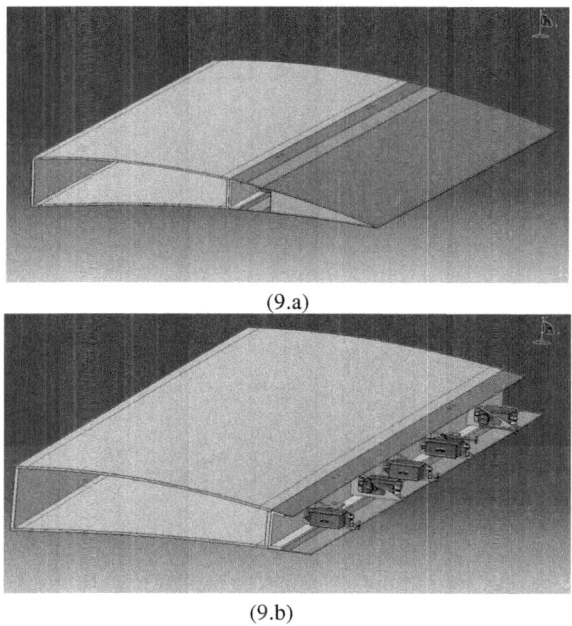

(9.a)

(9.b)

Fig. 9. The CHANGE trailing edge of METU design (a) control surface, (b) actuators [12, 16].

The transverse displacement patterns for NACA6510 to NACA 3510 motion, as an example, is given in Fig. 10 for in-vacuo condition. The same structure is then analyzed under aerodynamic loading which represents the intended flight conditions of the UAV at which the developed wing is attached to. As expected the aerodynamic load had provided additional lift and the resultant decamber became larger. Figure 11 gives the same patterns under aerodynamic loading.

Currently the studies are concentrated for developing structural concepts in order to simultaneously increase the chord and the camber of the trailing edge of the wing. A preliminary concept is shown in Fig. 12. The link A_0A will be given an input and the point B will follow a prescribed path. A mechanism will be synthesized for this concept.

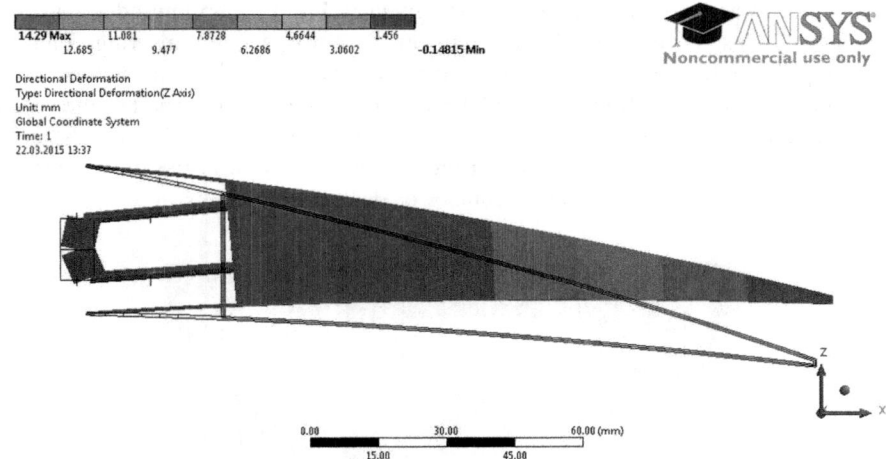

Fig. 10. Displacement contours (z direction) for NACA 6510 to NACA 3510 morphing in in-vacuo [12].

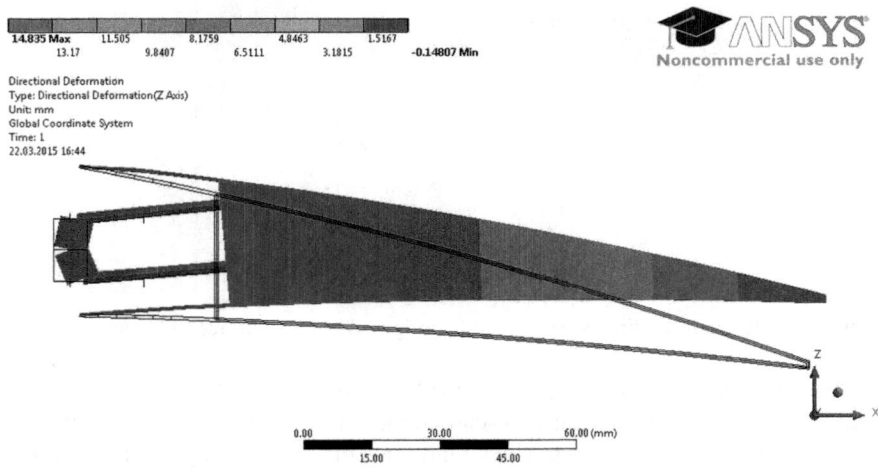

Fig. 11. Displacement contours (z direction) for NACA 6510 to NACA 3510 morphing under aerodynamic loading [12].

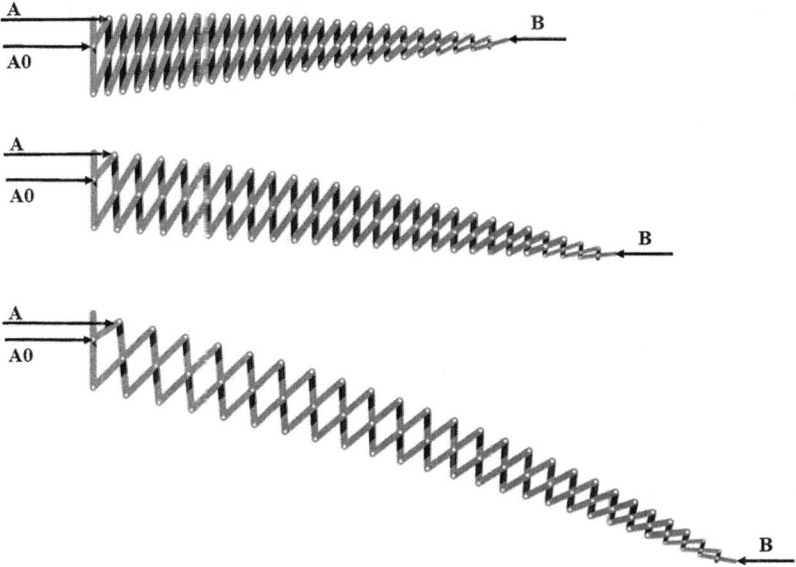

Fig. 12. The proposed simultaneous chord and camber increase mechanism

Acknowledgements. Part of the work presented herein has been partially funded by the European Community's Seventh Framework Programme (FP7) under the Grant Agreement 314139. The CHANGE project ("Combined morphing assessment software using flight envelope data and mission based morphing prototype wing development") is a L1 project funded under the topic AAT.2012.1.1-2 involving nine partners. The project started in August, 1, 2012.

References

1. Ajaj R, Keane A, Beaverstock C, Friswell MI, Inman DJ (2013) Morphing aircraft: the need for a new design philosophy. In: 7. Ankara international aerospace conference, METU, Ankara, Paper AIAC-2013-054
2. Arslan P, Kalkan U, Tıraş H, Tunçöz İO, Yang Y, Gürses E, Şahin M, Özgen S, Yaman Y (2014) A hybrid trailing edge control surface concept In: DeMEASS-VI: 6th conference for design, modelling and experiments of advanced structures and systems, Ede-Wageningen, Holland
3. Arslan P, Kalkan U, Tıraş H, Tunçöz İO, Yang Y, Gürses E, Şahin M, Özgen S, Yaman Y (2014) Structural analysis of an unconventional hybrid control surface of a morphing wing. In: ICAST 2014: 25th international conference on adaptive structures and technologies, The Hague, Holland
4. Barbarino S, Bilgen O, Ajaj RM, Friswell MI, Inman DJ (2011) A review of morphing aircraft. J Intell Mater Syst Struct 22:823–877
5. Canfield B, Westfall J (2008) Distributed actuation system for a flexible in plane morphing wing. In: Proceedings of the advanced course on morphing aircraft, mechanisms and systems, Lisbon, Portugal

6. Gandhi N, Ward D, Jha A, Monaco J, Seigler T, Inman DJ (2007) Intelligent control of a morphing aircraft. In: AIAA/ASME/ASCE/AHS/ASC Structures, structural dynamics and materials conference, vol 1, pp 166–182
7. Gürses E, Tunçöz İO, Yang Y, Arslan P, Kalkan U, Tıraş H, Şahin M, Özgen S, Yaman Y (2016) Structural and aerodynamic analyses of a hybrid trailing edge control surface of a fully morphing wing. J Intell Mater Syst Struct. doi:10.1177/1045389X16641200
8. NOVEMOR FP7 Project (2012–2015). http://www.novemor.eu/
9. Ricci S (2008) Adaptive camber mechanism for morphing-experiences at DIA PoliMi. In: Proceedings of the advanced course on morphing aircraft, mechanisms and systems, Lisbon, Portugal
10. SADE FP7 Project (2008–2012). http://www.sade-project.eu/
11. Skillen MD, Crossley WA (2007) Modeling and optimization for morphing wing concept generation. NASA/CR-2007-214860
12. Tunçöz İO, Yang Y, Gürses E, Şahin M, Özgen S, Yaman Y (2015) A hybrid trailing edge control surface capable of camber and decamber morphing. In: SMART 2015: 7th ECCOMAS thematic conference on smart structures and materials, Ponta Delgada, Azores, Portugal
13. Tunçöz İO, Yang Y, Gürses E, Şahin M, Özgen S, Yaman Y (2015) Design and analyses of an unmanned aerial vehicle hybrid trailing edge control surface having camber and decamber capabilities. In: AIAC 2015: 8th Ankara international aerospace conference, AIAC-2015-131. ODTÜ, Ankara
14. Tunçöz İO, Yang Y, Gürses E, Şahin M, Özgen S, Yaman Y (2016) A hybrid morphing trailing edge designed for camber change of the control surface. In: SciTech 2016, AIAA science and technology forum 2016, AIAA-2016-0316, San Diego, USA, 04–08 January 2016
15. Yaman Y (2013) Combined morpHing Assessment software usiNG flight Envelope data and mission based morphing prototype wing development, CHANGE; an overview presentation. In: 3rd EASN international workshop on aerostructures, Milano, Italy
16. Yaman Y, Tunçöz O, Yang Y, Arslan P, Kalkan U, Tıraş H, Gürses E, Şahin M, Özgen S (2015) Decamber morphing concepts by using a hybrid trailing edge control surface. Aerosp. Open Access J. 2(3):482–504. doi:10.3390/aerospace2030482
17. Yang Y, Özgen S, Yaman Y, Ciarella A, Hahn M, Beaverstock C, Friswell MI (2016) MDAO for aerodynamic assessment of a morphed wing for the loiter segment of a UAV flight mission. In: SciTech 2016, AIAA science and technology forum 2016, AIAA-2016-0314, San Diego, USA, 04–08 January 2016

Designing Human Powered Balers for Straw Bale Construction in Developing Countries: The Case of Haiti

Carlo Ferraresi, Walter Franco[(✉)], and Giuseppe Quaglia

Politecnico di Torino, Turin, Italy
`walter.franco@polito.it`

Abstract. Straw bale constructions are appropriate for the improvement of the housing conditions in developing Countries. The paper presents, starting from the analysis of the context to the fabrication of prototypes, the design process of several human powered balers useful for the production of straw bales for straw bale construction in poor Countries.

Keywords: Straw bale construction · Human powered baler · Hand-operated machine, humanitarian mechanical engineering

1 Introduction

In January 2016, thanks to the efforts of United Nations, Countries have adopted a new sustainable development agenda, with the aim to end poverty and hunger, to protect the planet from degradation, and that all human beings can enjoy prosperous and fulfilling lives [12]. The agenda is organized in seventeen goals; the eleventh goal provides to make cities and human settlements inclusive, safe, resilient and sustainable. Today in the World, 828 million people live in slums and the number keeps rising; in addition one in five people in developing regions still live on less than $1.25 a day [12]. How to ensure access for all to adequate, safe and affordable housing?

The problem is very complex. Typically the appropriate solutions provide the use of widespread, easily available and low cost construction materials, such compressed earth block, [4, 11], straw [6–8] or timber [1]. The solution, however, can only be found from a detailed analysis of the contexts. The present work concerns the development of straw presses for producing straw bale constructions aimed at mitigating the tremendous housing crisis in Haiti.

Haiti is one of the poorest countries of the world and the poorest in the Americas region, with corruption, poor infrastructure, lack of health care and lack of education. Fifty-four percent of the population lives with less than one dollar a day. The Human Development Index is equal to 0.48. In addition Haiti has had in recent years two significant emergencies in 2010 the earthquake, with 230.000 victims, 600.000 evacuees, 250.000 homes destroyed; in 2016 the hurricane Matthew, with 20.000 evacuees, in particular in the Jéremie region. Both natural disasters have increased a strong housing emergency.

The question is: how to mitigate the housing emergency in this situation? Which construction techniques may be appropriate in this context? In 2011, ASF Piemonte, in an international cooperation project, has built a school centre using wood as construction material [1]. The experience has not proved effective: in Haiti, the wood is hard to find, expensive and of low quality. Haiti is quite completely deforested. The study of the context allowed identifying in the traditional rice cultivation the possible solution of the problem. In fact a by-product of the rice cultivation is the straw, which currently is considered a waste, and is burnt in the field at the end of the threshing. Every farmer, in an average plot of 0.25 ha, produces 1.6 tons of straw per year that can be used to realize one hundred fifty straw bales suitable for straw bale construction [2].

There are two basic styles of straw bale construction: the first is non-load-bearing, or post-and-beam, or infill style, in which bales are used as infill panels between or around a structural frame; the second one is load-bearing or Nebraska-style in which the bale wall carries vertical load.

The choice of the appropriate construction technique depends on the local situation. Infill constructions requires to build wood frames, which can be problematic when wood is scarce; on the other hand they have the advantage of being easily designable from the structural point of view; finally are sufficient medium density straw bales (90 kg m^{-3}). Load bearing constructions require higher density straw bales (120 kg m^{-3}), but they have the advantage of presenting high resistance to dynamic loads of earthquakes [3, 9, 10] and not require wood for the structures.

In all cases, it is necessary a suitable equipment for the production of the straw bales. At present, where do not exist balers for agricultural use, it is required to rigorously design appropriate presses, improved over the traditional local solutions. This paper presents several solutions designed by the authors, some of which have been realized and tested on the field.

First, the design specifications of this kind of balers are presented. Then is described the architecture of possible presses and actuation mechanisms are discussed. Concluding, the different solutions are compared.

2 Design Specifications of the Balers

The presses must be able to produce bales of defined and constant dimension (in the specific case the dimensions chosen were 0.30 ÷ 0.36 × .45 × .90 m), with a density of between 90 ÷ 120 kg m^{-3}, depending on the style construction.

The press must be able to operate also without fuels or electricity, in order to reduce the running cost and being available and usable in low-income communities. For this reason must be manually actuated.

Since one of the aims of the project was to involve local communities and to engage a self-constructing and sustainable technology process at a local level, the press has to be simple, ergonomic and easy to self-build; it has to be made using metal products and other materials that can be found directly on site.

Finally the presses must be manufactured using simple tools available locally: metal circular saw, angle grinder, welding machine, drill (Fig. 1).

Fig. 1. Simple tools locally available.

3 Concept Design of the Balers

To produce a $0.3 \times 0.45 \times 0.9$ m straw bale, with final density ρ_f of 90 kg m^{-3}, it is required a mechanical compression work of about 2500 J [6]. The maximum work doable by a human operator, acting with a force F_{cp} of about 200 N on an operating lever of 2 m length, rotating it of 90°, is approximately 600 J. A human powered press must then be able to produce a bale in more compression cycles.

Following, first some possible architecture and strategies for the production of the bales are described, and then some simple actuation mechanisms are presented.

3.1 Types of Presses

A first type of baler is characterized by a closed compression chamber (Fig. 2). A certain mass of straw m_c, with initial density ρ_o, is introduced in a closed compression chamber with section A and initial length l_o (Fig. 2a). The compression plate is moved reaching the stroke y_c, necessary to obtain the desired final density of the straw ρ_f (Fig. 2b). Afterwards, the compression plate is back in the initial position (Fig. 2c). Then the sliding-lockable end of the compression chamber is translated by a quantity l_f equal to the length of the compressed straw in the previous compression cycle, and a new compression cycle start (Fig. 2d). Once formed the whole bale in a congruous number of compression cycles, it must be tied and extracted from the compression chamber.

In a second type of baler, called with continuous production, a certain mass of straw m_c, with initial density ρ_o, is introduced in the compression chamber at each compression cycle (Fig. 3a). The being formed straw bale, whose final density ρ_f has been obtained during the precedent compression cycle, takes place in the final part of the compression chamber, and constitutes its end. After this, a complete formed bale is forced to pass through a vertically restricted opening, that crushes the bale. Moving the compression plate, in a first phase the bale already formed is in static friction condition, so it works by an end of the compression chamber, and the straw introduced in the current cycle is compressed (Fig. 3b), until reaching the desired final density ρ_f.

Adjusting the transverse crushing of the bale already formed, it is possible to regulate the friction force, and then, the density of the being formed straw bale. In fact, at the desired density ρ_f, the force applied by the compression plate reaches the

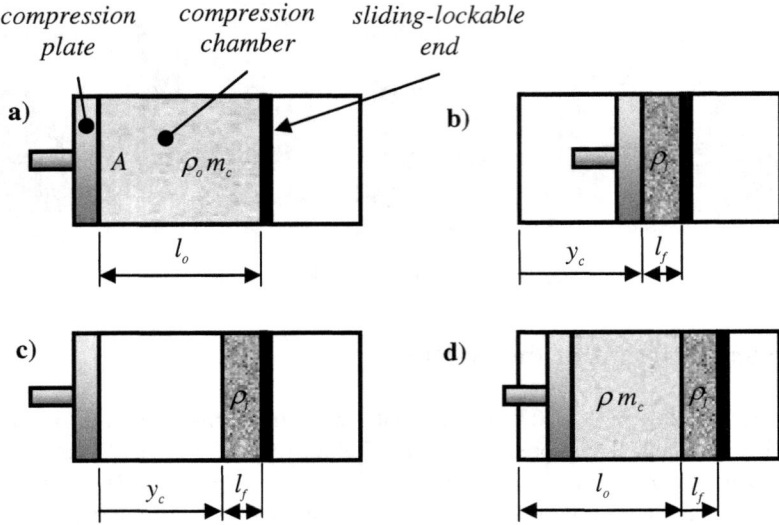

Fig. 2. Bale production storyboard: closed compression chamber press.

dynamic friction force and all the straw translates of y_b (Fig. 3c). The compression cycle is ended when the whole bale moved of l_f, i.e. the length of the straw mass m_c compressed ad the density ρ_f (Fig. 3d). At this point the compression plate is brought back and a new compression cycle is repeated. With this solution the formed bale is ejected gradually, and the bale production is continuous, with possibility to adjust the density.

3.2 Actuation Mechanisms

Apart from the type of press, the actuation mechanism that moves the compression plate should be simple. Considering that, in order to increase the straw density, it is necessary to increase the pressure on the piston surface (and the resultant force), it would be useful an actuation mechanism with variable transmission ratio, in order to require almost constant operating force, while exerting increasing force on the compression plate.

In all cases, for ergonomic reasons, it is considered appropriate an input actuation operated by a lever. The lever, in order to be easily grasped even in vertical position, must have a maximum length of 2 m; the operator, during a compression cycle, should rotate the lever to almost $\alpha_c = 90°$, from a vertical position to a horizontal one, optimizing the application of the force.

In the type synthesis of the actuation mechanism, it is taken into account, first of all, the specification of simplicity and constructability, rather than the requirement of optimization of the transmission ratio. For this reason, simple planar link mechanisms have been chosen instead of cam mechanisms.

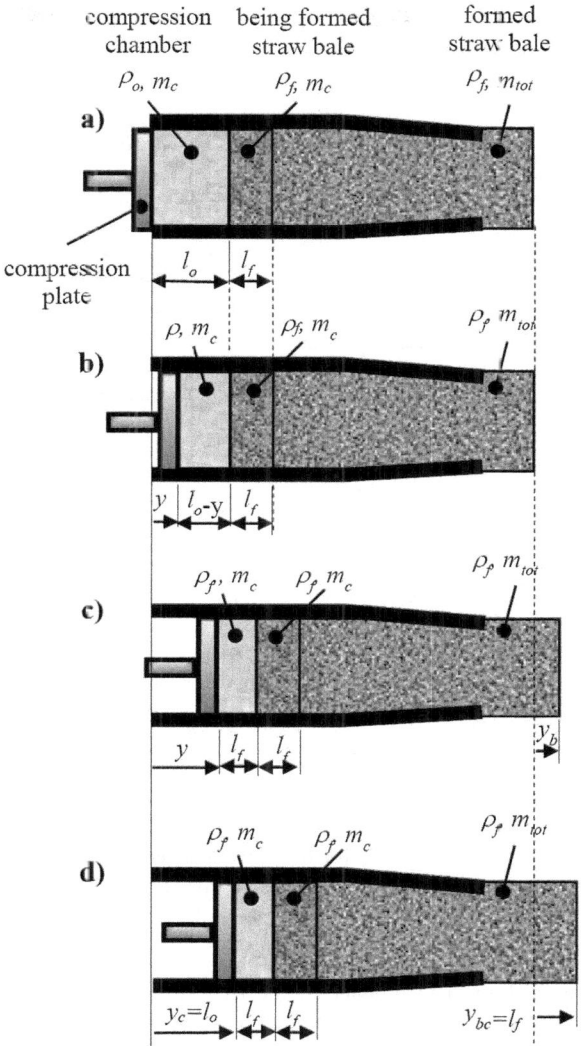

Fig. 3. Bale production storyboard: continuous production press with density control.

A first actuation system proposed is a centred slider crank mechanism (Fig. 4). The actuation lever is rigidly connected to the crank. Making the dimensional synthesis of the mechanism, as described in detail in [6, 7], it must be defined the proper length of the crank m, of the connecting rod b, and the optimum initial angle of the crank α_o, such that the compression plate can be moved of desired stroke y_c, minimizing the operating force F_{op}.

For example, Fig. 5 shows the trend of the operating force F_{op} versus the operating lever angular position α, in a closed compression chamber press, with a compression

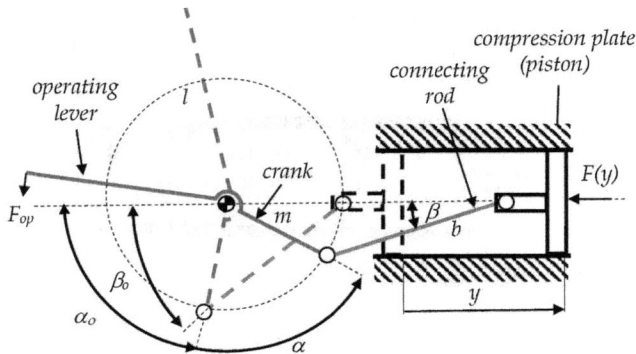

Fig. 4. Baler slider crank actuation mechanism.

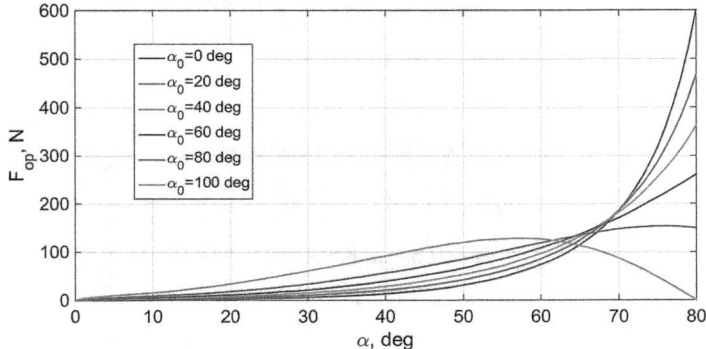

Fig. 5. Operating force F_{op} versus the rotation of the lever α in the case of a slider crank mechanism actuation ($\rho_o \approx 30$ kg m^{-3}; $\rho_f = 90$ kg m^{-3}; $\alpha_c = 80°$; $l = 2$ m; $m = 0.3$ m; $b = 0.7$ m).

plate stroke y_c of 0.31 m. The density of the rice straw is increased to a final value $\rho_f = 90$ kg m^{-3} starting from an initial value $\rho_o \approx 30$ kg m^{-3} [7].

Choosing an initial angle of the crank α_o of about 80÷100°, it is possible to reduce the maximum of the operating force. Unfortunately, the operating force changes significantly during the rotation of the lever. To overcome this drawback, a second actuation mechanism is proposed (Fig. 6). It uses a pantograph mechanism in series to a double slider crank mechanism. The operating lever is connected to the central link of the mechanism (in red in Fig. 6) of length equal to 2d. Thanks to the higher number of design parameters, and the adjustable phase shift between the two mechanisms in series, it is possible to adapt the mechanism to the required performances, to have more regular operator force, and to avoid normal forces on the sliding plate, that introduces friction and lower efficiency of the press.

Designing the mechanism, the length of the members a, c, d, and the initial angle α_o of the member d must be defined, as discussed in [5]. Figure 7 shows the trend of the

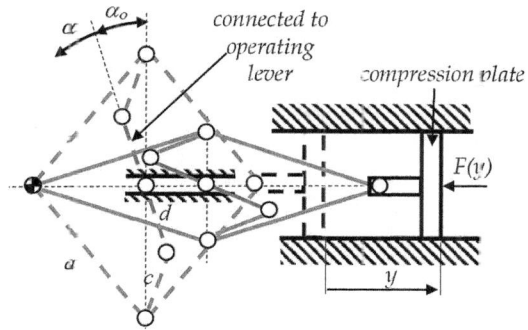

Fig. 6. Baler pantograph actuation mechanism.

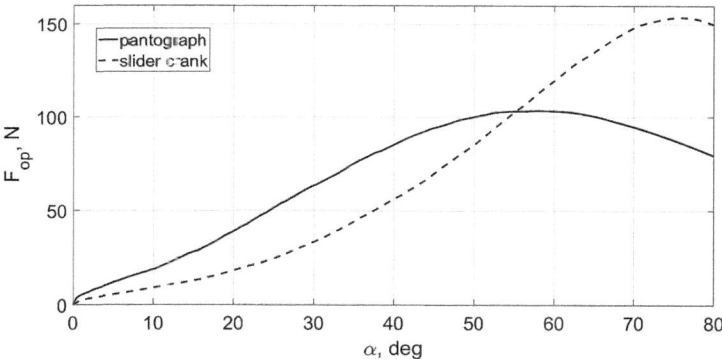

Fig. 7. Operating force F_{op} versus the lever rotation α in the case of a pantograph mechanism (a = 0.35 m, c = 0.2 m, d = 0.131 m, α_o = 30°) and a slider crank mechanism (α_o = 80°)

operating force F_{op} versus the rotation angle of the lever α in the case of a pantograph mechanism, compared to the slider crank mechanism with the same stroke y_c = 0.31 m. In both cases, the density of the rice straw, starting from an initial value $\rho_o \approx 30$ kg m^{-3}, is increased to the final value ρ_f = 90 kg m^{-3}.

4 Detailed Design of the Presses and Prototypes

Starting from the considerations presented in the previous sections, the detailed design of different balers have been developed, some of which have been realized.

A first press adopt a closed chamber solution, with a slider crank actuation mechanism (Fig. 8). The straw is charged into the loading chamber. The sliding/lockable end of the compression chamber must be moved cycle after cycle, in order to restore the chamber dimension. The bale formed must be extracted by the unload door.

18   C. Ferraresi et al.

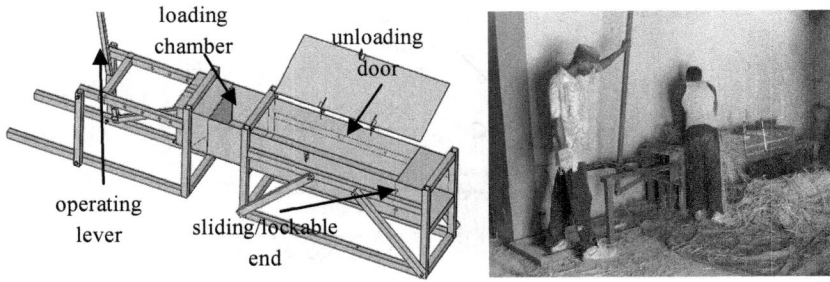

Fig. 8. Closed compression chamber baler with slider crank actuation mechanism: detailed design and prototype

A second kind of baler permit to produce bales continuously, with the adjustment of the density. The actuation transmission is still a centred slider crank mechanism (Fig. 9). The bale formed is progressively expulsed.

Fig. 9. Press with continuous bale production and slider crank actuation mechanism: detailed design and prototype

A last solution proposed (and not realised) is equipped with the continuous production system, and actuated by a pantograph mechanism (Fig. 10).

In all cases, the revolute joints and the prismatic joints of the compression plate are plain bearings. Very simple solutions, as shown in Fig. 11, have been adopted. In order to reduce the friction force, lubricating grease was used.

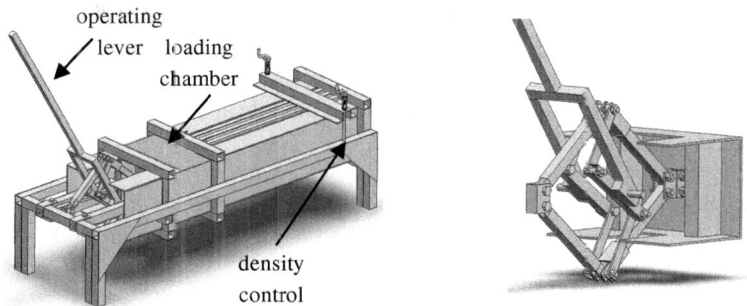

Fig. 10. Press with continuous bale production and pantograph actuation mechanism: detailed design and particular of the actuation mechanism

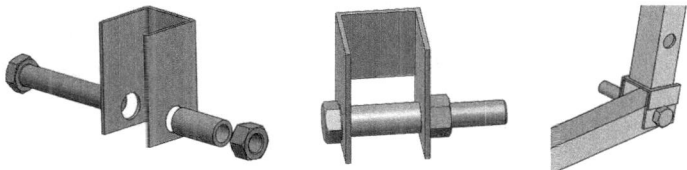

Fig. 11. Constructive solution of the revolute joints of the actuation mechanism.

5 Conclusions

Different solutions of human powered balers for production of bales for bale construction have been proposed. The presses differ both for the architectures and the bale production strategies (closed compression chamber, continuous production), both for the actuation mechanisms (slider crank or pantograph).

Regarding the *closed chamber presses*, they have the disadvantage of having to reposition, at each compression cycle, the sliding lockable-end; in addition the bale density adjustment can only be done by weighing the mass of straw inserted in the bale at each compression cycle; finally the extraction of the entire bale is difficult.

Regarding the *continuous production presses*, they have the significant advantage of being able to control the density of the straw, independently from the mass of straw inserted in the press at each compression cycle; in addition they have the convenience of having the progressive expulsion of the bale being formed. In this case a set up phase of the press at the beginning of straw bale production is required, regulating the friction on the bale adjusting the transverse crushing of the bale, in order to have the proper final density.

Regarding the *actuation mechanisms*, the *slider crank mechanism* is simple and easy to be realized, but has some limitations from the point of view of the optimization of the trend of the operating force. The *pantograph mechanism* is more complex, but it allows to optimize the trend of the operating force.

In general, the solutions proposed have proved to be appropriate for the production of bales for straw bale construction. In fact, two prototypes were constructed, and effectively used for the construction of an infill straw bale warehouse for a rice farmer cooperative of Haiti, and of a load bearing straw module at the Politecnico di Torino, as part of the Anpilpay 2.0 student workshop.

Acknowledgments. Our gratitude goes to eng. Federico Iarussi for the collaboration in the detailed design of the press Anpilpay 1.0; the Architettura Senza Frontiere Piemonte (ASF) association, in the person of Valeria Cottino, Annalisa Mosetto and Veronica Brugaletta for the realization of the Anpilpay1.0 prototype; the Politecnico di Torino student team Anpil Pay 2.0 for the detailed design of the second prototype of baler.

References

1. ASF Piemonte (2013) Haiti |1| Foyer d'accueil aux enfants démunis d'Haiti, III Congress of the University Network for Development Cooperation, CUCS Torino 2013, Torino, 19–21 September 2013, p 317
2. ASF Piemonte (2013) Haiti |2| Re-start from straw, III Congress of the University Network for Development Cooperation, CUCS Torino 2013, Torino, 19–21 September 2013, p 318
3. Bonoli A, Rizzo S, Chiavetta C (2015) Straw as construction material for sustainable buildings: life cycle assessment of a post-earthquake reconstruction. In: vernacular architecture: to-wards a sustainable future. CRC Press, pp 143–146
4. Ferraresi C, Franco W, Quaglia G (2011) Human powered press for raw earth blocks. In: ASME IMECE 2011 international mechanical engineering congress & exposition, Denver, USA, November 2011
5. Ferraresi C, Franco W, Quaglia G (2017) Synthesis of the actuation mechanisms of a human powered straw baler for straw bale constructions in developing countries. In: ASME IMECE 2017 international mechanical engineering congress & exposition, Tampa, USA, November 2017. Under review
6. Franco W, Iarussi F, Quaglia G (2016) Human powered press for producing straw bales for use in construction during post-emergency conditions. Biosys Eng 150:170–181
7. Franco W, Quaglia G, Ferraresi C (2017) Experimentally based design of a manually operated baler for straw bale construction. In: Proceedings of the 1st international conference of IFToMM ITALY, Vicenza, Italy, 1–2 December 2016. Mechanism and machine science, vol 47, pp 307–314
8. Franco W, Anastasio D, Ferraresi C, Gondino F, Quaglia G, Soprana L (under review) Anpilpay 2.0: a new human powered press for producing straw bales for load bearing constructions. Agric Eng Int CIGR J
9. King B (2006) Design of straw bale buildings: the state of art. Green Building Press, San Rafael
10. Pakistan Straw Bale and Appropriate Building. www.paksbab.org. Accessed 30 Dec 2016
11. Sassu M, Romanazzi A, Giresini L, Franco W, Ferraresi C, Quaglia G, Orefice E (under review) Mechanical characterization of compressed earth blocks (CEB) made by float ram 1.0 press. J Mater Civ Eng
12. United Nations. http://www.un.org/sustainabledevelopment/. Accessed 16 Dec 2016

Guiding Linkages with Remote Centre of Rotation for Thermal Cutting Processes

Carsten Teichgräber[1(✉)], Jörg Müglitz[2], and Maik Berger[1]

[1] Technische Universität Chemnitz, Chemnitz, Germany
{carsten.teichgraeber,maik.berger}@mb.tu-chemnitz.de
[2] ZIS Industrietechnik GmbH, Meerane, Germany
info@zis-meerane.de

Abstract. This paper offers some thoughts on the synthesis of planar guiding mechanisms for thermal cutting. One of the links, namely the cutting torch holder, performs a rotation about a virtual axis that is not carried out as a real joint. The considerations are presented with respect to the boundary conditions of the process, mainly the design and collision space restrictions and the demands on the available tilting angle. As it will be shown, many of the existing solutions in the literature and industry are variants of – or derived from - one 6-bar linkage based on a parallel crank mechanism. Additionally one solution is given, that is not related to the 6-bar linkage.

Keywords: Remote centre of rotation · Guiding linkage · Thermal cutting

1 Introduction

This paper deals with the synthesis of mechanisms with one link performing a pure rotation about a virtual axis – i.e. an axis not being physically designed as a revolute joint. This feature is often referred to as a 'remote center of rotation' in the literature when applied e.g. on robots that perform rotations about their tool centre point (TCP) mechanically.

Firstly the advantages of a TCP-fixed motion for thermal cutting processes are outlined. At CNC cutting machines, the bevel unit – which performs the tilting operation of the tool – is attached the Cartesian X-, Y- and Z-axes. The technology demands the cutting beam – oxygen, plasma, laser or waterjet – to enter the workpiece at a specific point. The motion of this point on the workpiece together with the beam direction then generates the final shape of the part (at least as a first approximation).

When a chamfered plate is produced out of sheet metal, a re-orientation of the beam – e.g. at corners or rounding – must not lead to a change of the cutting point's position. That is why in the general case, the Cartesian axes perform an offset motion during the orientation change in order to keep the TCP fixed with respect to the part (compare Fig. 1a). That principle is mechanically simple, as all machine axes are serially arranged. Nevertheless the machine accuracy can be effected negatively, e.g. when the axis controllers possess unequal loop errors. TCP-fixed bevel units do not face those problems. As rotations about the TCP are mechanically decoupled, the inaccuracies of translational axes and bevel unit can be coped with independently.

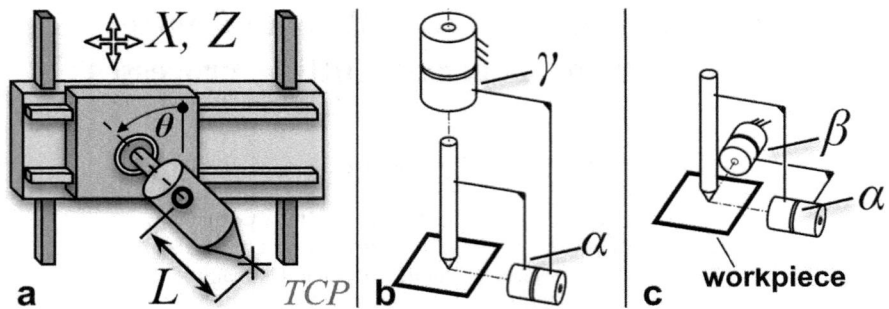

Fig. 1. (a) General arrangement requires an offset motion of the Cartesian axes decoupled rotation of the torch with (b) polar and (c) Cardanic configuration

2 Technology Overview and State of the Art

When starting the development of a mechanism with a virtual remote rotation axis, it is logical to define the constraints that must not be violated already by the kinematic draft (see Sect. 2.1) and to perform a thorough analysis of previous solution for related tasks (Sect. 2.2). The analysis lead to a few basic mechanisms being found in many applications with different kinematic dimensions.

2.1 Technological Constraints

In the case of automated cutting processes it suffices to become aware of three main facts.

Firstly one must realize the two structural variants that differ in the order of actual rotations of the cutting tool – see Fig. 1b and c and [2, 5]. A third rotation would be needless due to the beam's rotational symmetry. Both figures show the axes of rotation intersect at the TCP.

When imagining the cutting process even the polar configuration – first rotation about X-axis, then about Z-axis – leads to a collision of the revolute joint with the workpiece, which is the second fact. To handle the design space restrictions, virtual joints need to execute the rotations of angles α and β.

The third issue concerns the reaction forces on the guiding mechanism, which are very small. Only the weight and damping of the consumables pipe add non-linear forces to the system – the remaining loads can be estimated quite accurately as the process forces itself are negligible. The range of rotation is in most cases demanded as the ability to set a tilting angle of ψ_{max} about a horizontal axis and allow this to be rotated within $\varphi \in [0°, 360°]$ about the vertical direction. Usual demands on ψ_{max} reach from 30° up to 52° or even 68°. For the torch vector the coordinates φ and ψ are of course only one parameterization. The choice of rotational representation is independent of the mechanical axes' actual configuration.

2.2 Some Exemplary Solutions

The first application of a planar mechanism for a TCP-fixed robot wrist – that could be found by the authors – is shown in a patent from Mosher [4]. The system presented is able to tilt by 120° in one direction but can be fully rotated by a vertical (polar) revolute joint – see Fig. 2b.

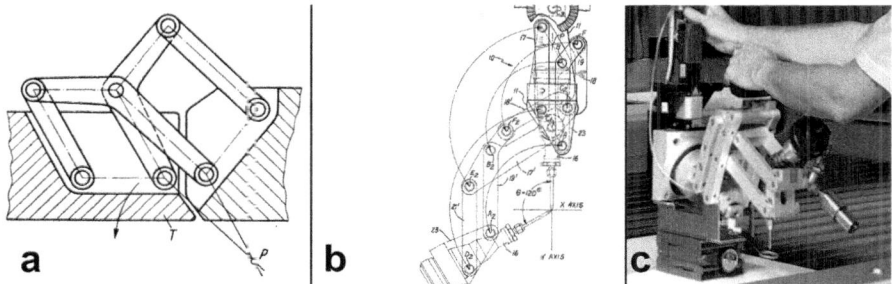

Fig. 2. (a) Aircraft door mechanism [6] (b) robot wrist patent [4] (c) waterjet head [1]

Even before the same well-known 6-link mechanism has been in use for a door being able to close an aircraft shape without overlap (Fig. 2a). Until now the mechanism can be found as a bevel unit for 5-axis cutting applications (e.g. at the waterjet machine in Fig. 2c). The first two examples show quite different link lengths and the latter case even uses an overconstrained design. Yet the basic principle is the same.

Another effective means to perform the TCP-fixed rotation is an arc guide rail. It can be used both in the polar (Fig. 3) and the Cardanic configuration (Fig. 4). The planar mechanism shown in Fig. 4a is another example of the same basic mechanism appearing in a totally different design – being discussed in the next chapter.

Fig. 3. (a) Arc guiding rail in a polar arrangement (b) and (c) show a plasma process [5]

3 Basic 6-Link Guiding Mechanism

It is the intention of this paper to outline the geometric constraints that lead to the basic 6-link guiding mechanisms and some related linkages, that can be directly derived from it. As mentioned before, the purpose is to guide the cutting torch. Thus the area below

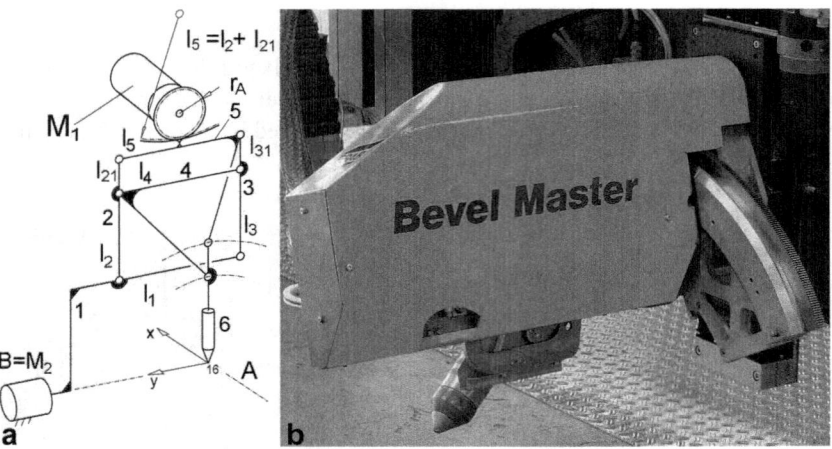

Fig. 4. Cardanic arrangement (a) kinematic scheme of mechanisms for virtual A-axis (b) TCP-fixed wrist for plasma cutting [5]

has to be seen as the workpiece region and must neither be used as design space nor moving zone due to the danger of collisions.

To start the considerations the 4-link parallel crank mechanism in Fig. 5a is analysed. Its four kinematic parameters are the crank length L_{24}, the coupler length L_{13} and the offset $[x_C \; y_C]^T$ of the coupler point in a local coordinate frame on the coupler (see upper sketch of Fig. 5c). The coupler curve k_C of point C in Fig. 5a is of circular shape with radius L_{24}, the arcs centre is located at $[-x_C \; -y_C]^T$ starting from the base joint A_0. The use of the identical link lengths in the way depicted ensures, that the point guiding is circular.

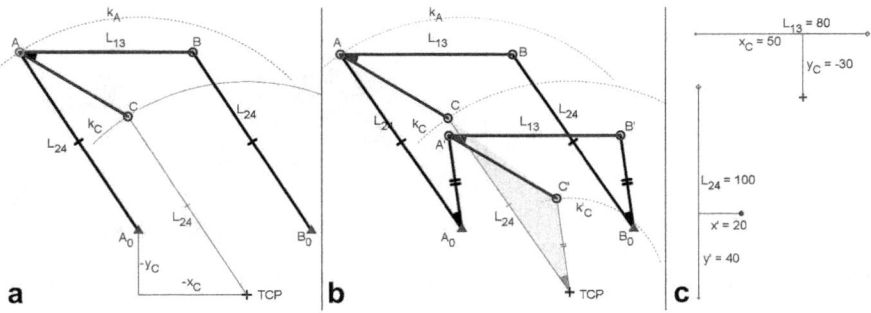

Fig. 5. (a) Parallel crank mechanism with coupler point (b) double-closure parallel crank, end-effector link is guided by two pin joints (overconstrained) (c) six independent parameters

The next step is to add two links, which will be performing the rotation about the virtual axis. Firstly two revolute joints A', B' are added to the cranks at coordinates

$[x'y']^T$ with respect to a local coordinate system on each crank (see lower sketch of Fig. 5c). Secondly an identical coupler is mounted between the new revolute joints A', B' leading to the circular path k'_C of the point C'. Notice the different location of k'_C. While

$$r' = \sqrt{x'^2 + y'^2} \tag{1}$$

defines the arc radius,

$$\Delta\psi' = \arctan\left(\frac{y'}{x'}\right) \tag{2}$$

equals its phase shift about the TCP with respect to k_C (see Fig. 5b).

It is now easy to understand, that a 6th link – namely the torch holder – completes the mechanism to fulfil its task. The geometry of this link correlates to the two cranks. The triangles A_0 A A' are congruent to TCP C C' and of course B_0 B B'. The joint angles of the cranks are equal to one another and to the angle of the torch holder as well – thus resulting in the beneficial constant transmission ratio of $i = 1$.

The scheme in Fig. 5b is structural similar to the "Bevel Master" solution in Fig. 4a – yet the differences are worth discussing. The first thing to notice is the innovative actuation of the mechanism via the gear segment. It allows to place the drive above the linkage and quite distant to the location of the cutting. Secondly the arrangement of the revolute joints in a straight line is a special case and may seem as an unnecessary condition.

However the mechanism is twice overconstrained. The determination of the second coupler between A' and B' is redundant as well as the connection of C and C' with the torch holder link. As the production and assembly of the parts is only possible within certain tolerances, there will always be prestress and deformation between the links – as the exact kinematic dimensions cannot be realized. On the one hand, the mechanism gains stability and stiffness with every redundant binding, yet an adjustment is getting hardly possible.

Another notable issue comes up, if the TCP is wanted at a greater vertical distance $|y_C|$ from the base joints A_0 and B_0. Then in the practical case, the ternary links ABC and $A'B'C'$ will increasingly suffer from bending loads and might show an insufficient stiffness. This point will be addressed again later.

3.1 Simplification and Variation

This section is intended to show some variations of the solution presented above. The problem of overconstraints can be solved quite easily. Looking again at the 4-link mechanism in Fig. 5a, the position of point C of the torch holder is already assured. Merely the correct orientation of the link needs to be taken care of, e.g. using gears or a belt drive with constant ratio $i = 1$ between the crank motion at joint A_0 and the torch holder motion at joint C – compare Fig. 6b.

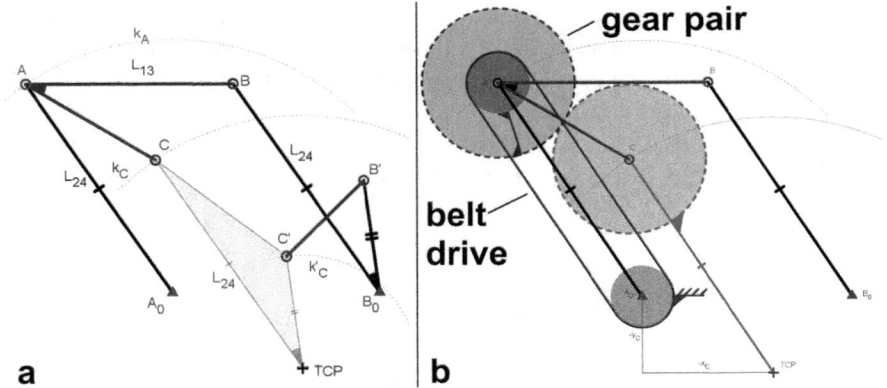

Fig. 6. (a) Non-overconstrained 6-link mechanism (b) implementing gears or a belt/chain drive

A more elegant way to define the end effector's orientation is the use of the crank-point B' already introduced in Fig. 5b. As the lower coupler in that image is of course a rigid body, points B' and C' remain in a constant distance when moving on the circular curve. Therefore a rigid link between those two points can be added to guarantee the correct torch angle (Fig. 6a). The resulting mechanism is free of redundant constraints but not different in behaviour. The link $B'C'$ is sliding along the arc k'_C – like the coupler ABC slides along the arc k_C.

3.2 Derivation of the 8-Link Mechanism

The exchange of the lower ternary coupler from the mechanism in Fig. 5b leads to the simplified one in Fig. 6a. The latter linkage is easier to produce and assemble, and even a mechanical adjustment becomes more convenient.

Considering the need for lightweight design, it also might be helpful to distribute the end-effector forces on tension and compression rather than bending loads. Following this thought, an end-effector, that is supported by three RR-links (each with two revolute joints), will be hold only by normal forces. In this context there is another related mechanism, shown in Fig. 7, which is worth investigating.

Starting with the non-overconstrained 6-bar mechanism in Fig. 6 another revolute joint A'' is added to the left crank at local coordinates $[x''\ y'']^T$ – see Fig. 7a and b. The corresponding joint C'' is found by moving the same amount in local coordinates from the *TCP* on the torch holder. Between them, the new link $A''C''$ is placed.

In the second step the common joint C between the ternary coupler and the end-effector is disconnected and another pin joint D on the coupler is introduced at the local coordinates $[x_D\ y_D]^T$ (see Fig. 7a and b). Between them an additional RR-link DC is mounted, thus completing the 8-link mechanism and the threefold suspension of the torch holder by RR-links. All those links $A''C''$, $B'C'$ and DC translate on circular curves without changing their orientation – and of course link ABD also does. The

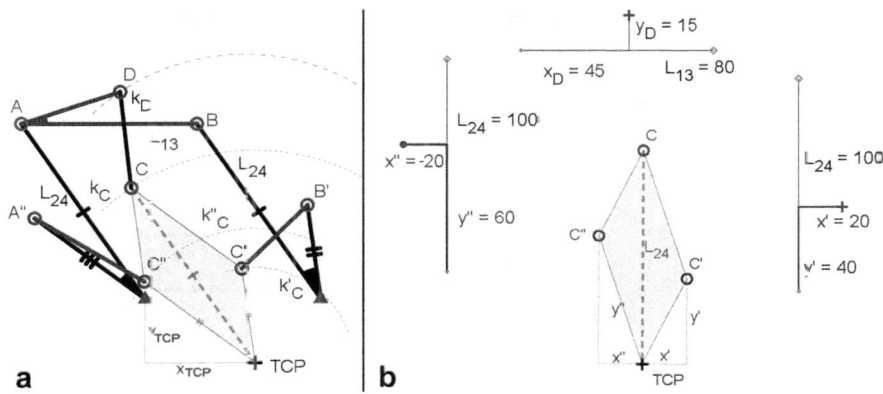

Fig. 7. (a) Kinematic scheme of the generalized 8-link mechanism with virtual rotation axis (b) overview of the kinematic parameters of end-effector, coupler and cranks

forces within the links will either be tensile or compressive, while moments are excluded.

Figure 7a and b show the 10 independent design parameters of the 8-link mechanism. The coordinates $[x_{TCP}\ y_{TCP}]^T$ were introduced, as the location of point C is no longer determined on a rigid link (although is stays the same – compare with Fig. 5a). The link lengths $|A''C''|$, $|B'C'|$ and $|DC|$ result from the choice of the other lengths as follows

$$|A''C''| = \sqrt{x_{TCP}^2 + y_{TCP}^2} \qquad (3)$$

$$|A'B'| = \sqrt{(L_{13} - x_{TCP})^2 + y_{TCP}^2} \qquad (4)$$

$$|DC| = \sqrt{(x_D - x_{TCP})^2 + (y_D - y_{TCP})^2} \qquad (5)$$

The alteration of the 10 independent parameters (see Fig. 7) will change the properties of the system strongly – like collision space, the available tilting angle and the static stiffness – so they have to be evaluated in a separate step. But the feature of the remote center of rotation will remain. Interestingly, if the distance $[x_{TCP}\ y_{TCP}]^T$ of base joints and virtual rotation axis is increased, only the three RR-links will gain length and none of the ternary links is affected. This fact leads to yet another astonishing variation.

As the 8-link mechanisms permits the free positioning of the TCP in the base plane, the points A'', B' and D (Fig. 7a) may be used to guide a second end-effector (Fig. 8a). Even cases with multiple torches (Fig. 8b) are possible and each *TCP* can be placed freely on the base plane.

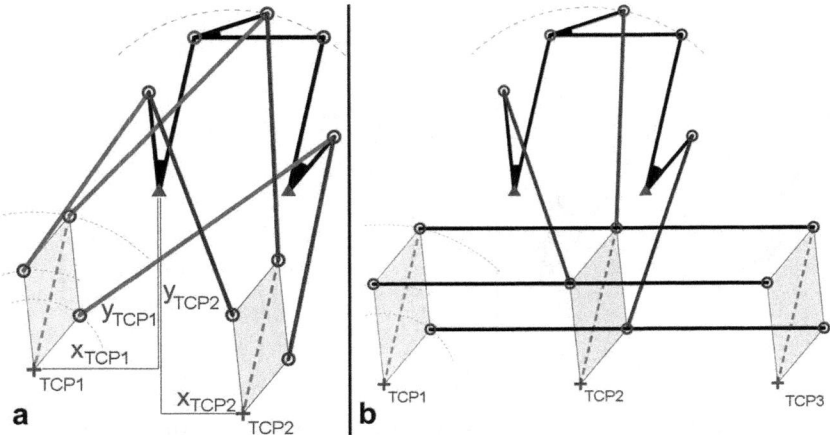

Fig. 8. (a) Two end-effectors guided by one mechanism at the top (b) several torches interconnected

4 A Different Approach – Double-Slider Mechanism

Referring to Fig. 2a the German textbook on mechanism theory [6] says:

> "Guiding a link about a fixed axis is – except by the revolute joint – only possible with the 6-link mechanism [...]. An exact rotation is achieved only with parallel links, as shown by the door hinge example [...]."

As shown in the previous section, many different linkages can be deduced from the 6-link basic guiding mechanism in Fig. 6a – regardless of the change concerning the mere dimensions or even the structure. Yet the authors have found two further non-related mechanisms, of which one is presented briefly in this section.

The explanation starts with a look on the double-slider mechanism with orthogonal prismatic joints in Fig. 9a. A circular coupler curve k_P with radius r demands the coupler point P to be centrally arranged on the coupler.

Figure 9b shows a possible guiding mechanism about the virtual rotation axis through the *TCP*. Using the coupler 3 and the broadened slider 4 a circular translation

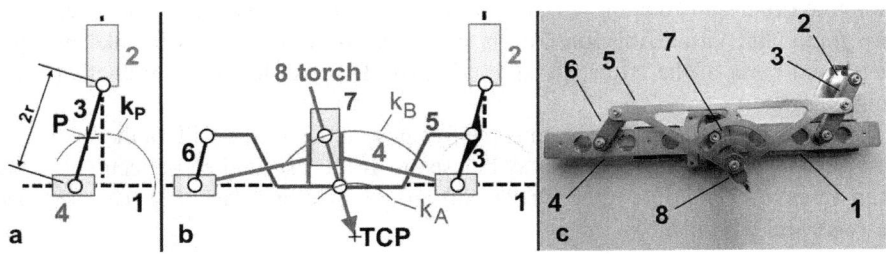

Fig. 9. (a) Double-slider mechanism (special dimensions) (b) kinematic scheme (c) functional model

Fig. 10. "3D-Link" unit from KOIKE Europe B.V. [7]

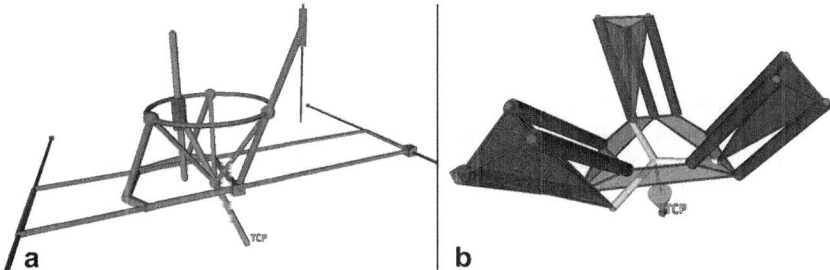

Fig. 11. Generalised spatial mechanism of both (a) double-slider and (b) parallel crank mechanism for a use a TCP-fixed bevel unit

along k_A is gained via the dyad 5 6 – where link 6 is of length r. The revolute joint on k_A connecting links 5 and 8 already defines the position of the torch 8. Its correct orientation finally is set by the vertical slider 7.

Again the parallel crank principle can be recognized at the links 3, 4, 5 and 6. The mechanism may be actuated at the horizontal slider 4 – which requires a non-linear transformation – or for instance between links 4 and 3 with a constant transmission ration of $i = -1$.

5 Conclusions and Prospect

This paper presents the derivation of planar guiding mechanism to realise a virtual rotation axis. One example is elaborated in detail, because it seems the one applied more often. Yet few designers might have realised the relationship of their design to the basic mechanism. To the authors no application of the second mechanism is known – but a working table-sized model was created (see Fig. 9c).

There are several issue to be addressed in further work. As pointed out above, from a technological point of view it is highly desirable to achieve a bevel unit with a Cardanic principle (see Fig. 1c). A planar mechanism covers only one rotation but there might be a generalised spatial mechanism, which guides both tilting directions. Several generalisations of the parallel crank mechanism (Fig. 5) can be found in the literature – of which the 8th chapter of Kiper's work [3] contains a helpful summary. The only application as a TCP-fixed wrist that was found by the authors is shown in Fig. 10. The bevel unit is of high dynamic, yet it is overconstraint and thus makes considerable demands on the manufacturing precision. Concerning the double-slider mechanism (Fig. 9b) a theoretical generalisation could be found by the authors – see the spatial mechanisms in Fig. 11, which will not be further discussed here. Yet a collision-free and practical implementation is still open for the latter.

References

1. Bystronic AG, Switzerland: 3D-cutting with Bystronic 3dcutting. bystronic.com
2. Chen L, Teichgräber C, Berger M (2015) A case study about spherical robot wrists with serial and parallel structure. In: Trc-IFToMM symposium on theory of machines and mechanisms, Izmir, Turkey, p 581–589. ISBN 978-605-84220-0-1
3. Kiper G (2011) Design methods for planar and spatial deployable structures. PhD thesis, Middle East Technical University, Ankara
4. Mosher RS (1985) Low cost articulating and rotating wrist mechanism. US Patent 4,551,058
5. Müglitz J, Berger M, Wegert E (2011) Roboterhandgelenke zum thermischen Schneiden – Robot wrists for thermal cutting (German). In: 9. Kolloquium Getriebetechnik – Chemnitz
6. Volmer J (ed) (1979) Getriebetechnik – Koppelgetriebe (dt.) – Mechanism theory (German). Verlag Technik, Berlin
7. Mercator Media Ltd. Cutting out second ops. www.engineeringcapacity.com/....../news101/process-news/forming-and-fabrication/cutting-out-second-ops

Structural Synthesis of 2R1T Type Mechanisms for Minimally Invasive Surgery Applications

Abdullah Yaşır and Gökhan Kiper[✉]

Izmir Institute of Technology, Izmir, Turkey
{abdullahyasir,gokhankiper}@iyte.edu.tr

Abstract. Assistive and operative manipulators allow easier and more precise operations for minimally invasive surgery. Such manipulators often have a pivot point at the incision port on the patient's body, so the manipulator should have a remote center of motion. This study presents the structural synthesis of a non-parasitic 3-dof manipulator with 2R1T motion pattern to be used as a remote center of motion mechanism for minimally invasive surgery applications. The manipulators of various kinematic structure are evaluated considering criteria such as possibility of construction of the mechanism for remote center of motion, ease of dynamic balancing, number of links, structural symmetry, the number of actuators connected to the base and decoupling of the joint inputs and the output motion of the platform.

Keywords: Minimally invasive surgical manipulator · Structural synthesis · Non-parasitic motion · Remote center of motion

1 Introduction

The use of robotic devices in surgical applications is becoming widespread day by day [2]. Some of these devices are used directly in operations while others are being developed as assistive devices. Compared with manual surgical operations, robot operated surgeries appear to be more precise [4]. Surgical robots are usually used in minimally invasive surgery (MIS) applications. MIS is performed with a surgical tool inserted through a small hole (incision port) into the patient's body. Such operations can be completed in less time, with less pain and less blood loss compared to conventional surgery. The post-operative process also results in faster recovery and smaller surgical scars [6].

The necessity of moving through an incision port in MIS requires that the robot should have a pivot point at the port. Remote-center-of-motion (RCM) is a point which remains stationary with respect to the base of the manipulator without necessitating any joints at that point such that the end-effector of the manipulator can rotate about and slide through this point. The RCM of a manipulator can be obtained by properly designing the mechanical structure, or by using a kinematically redundant manipulator and ensuring RCM by control. Mechanical RCMs are more reliable and considered more suitable for clinical practice compared to non-mechanical ones [7].

© Springer International Publishing AG 2018
M.I.C. Dede et al. (eds.), *Mechanisms, Transmissions and Applications*,
Mechanisms and Machine Science 52, DOI 10.1007/978-3-319-60702-3_4

In MIS, a surgical tool can have at most four degrees-of-freedom (dof) through the incision port: yaw, pitch, roll and heave [7]. However, yaw, pitch and heave movements are sufficient for endoscope movements [8]. This study follows Kong and Gosselin's [3] structural synthesis methods in order to classify manipulators, end-effectors of which are capable of non-parasitic yaw, pitch and heave motions, i.e. 2R1T (two rotational, one translational dof) motion. The results are comparable with Li and Hervé's [5] results for classifying non-parasitic 1T2R manipulators, which are the kinematic inversion of the 2R1T case. Recently Huang et al. [1] also worked on RCM mechanisms with 2R1T motion, where they have similar results for synthesizing the legs, but the manipulators they obtained are few in number. After the classification, proper mechanical structures are compared according to design criteria for MIS applications with RCM.

2 Structural Synthesis

A manipulator with 2R1T motion can have various kinematic structures. The simplest kinematic structure would be an RRP (R: revolute, P: prismatic) serial manipulator where the R axes intersect each other and the P direction is perpendicular to the plane defined by the R axes. Also several types of hybrid kinematic structures can be used such as: type (1) the first R of the 2R1T motion is serially connected to a 2-dof parallel kinematic chain (PKC) for the RT motion; type (2) 2R motion is obtained with a PKC while the T motion is connected serially; type 3) second R motion is obtained with a PKC while the first R and the T motion are connected serially. Examples for serial and hybrid kinematic structures for orienting (2R) manipulators are presented in [9].

In this section, the structural synthesis for 3-dof parallel manipulators (PM) for 2R1T motion pattern is performed. The virtual chain (VC) corresponding to the 2R1T motion pattern is **RRP**. The motion and constraints of a kinematic chain are respectively represented by twist systems and wrench systems in screw theory [3]. Then, the wrench system for an **RRP**= PKC is a $2F_0$-$1F_\infty$-system (Fig. 1a), where F_0 and F_∞ represent 0-pitch and ∞-pitch wrenches, respectively.

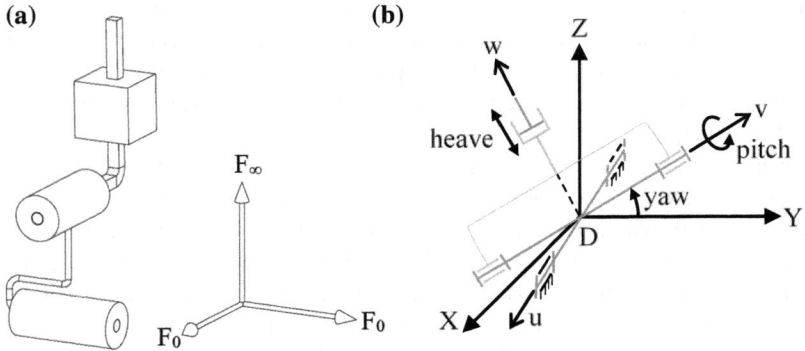

Fig. 1. a. $2F_0$-$1F_\infty$-system. b. Description of the **RRP** VC

An **RRP** VC is illustrated in Fig. 1b. u- and v-axes represent the R joint axes of the VC and intersect at the pivot point D. w-axis is along the direction of the P joint and passes through the pivot point. The u-axis lies on the XY-plane. u-axes attached to different legs may have different directions. v-axis is perpendicular to the u-axis, while w-axis is perpendicular to the v-axis. In Fig. 1b, when both yaw and pitch angles are equal to zero, u- and v-axes lie on the XY-plane and w-axis coincides with the Z-axis of the XYZ frame attached to the base.

2.1 Decomposition of the Wrench System of the PKC

Combinations of leg constraint degrees (c^1, c^2, c^3 for the three legs) for 3-legged 3-dof (M = 3) PKCs are given in Table 1. Δ is the total degree of overconstraint of the PKC such that $c^1 + c^2 + c^3 = \Delta + M$. The number of single dof joints, f, in a leg is calculated with the formula f = (6 − c) + M, where c is either of c^1, c^2, c^3.

Table 1. Combinations of ci for 3-legged 3-dof PKCs

Δ	6	5	4			3			2			1			0
c^1	3	3	3	3	3	3	2	3	3	2	3	2	2	3	2	1
c^2	3	3	3	2	3	2	2	2	1	2	1	2	1	0	1	1
c^3	3	2	1	2	0	1	2	0	1	1	0	0	1	0	0	1

The platform wrench system is a $2F_0$-$1F_\infty$-system, so the legs may have one of the following systems: $2F_0$-$1F_\infty$-, $1F_0$-$1F_\infty$-, $2F_0$-, $1F_0$-, $1F_\infty$- or 0-system. The wrench system of the moving platform is simply the linear combination of those of all the legs [3]. So, combinations of leg-wrench system which do not have two F_0- and one F_∞-system in total fail to produce the platform wrench system and are not feasible. Feasible combinations are listed in Table 2.

Table 2. Combinations of leg-wrench systems for 3-legged **RRP**= PKCs

Δ	6	5	4			3				
c^1,c^2,c^3	3,3,3	3,3,2	3,3,1	3,2,2	3,3,0	3,2,1			2,2,2	
$2F_0$-$1F_\infty$	3	2	2	2	2	1	1	1	2	1	1	1	1	0	0	0
$1F_0$-$1F_\infty$	0	1	0	0	0	2	0	1	0	1	1	0	0	3	2	1
$2F_0$	0	0	1	0	0	0	2	1	0	0	0	1	1	0	1	2
$1F_0$	0	0	0	1	0	0	0	0	0	1	0	0	1	0	0	0
$1F_\infty$	0	0	0	0	1	0	0	0	0	0	1	1	0	0	0	0
0	0	0	0	0	0	0	0	0	1	0	0	0	0	0	0	0

Δ	2			1			0		
c^1,c^2,c^3	3,2,0	3,1,1	2,2,1	3,1,0	2,2,0	2,1,1	3,0,0	2,1,0	1,1,1
$2F_0$-$1F_\infty$	1	1	1	1	1	0	0	0	0	1	1	0	0	0	0	0	1	0	0	0
$1F_0$-$1F_\infty$	1	0	0	0	0	2	0	2	1	0	0	2	1	1	0	1	0	1	0	0
$2F_0$	0	1	0	0	0	0	2	0	1	1	0	0	1	0	1	0	0	0	1	0
$1F_0$	0	0	2	0	1	1	0	0	1	0	1	0	0	2	0	1	1	0	1	0	2
$1F_\infty$	0	0	0	2	1	0	1	1	0	1	0	1	0	0	2	1	1	0	0	1	1
0	1	1	0	0	0	0	0	0	0	1	1	1	1	0	0	0	0	2	1	1	0

2.2 Type Synthesis and Assembly of the Legs

$2F_0$-$1F_\infty$-system: There should be $f = (6 - c) + M = (6 - 3) + 3 = 6$ joints in a virtual loop (VL). Since there are already 3 joints in the **RRP** VC, there should be 3 joints on the leg. In order to obtain a $2F_0$-$1F_\infty$-system, two coaxial ($3F_0$-$2F_\infty$-system) and one codirectional ($2F_0$-$3F_\infty$-system) compositional units (CUs) are combined (Table 3.3 in [3]). Coaxial or codirectional CUs are denoted by $()_L$, where the R or P joints are written inside the parenthesis. A coaxial CU can include one or more R joints with coincident axes. A codirectional CU can include one or more P joints with the same direction. Together with the **RRP** VC (or more specifically $(\mathbf{R})_L(\mathbf{R})_L(\mathbf{P})_L$ VC), there are no feasible solutions for a VL, because the leg would have multiple coaxial R joints or multiple codirectional P joints. However, the VC itself can be used as a leg. Leg structure alternatives are $(R)_L(R)_L(P)_L$ and $(R)_L(RR)_A$, where $()_A$ represents a parallel axis CU. Two or three legs of this type cannot be used in the PKC, because the centers of the universal (U) joints of multiple legs would necessarily be at the pivot point, hence not resulting in a RCM. So the columns in Table 2 with more than one $2F_0$-$1F_\infty$-system can be disregarded.

$1F_0$-$1F_\infty$-system: A loop is composed of a coaxial CU and a planar CU (denoted by $()_E$). There are $f = (6 - c) + M = (6 - 2) + 3 = 7$ joints in the loop, 4 of which are on the leg. A planar CU has at least two joints, at least one of which is an R joint and all links move along parallel planes. There can be at most two P joints and more than one coaxial R joints is not allowed in the leg, because otherwise the leg has internal mobility. Since the R axes in the VC are orthogonal, the coaxial unit should definitely be a part of the VC. Also a part of the coaxial chain needs to be in the leg as well, because otherwise we have a coplanar 4-joint leg, which has internal mobility. So, the only possible structure for the VL is $((\mathbf{RR})_L\mathbf{RP}E)_E$, where E stands for a planar joint. The VC joints are represented with bold letters in the VL in order to distinguish them from the joints of the leg. In practice, the E joint may be constructed as a planar RRR, PRR, RPR, RRP, PPR, PRP or RPP chain. $(R)_L$ part being inside the $()_E$ parenthesis means that the coaxial R joint can be positioned arbitrarily in between any two joints of the planar chain. Practically we do not prefer an unactuated P joint, so we disregard the solutions with multiple P joints in a leg. The feasible legs are $((R)_LRRR)_E$, $((R)_LRRP)_E$, $((R)_LRPR)_E$ and $((R)_LPRR)_E$. The first three are equivalent to URR, URP and UPR legs, respectively. For $((R)_LPRR)_E$, if the $(R)_L$ axis is along the P direction, this specific case corresponds to a CRR chain.

$2F_0$-system: This wrench system is composed of a spherical CU (denoted by $()_S$) and a codirectional CU. There are $f = (6 - c) + M = (6 - 2) + 3 = 7$ joints in the VL. A spherical CU includes at least two R joints and a codirectional CU includes at least one P joint. Two or more codirectional P joints in the leg are not feasible in practice. With these conditions, the only possible structure for the VL is $(\mathbf{SRR})_S(PP)_L$, where S stands for spherical joint. So the leg has a SP structure, which can have $(RRR)_S(P)_L$ form in practice.

$1F_0$-system: There are $f = (6 - c) + M = (6 - 1) + 3 = 8$ joints in the VL of this wrench system. In Table 3.4 in [3], six different ways are given to obtain a $1F_0$ system: (a) a planar CU + a spherical CU, (b) two spherical CUs with distinct centers, (c) inserting two coaxial CUs into a single-loop kinematic chain (SLKC) composed of a planar CU, (d) inserting two coaxial CUs into a SLKC composed of a spherical CU, (e) inserting a coaxial CU and a codirectional CU into a SLKC composed of a spherical CU and (f) inserting two codirectional CUs into a SLKC composed of a spherical CU. In case (a), the only possible SLKCs are in the form of $(RRR)_S(RPE)_E$. So, leg structures can be in the form of UE. The center of the U joint should be on the u-axis and the plane of the U joint should not be parallel to the u-axis. Plane of the E joint should be perpendicular to the v-axis. (b) and (d) cases are not feasible because of the **P** joint in the VC. The only possible structure of VL for case (c) is $((R)_L(RR)_L RPE)_E$. So, leg structure is $((R)_L(R)_L E)_E$. The axis of the 1st $(R)_L$ is along the u-axis; the axis of the 2nd $(R)_L$ is along the virtual P direction; plane of the E joint should be perpendicular to the v-axis; $(R)_L$'s can be distributed in the E joint. For case (e), the possible structures for VL are $(RRRRR)_S(PP)_L(R)_L$ and $((RR)_L R(PP)_L RRR)_S$. Thus, the leg structure can be $S(R)_L(P)_L$, where the center of the S joint is on the pivot point, the axis of the $(R)_L$ does not pass through the pivot point and direction of $(P)_L$ is along the w-axis. Or alternatively the leg structure can be $((R)_L S(P)_L)_S$, where the axis of $(R)_L$ is along the u-axis, the center of the S joint is on the v-axis and direction of $(P)_L$ is along the w-axis. There are two possibilities for case (f): $(SRR(PP)_L(P)_L)_S$ (corresponds to leg structure $(RRR(P)_L(P)_L)_S$, which is not desirable due to two P joints in the leg) and $(SRRR(P)_L(P)_L)_S$ (corresponds to leg structure $(SR(P)_L)_S$, which is not feasible due to internal mobility of the spherical 4R in the leg).

$1F_\infty$-system: In this system, there are $f = (6 - c) + M = (6 - 1) + 3 = 8$ joints in the VL. For obtaining such a wrench system, we have two cases: (a) combining two parallel axis CUs or planar CUs or (b) inserting a coaxial CU into the SLKC composed of a parallel axis CU or planar CU. For case (a), using planar CUs is favorable over using parallel axis CUs due to constructional ease, so only planar CUs will be considered. Possible structures for the VL with two planar CUs are $(..R)_E(RP...)_E$ and $(...R)_E(RP..)_E$, where each dot represents an R or P joint. Plane of the first $()_E$ is perpendicular to the u-axis and plane of the second $()_E$ is perpendicular to the v-axis. So the leg structure is $(..)_E E$ or $E(..)_E$ where .. may be RR, RP or PR. For case (b), possible structures for the VL have the form of $((RR)_L RP....)_A$. So the leg can be $((R)_L RRRP)_A$, $((R)_L RRPR)_A$, $((R)_L RPRR)_A$ or $((R)_L PRRR)_A$. $(R)_L$ is coaxial with the u-axis and the remaining R joint axes are all parallel to the v-axis.

0-system: In this system, there are no constraints for the dof of the leg and any 6-dof chain is suitable. SPU can be an example.

Possible leg configurations are summarized in Table 3.

RRP = PKCs can be generated by assembling three of the alternative leg structures given in Table 3. For each six alternative leg-wrench systems in Table 3, a representative leg is selected and possible assemblies according to Table 2 are constructed in a CAD program. Some of the assemblies result in 4-dof platform motion due to dependency of the leg wrench systems, so they are disregarded. For the assemblies that

Table 3. Leg alternatives for each system

System	Leg Structure Alternatives
$2F_0$–$1F_\infty$	$(R)_L(R)_L(P)_L$; $(R)_L(RR)_A$
$1F_0$–$1F_\infty$	$(R)_L E$ ($(R)_L$ is co-axial with 1st R joint axis of the VC): $((R)_L RRR)_E$ or URR; $((R)_L PRR)_E$ (may be CRR); $((R)_L RPR)_E$ or UPR; $((R)_L RRP)_E$ or URP
$2F_0$	SP (The center of S joint should be coincident with the pivot point D)
$1F_0$	(a) UE (center of U joint is on u-axis; plane of U joint is not parallel to u-axis; plane of E joint is perpendicular to v-axis): $S(RR)_E$, $S(RP)_E$, $S(PR)_E$, $U(PRR)_E$ (c) $((R)_L(R)_L E)_E$ (axis of 1st $(R)_L$ is along u-axis; axis of 2nd $(R)_L$ is parallel to w-axis; plane of joint E is perpendicular to v-axis; $(R)_L$'s can be distributed in E joint) (e) $S(R)_L(P)_L$ (center of S joint is on pivot point; axis of $(R)_L$ does not pass through pivot point; direction of $(P)_L$ is along w-axis) or $((R)_L S(P)_L)_S$ (axis of $(R)_L$ is along u-axis; center of S joint is on v-axis; direction of $(P)_L$ is along w-axis)
$1F_\infty$	(1) $(..)_E E$ (plane of $()_E$ is perpendicular to u-axis; plane of E joint is perpendicular to v-axis): $(RR)_E(RRR)_E$, $(RR)_E(RRP)_E$, $(RR)_E(RPR)_E$, $(RR)_E(PRR)_E$; $(RP)_E(RRR)_E$; $(PR)_E(RRR)_E$; $E(RR)_E$: $(RRR)_E(RR)_E$, $(RRP)_E(RR)_E$, $(RPR)_E(RR)_E$, $(PRR)_E(RR)_E$; $(RRR)_E(RP)_E$; $(RRR)_E(PR)_E$ (2) $((R)_L RRRP)_A$, $((R)_L RRPR)_A$, $((R)_L RPRR)_A$, $((R)_L PRRR)_A$
0	Example: SPU

satisfy the 2R1T motion which can be used as an RCM mechanism, the options for the three joints to be actuated are evaluated. Preferably, the actuated joints should be connected to the base.

3 Design Evaluation

For some of the assemblies obtained in Sect. 2 the legs share the first R axes on the base, hence they actually have a hybrid type 1 kinematic structure. All the manipulators obtained in Sect. 2 are evaluated along with hybrid types 2 and 3 and serial assemblies. All alternatives are compared with each other considering the following evaluation criteria with their weights, w: ease of dynamic balancing (w = 3); number of links (w = 1); structural symmetry (w = 1); decoupling of the inputs and the output 2R1T motion (w = 2); and the number of actuators connected to the base (w = 2). For grading, the following considerations were taken into account: For ease of dynamic balancing, grade of legs with prismatic joints (if un-avoidable) and/or spherical sub-chains was kept low. Less number of links was considered as an advantage. Structural symmetry grade is determined based on whether there are same types of legs and the legs can be positioned in opposite sides. Decoupling of the inputs and the output 2R1T motion was graded based on how many inputs directly correspond to an R or T motion of the end-effector. Finally, more number of actuators that can be connected to the base was considered to be in favor.

The evaluation chart is too big to be presented here. The kinematic structure with the highest grade in the evaluation chart has a 2 ($1F_0$-system) and 1 ($1F_0$-$1F_\infty$-system)

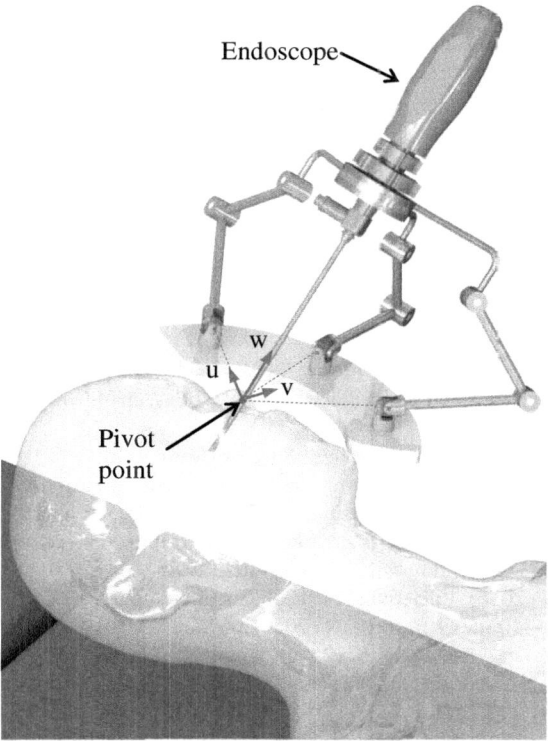

Fig. 2. Conceptual CAD model of an RCM manipulator for transsphenoidal surgery

leg structure. A CAD model of such a system designed as an RCM manipulator for transsphenoidal surgery is depicted in Fig. 2. The PM in Fig. 2 has 2URRR-URR kinematic structure. For this PM, the R joint axes on the base should be intersecting at the pivot point, but the angles between the axes are arbitrary. The R axes of the URRR legs connected to the endoscope should be concurrent along the w-axis. Although all legs in Fig. 2 seem to have identical structure, one of the legs has one less R joint (about the w-axis), hence the part of the leg after the second R joint is rigidly connected to the endoscope.

4 Conclusions

This study deals with the structural synthesis of a non-parasitic 3-dof manipulator with 2R1T motion pattern to be used as an RCM mechanism for MIS applications. For the synthesis of the PMs, the constraint based type synthesis method of Kong and Gosselin [3] is used. The resulting PMs together with manipulators with hybrid and serial kinematic structure are evaluated. The key evaluation criteria are possibility of construction of the mechanism for RCM and ease of dynamic balancing. The best option is evaluated to be a PM with 2 ($1F_0$-system) and 1 ($1F_0$-$1F_\infty$-system) leg structure which has a total degree of overconstraint of $\Delta = 1$. The future studies include optimization of the dimensions of the PM for a desired dexterous workspace, dynamic balancing and constructional design.

Acknowledgments. This study is funded by The Scientific and Technological Research Council of Turkey (grant number 115E726).

References

1. Huang L, Guang C, Yang Y, Su P (2017) Type synthesis of parallel 2R1T remote center of motion mechanisms based on screw theory. In: MATEC Web of Conference, 95. doi:10.1051/matecconf/20179508009
2. Kim KC (ed) (2014) Robotics in General Surgery. Springer
3. Kong X, Gosselin C (2007) Type Synthesis of Parallel Mechanisms. Springer
4. Kuo CH, Dai JS (2009) Robotics for minimally invasive surgery: a historical review from the perspective of kinematics. In: International symposium on history of machines and mechanisms, pp 337–354. doi:10.1007/978-1-4020-9485-9
5. Li Q, Hervé JM (2010) 1T2R parallel mechanisms without parasitic motion. IEEE Trans Robot 26(3):401–410. doi:10.1109/TRO.2010.2047528
6. Li J, Xing Y, Liang K, Wang S (2015) Kinematic design of a novel spatial remote center-of-motion mechanism for minimally invasive surgical robot. J Med Devices 9 (1):011003. doi:10.1115/1.4028651
7. Liu S, Chen B, Caro S, Briot S, Harewood L, Chen C (2016) A cable linkage with remote centre of motion. Mech Mach Theory 105:583–605. doi:10.1016/j.mechmachtheory.2016.07.023
8. Taniguchi K, Nishikawa A, Sekimoto M, Kobayashi T, Kazuhara K, Ichihara T, Kurashita N, Takiguchi S, Doki Y, Mori M, Miyazaki F (2010) Classification, design and evaluation of endoscope robots. In: Baik SH (ed) Robot surgery. InTech. doi:10.5772/6893
9. Teichgräber C, Müglitz J, Berger M (2018) Guiding linkages with remote centre of rotation for thermal cutting processes. In: Dede MIC et al. (eds.) Mechanisms, transmissions and applications, mechanisms and machine science, vol 52, pp 21–30. Springer. doi:10.1007/978-3-319-60702-3_4

3R1H Pseudo-Rigid-Body Model for Compliant Mechanisms with Inflection Beams

Yue-Qing Yu[✉] and Shun-Kun Zhu

Beijing University of Technology, Beijing, China
yqyu@bjut.edu.cn, takatol888@126.com

Abstract. A 3R1H Pseudo-Rigid-Body Model (PRBM) is proposed to simulate the flexible beam with an inflection point in compliant mechanisms. The characteristic radius factors of the PRBM are determined via optimization. The spring stiffness coefficients are obtained using a linear regression technique. The numerical simulation shows that both the tip locus and inflection point of the flexural beam with single inflection point can be simulated using the model proposed in this paper.

Keywords: Compliant mechanism · Pseudo-rigid-body model (PRBM) · Inflection point · Flexible beam

1 Introduction

Compliant mechanisms gain the mobility from the deflection of their flexible members. Such mechanisms offer many advantages such as increased precision, reduced backlash and parts number, ability to store energy [1].

The pseudo-rigid-body model (PRBM) method is a simple and accurate approach enabling one to use the well-developed knowledge of rigid-body mechanisms to study the compliant mechanisms [2, 3]. Howell and Midha [4], and Howell et al. [5] proposed the PRBM with two rigid segments connected by a revolute joint. It is named as the 1R PRBM in this paper as "R" represents the revolute joint and "1" indicates the number of joints. A torsional spring is placed at the joint to simulate the resistance to deflection of the compliant beam. However, the accuracy is guaranteed only in a limited range of the slope angle. Su proposed a 3R PRBM [6] whose parameters are independent to the load. The accuracy of 3R PRBM is relatively high when there is no inflection point in the flexible beam. In Ref. [7], a 2R PRBM was proposed to improve the simulating accuracy of the 1R PRBM and simplify the iterative process of the 3R PRBM. In Ref. [8], a PR PRBM was proposed, where P indicates the prismatic pair and R represents the revolute joint for the PR. Because the axial deflection of the flexible beam can be simulated by the PR PRBM, the simulating precision is higher than the 2R PRBM. Furthermore, a PRR PRBM was proposed in Ref. [9], where P also indicates the prismatic pair and R represents the revolute joint too for the PRR. This model has 3 degree of freedom as same as the 3R PRBM and it can also simulate the axial deflection of the flexible beam. Therefore, the PRR PRBM may be the most accurate model in these PRBMs.

It can be seen that the PRBM method has been widely used in the analysis and designing of compliant mechanisms. However, there are limited PRBMs that can be applied to analyze flexible beam with inflection points. Kimball and Tsai [10] used PRBM method to model the large deflection of a cantilever beam subjected to arbitrary end load. However, the model is oversimplified and it cannot ensure the accuracy. Midha et al. [11] develop a model to analyze fixed-guided compliant beam with an inflection point based on the PRBM concepts. The results obtained from this method were compared with those from FEM and elliptic integral solutions. A further study on the PRBM for the compliant mechanisms with inflection beams should be continued.

In this paper, a new type of 3R1H PRBM is proposed to predict both the locations of end tip and the inflection point of flexural beams in compliant mechanisms. The characteristic factors of the 3R1H PRBM are determined using optimization and linear regression techniques. The numerical simulation shows the effectiveness and superiority of the 3R1H PRBM.

2 3R1H Pseudo-Rigid-Body Model

A large deflection cantilever flexural beam with an inflection point is shown in Fig. 1. θ_0 is the deflection angle of the beam, a and b are the horizontal and vertical coordinates of the tip point of the beam, respectively. The flexible beam is subjected to a tip load combined with end force F_0 and clockwise torque M_0. The end force F_0 can be divided into a vertical force P and a horizontal force nP. i.e.

$$F_0 = \begin{bmatrix} nP \\ P \end{bmatrix}, \; n = \frac{F_{0x}}{F_{0y}}$$

Where, n is the force ratio. Φ is the angle of the force load F_0. A flexible beam in such load condition may have an inflection point.

In order to simulate the flexural beam with an inflection point, a 3R1H PRBM is proposed here, as shown in Fig. 2. It consists of five rigid links jointed by three revolute (R) pairs with torsional springs and one hinge (H) without spring. It is well known that the moment at the inflection point in the flexural beam is zero, so a free hinge without spring at point Q_i in the PRBM can be used to represent the inflection point of the flexural beam.

The whole 3R1H PRBM can be divided into two parts from the hinge Q_i. The segment I is simulated by a 2R PRBM [7] and the segment II is simulated a 1R PRBM [4], as shown in Fig. 3(a) and (b).

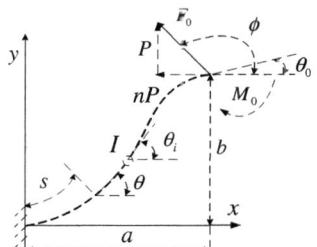

Fig. 1. Flexural beam with an inflection point

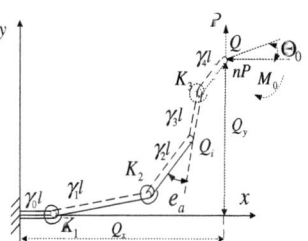

Fig. 2. 3R1H PRBM

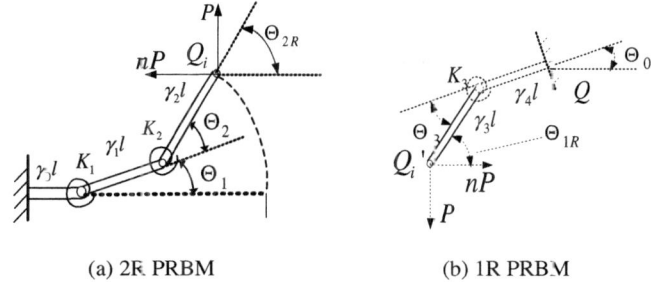

(a) 2R PRBM   (b) 1R PRBM

Fig. 3. Two parts of the 3R1H PRBM

3 Characteristic Parameters

The characteristic parameters [1] of the 3R1H PRBM include the characteristic radius factors γ_i ($i = 0, 1, 2, 3, 4$) representing the link length factor of each links of the model and stiffness coefficients k_i ($i = 1, 2, 3, 4$) representing the ability to resist deflection. Finding the optimal characteristic radius factors is the key to build the PRBM.

The kinematic analysis is necessary before the optimal radius factors γ_i ($i = 0, 1, 2, 3, 4$) of the 3R1H PRBM can found.

The location of hinge Q_i in the 2R PRBM in Fig. 3(a) can be written as

$$Q_{ix} = [\gamma_0 + \gamma_1 \cos\Theta_1 + \gamma_2 \cos(\Theta_1 + \Theta_2)]l \tag{1}$$

$$Q_{iy} = [\gamma_1 \sin\Theta_1 + \gamma_2 \sin(\Theta_1 + \Theta_2)]l \tag{2}$$

In the 1R PRBM in Fig. 3(b), the point Q can be seen as a fixed end because it is subjected to a force and a torque load. The location of Q can be written as

$$Q_x = Q_{ix} + [\gamma_4 \cos\Theta_0 + \gamma_3 \cos(\Theta_0 + \Theta_3)]l \tag{3}$$

$$Q_y = Q_{iy} + [\gamma_4 \sin\Theta_0 + \gamma_3 \sin(\Theta_0 + \Theta_3)]l \tag{4}$$

The joint displacement of each link for the 3R1H PRBM can be determined through inverse kinematics analysis

$$x = Q_{ix} - \gamma_0 l \tag{5}$$

$$y = Q_{iy} \tag{6}$$

$$\Theta_2 = \cos^{-1}\left(\frac{x^2 + y^2 - \gamma_1^2 l^2 - \gamma_2^2 l^2}{2\gamma_1 \gamma_2 l^2}\right) \tag{7}$$

$$\Theta_1 = \tan^{-1}\left(\frac{(\gamma_1 l + \gamma_2 l \cos \Theta_2)y - \gamma_2 l \sin \Theta_2 x}{(\gamma_1 l + \gamma_2 l \cos \Theta_2)x + \gamma_2 l \sin \Theta_2 y}\right) \tag{8}$$

$$\Theta_3 = \tan^{-1}\left(\frac{Q_{iy} - Q_y + \gamma_0 l \sin \Theta_0}{Q_{ix} - Q_x + \gamma_0 l \cos \Theta_0}\right) - \Theta_0 \tag{9}$$

The slope of right end of the 2R PRBM shown as Θ_{2R} in Fig. 3(a) can be written as

$$\Theta_{2R} = \Theta_1 + \Theta_2 \tag{10}$$

For the slope of left end of 1R PRBM shown as Θ_{1R} in Fig. 3(b), we have

$$\Theta_{1R} = \Theta_3 + \Theta_0 \tag{11}$$

After the kinematics analysis, the characteristic radius factors of the 3R1H PRBM can be obtained through optimization. Assume that the flexible beam with an inflection point deflects to a certain position. If the hinge and free end of a 3R1H PRBM are forced to be coincided with the inflection point and tip of the flexible beam, the configuration of the model can be determined. It is known that the curvature of inflection point of a flexible beam is zero. In the 3R1H PRBM, the hinge is used to represent the inflection point of a flexible beam. In order to reduce the difference between the configuration of the flexible beam and that of the PRBM, the angle of two links jointed by the hinge without spring, e_a, as shown in Fig. 2, should be as small as possible. However, for different value of γ_i ($i = 0, 1, 2, 3, 4$), the value of e_a may differ. The objective is to find a set of characteristic radius factors that minimize the value of e_a. Therefore, the objective function can be established as the average absolute value of e_a.

To determine the objective function in this study, the force ratio n and the load ratio index $\kappa = \frac{M_0}{EI}$ [1] is set firstly. Secondly, the maximum value of end slope of the flexible beam, θ_{0max}, is determined as

$$\theta_{0max} = \phi - \cos^{-1}(1 - \kappa) \tag{12}$$

where

$$\phi = \frac{\pi}{2} + \tan^{-1}(n) \tag{13}$$

and the initial value of θ_o in each step is determined as

$$\theta_{0j} = \theta_{0j-1} + j\Delta\theta_c \tag{14}$$

where $\Delta\theta_0 = \theta_{Cmax}/m$, $\theta_{01} = 0$. Where, m is the step number. Thirdly, the coordinates of tip Q (Q_x, Q_y) and inflection point Q_i (Q_{ix}, Q_{iy}) are then obtained by elliptic integral solution which is commonly used to solve large-deflection problems and details can be seen in Ref. [12]. Set $(Q_x, Q_y) = (a/l, b/l)$, $(Q_{ix}, Q_{iy}) = (a_i/l, b_i/l)$, $\Theta_0 = \theta_0$. With a set of value of characteristic radius factors, the pseudo-rigid-body angles Θ_1, Θ_2, Θ_3, Θ_{1R}, Θ_{2R} can be determined through kinematic analysis using Eqs. (5)–(11). If one of the angles Θ_1, Θ_2, Θ_3 is not real, the values of characteristic radius factors are not feasible. To avoid such infeasible values of characteristic radius factors, the objective function is added with a relatively big value, M, as penalty factor. We have

$$\begin{cases} e_a = |\Theta_{2R} - \Theta_{1R}| \\ e_a = M \end{cases} \tag{15}$$

Then, the objective function E_a can be determined as

$$E_a = \sum_{j=1}^{m} e_{aj} \tag{16}$$

The optimization for the characteristic radius factors of the 3R1H PRBM can be described as

Minimize Fitness $= E_a$

Subject to $\sum_{i=0}^{4} \gamma_i = 1$ $\gamma_i > 0$ $(i = 0, 1, 2, 3, 4)$

In this paper, a genetic algorithm is used to obtain the optimization solution and the exact configuration of flexible beams is determined via the elliptic integral solution [12]. The process to determine E_a is shown in the flow flowchart in Fig. 4.

The optimal characteristic radius factors of the 3R1H PRBM in two load cases of different value of force ration index $n = 1$ and $n = 0$ are obtained in this study, respectively. In the case of $n = 1$, the optimal characteristic radius factors of the 3R1H PRBM are

$$\gamma_0 = 0.070, \ \gamma_1 = 0.342, \ \gamma_2 = 0.095, \ \gamma_3 = 0.313, \ \gamma_4 = 0.180$$

For $n = 0$, the optimal characteristic radius factors are

$$\gamma_0 = 0.047, \ \gamma_1 = 0.294, \ \gamma_2 = 0.235, \ \gamma_3 = 0.317, \ \gamma_4 = 0.107$$

The torsional spring stiffness coefficients describe the ability of flexible beam to resist deflection. After the characteristic radius factors are obtained, the join torques of spring in the 3R1H PRBM can be obtained through statics analysis as follows.

For the 3R1H PRBM, define k_1, k_2, k_3 as non-dimensional coefficients of stiffness constants of torsion spring K_1, K_2, K_3 satisfying

$$\begin{pmatrix} K_1 \\ K_2 \\ K_3 \end{pmatrix} = \frac{EI}{l} \begin{pmatrix} k_1 \\ k_2 \\ k_3 \end{pmatrix} \qquad (17)$$

The static equations can be presented as

$$k_1 \Theta_1 = \alpha^2 (n(\gamma_2 \sin(\Theta_1 + \Theta_2) + \gamma_1 \sin \Theta_1) + \gamma_2 \cos(\Theta_1 + \Theta_2) + \gamma_1 \cos \Theta_1) \qquad (18)$$

$$k_2 \Theta_2 = \alpha^2 \gamma_2 (n \sin(\Theta_1 + \Theta_2) + \cos(\Theta_1 + \Theta_2)) \qquad (19)$$

$$k_3 \Theta_3 = \alpha^2 (\gamma_3 \cos(\Theta_3 + \Theta_0) + n\gamma_3 \sin(\Theta_3 + \Theta_0)) \qquad (20)$$

Where, $\alpha^2 = Pl^2/EI$ is called nondimensionalized load index [1].

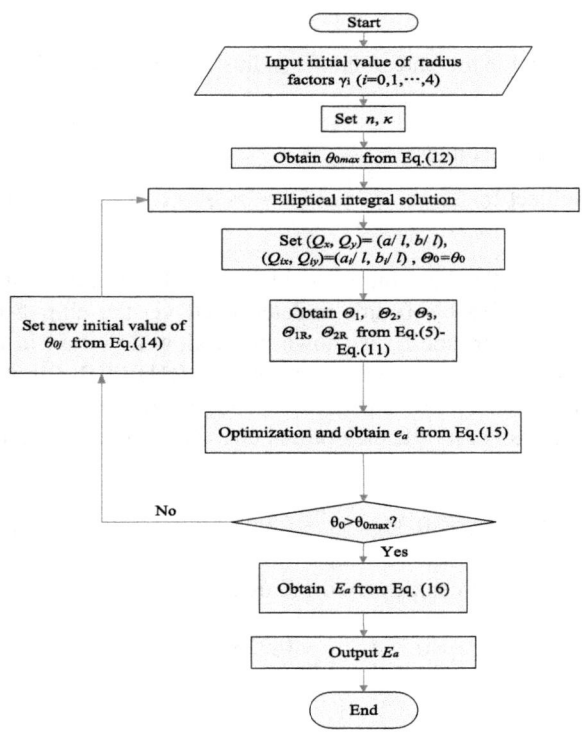

Fig. 4. Flow chart of determining the objective function E_a

The stiffness coefficients of the 3R1H PRBM can be obtained using the linear regression method [13]. For $n = 1$, the stiffness coefficients are obtained as

$$k_1 = 4.973, \ k_2 = 2.745, \ k_3 = 4.183$$

For $n = 0$, the stiffness coefficients are

$$k_1 = 6.106, \ k_2 = 3.229, \ k_3 = 3.835$$

4 Numerical Examples

Numerical examples are presented in this section to show the effectiveness and superiority of the 3R1H PRBM.

In the numerical examples, the force ratio n are presented as two case as $n = 0$ and $n = 1$. The load ratio index κ are selected as $\kappa = 0.2$ and $\kappa = 0.4$ for $n = 0$, and $\kappa = 0.3$ and $\kappa = 0.9$ for $n = 1$. The numerical results are presented Table 1. The configurations of the 3R1H PRBM and the flexural beams in the case of $\kappa = 0.3$ and $\kappa = 0.9$ for $n = 1$ are shown in Fig. 5.

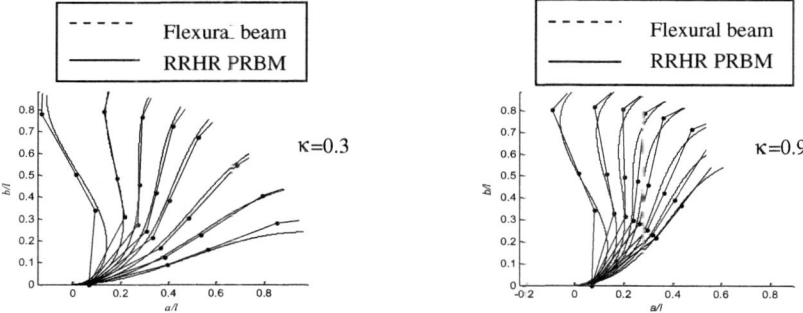

Fig. 5. Configuration of flexural beam and two PRBMs for various load indices κ

In order to present the simulation accuracy of the 3R1H PRBM, the average error e_{ia} and the maximum error e_{im} of inflection point and that of tip end e_{ta} and e_{tm} are taken into account. These errors of the 3R1H PRBM are shown in Table 1. The average errors and the maximum error of 3R PRBM proposed by Su [6] are shown in Table 2 for comparison. Because the errors of locus of inflection point were not discussed in Ref. [6], only errors of tip locus are presented here.

It can be seen that the errors of the 3R1H PRBM are small in simulating the tip end and inflection locus of flexible beams. The average errors of tip locus of the 3R1H PRBM are less than the 3R PRBM in all examples. The maximum errors of tip locus of the 3R1H PRBM are slightly bigger than the 3R PRBM only in cases of $n = 1$, $\kappa = 0.3$. In general cases, the configuration of the 3R1H PRBM are closed to those of the

Table 1. Errors of the 3R1H PRBMs

n	k	e_{ia}	e_{im}	e_{ta}	e_{tm}
1	0.3	5.24%	9.33%	1.76%	5.59%
1	0.9	5.85%	9.41%	2.36%	7.73%
0	0.2	3.83%	7.61%	0.73%	1.29%
0	0.4	2.97%	7.64%	0.63%	1.99%

Table 2. Errors of 3R PRBM

n	κ	e_{ta}	e_{tm}
1	0.3	3.82%	5.08%
1	0.9	9.95%	8.52%
0	0.2	1.65%	2.50%
0	0.4	1.88%	2.97%

flexible beam. Therefore, it can be concluded that the 3R1H PRBM is accurate enough in simulating the flexible beam with an inflection point.

5 Conclusions

A new 3R1H pseudo-rigid-body model has been proposed in this study to simulate the large deflection of flexural beam with an inflection point. The characteristic parameters of the model have been determined via optimization and linear regression technique. The numerical example shows the validation and superiority of the new PRBM proposed in this study in predicting both the tip locus and inflection point of large deflection beams with an inflection point. This model can be used for different force ratio index n and load indices κ of inflection beams in compliant mechanisms.

Acknowledgments. The financial support of this study was from the National Natural Science Foundation of China (Grant No. 51575006).

References

1. Howell LL (2001) Compliant mechanisms. Wiley, New York
2. Saxena A, Ananthasuresh GK (2001) Topology synthesis of compliant mechanisms for nonlinear force-deflection and curved path specifications. J Mech Des 123(1):33–42
3. Holst GL, Teichert GH, Jensen BD (2011) Modeling and experiments of buckling modes and deflection of fixed-guided beams in compliant mechanisms. J Mech Des 133(5):623–635
4. Howell LL, Midha A (1995) Parametric deflection approximations for end-loaded, large-deflection beams in compliant mechanisms. J Mech Des 117(1):156–165
5. Howell LL, Midha A, Norton TW (1995) Evaluation of equivalent spring stiffness for use in a pseudo-rigid-body model of large-deflection compliant mechanisms. J Mech Des 118(1):126–131

6. Su HJ (2009) A pseudo rigid-body 3R model for determining large deflection of cantilever beams subject to tip loads. J Mech Rob 1(2):795–810
7. Yu YQ, Feng ZL, Xu QP (2012) A pseudo-rigid-body 2R model of flexural beam in compliant mechanisms. Mech Mach Theory 55(9):18–33
8. Yu YQ, Zhou P, Xu QP (2015) A new pseudo-rigid-body model of compliant mechanisms considering axial deflection of flexural beams. Mech Mach Sci 24:851–858
9. Yu YQ, Zhu SK, Xu QP, Zhou P (2016) A novel model of large deflection beams with combined end loads in compliant mechanisms. Precis Eng 43:395–405
10. Kimball C, Tsai LW (2002) Modeling of flexural beams subjected to arbitrary end loads. J Mech Des 124(2):223–235
11. Midha A, Bapat SG, Mavanthoor A, Chinta V (2012) Analysis of a fixed-guided compliant beam with an inflection point using the pseudo-rigid-body model (PRBM) concept. J Mech Rob 64(4):147–154
12. Hancock H (1958) Elliptic integrals. Dover, New York
13. Weisberg S (2005) Applied linear regression, 3rd edn. Wiley, New York

A Spatial Four-Bar Linkage RSPS for Ball-Bar to Test R-pair

Delun Wang[1(✉)], Zhi Wang[1], Xiaopeng Li[1], Huimin Dong[1], and Shudong Yu[1,2]

[1] School of Mechanical Engineering, Dalian University of Technology,
Dalian 1106024, People's Republic of China
dlunwang@dlut.edu.cn
[2] Department of Mechanical and Industrial Engineering,
Ryerson University, Toronto, ON, Canada
syu@ryerson.ca

Abstract. A novel mechanism model of RSPS is firstly presented in the paper for testing the rotational ac-curacy of an actual R-pair, which is new in the literature. The spatial four-bar linkage RSPS is comprised of a ball-bar with three kinematic pairs S-P-S and an actual R-pair measured. The outputs of RSPS corresponds to the distance between two master balls' centers, and the assembly errors of the ball-bar instrument are concisely contained in the outputs of RSPS with different parameters. Hence, the rotational errors of an actual R-pair measured are readily identified by the measured data of the ball-bar instrument minus the theoretical outputs of RSPS.

Keywords: Spatial four-bar linkage · Kinematics · Ball bar · Testing · Error

1 Introduction

The rotational accuracy is one of key indexes in precision revolute pairs or R-pairs, such as spindles and rotary tables of machining tools. For testing and evaluating the rotational accuracy of R-pairs, the radial and axial run-outs of the rotary components [1, 2] are widely used as indexes, which cannot distinguish the error motions of the rotary from measured data since the geometrical errors of the rotary and the assembly errors of measuring instruments. Thus, some precise artifacts, such as the master ball and the cylinder, are used as the work-pieces to reduce the influences of geometrical errors [1–3], and the assembly errors of measuring instruments need to be eliminated and the displacements sensors properly have to be installed with a lot of time, which is not widely used in practice.

Recently, a ball-bar instrument, comprised of the double ball and one bar (ball-bar for short), shown in Fig. 1, is introduced to measure the rotational accuracy of an R-pair [4–6], since it is convenient to be installed and independent of the geometrical error of the work-pieces measured, which is believed to be a suitable instruments for testing the rotational accuracy of an actual R-pair.

For a ball-bar instrument, the length $s^{(i)}$ between two balls' centers S_A and S_B is the output of the ball-bar instrument, or the measured data $s^{(i)}$. The deviation Δ_S between

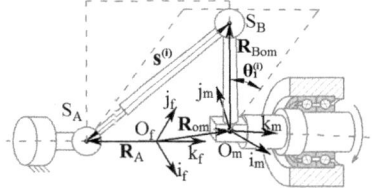

(a) The photo of ball-bar instrument (b) The schematic diagram

Fig. 1. The ball-bar instrument to test rotational accuracy of an actual R-pair

the output data $s^{(i)}$ and a constant is believed to eliminate the influences of assembly errors of the ball-bar from the measured results [7, 8] in the current way, but it is not true. It is impossible to precisely locate the exact locations of both the ball center S_B on the moving rotator and the ball center S_A on the fixed frame respectively when the ball-bar is installed. Therefore, what is the relationship between the measured data and the theoretical output of the ball-bar in kinematics and geometry? How to eliminate the influences of the assemble error from the measured data of an actual R-pair? All of these questions are not answered so far. A novel kinematic model of a spatial four-bar linkage RSPS is presented for testing the rotational accuracy of an actual R-pair and discussed in both kinematics and geometry, which are the theoretical base for a ball-bar instrument to test the rotational accuracy of an actual R-pair.

2 The Spatial Four-Bar Linkage RSPS

Based on Fig. 1, both an R-pair measured and a ball-bar instrument with two spherical pairs and a prismatic pair constitute a spatial four-bar linkage RSPS shown in Fig. 2, during testing the rotational accuracy of an R-pair, in which the rotational angular $\theta_1^{(i)}$ of the rotator (the link 1) will be the input displacement of RSPS and the length $s^{(i)}$ between two balls' centers S_A and S_B is the output displacement of RSPS. The spatial four-bar linkage RSPS has the number of link $n = 4$, the number of kinematic pair $g = 4$, the number of passive DOF $p = 1$ rotated around the line of two balls' centers, and the level of the kinematic pairs f_i double 3 and double 1 respectively, whose DOF [9] can be calculated by

$$F = 6(n - g - 1) + \sum_{i=1}^{g} f_i - p \qquad (1)$$

Hence, the spatial four-bar linkage RSPS has a single degree of freedom.

To describe the motions of the spatial four-bar linkage, a moving frame $\{O_1; \mathbf{i}_1, \mathbf{j}_1, \mathbf{k}_1\}$ attached to link 1 and a fixed frame $\{O_0; \mathbf{i}_0, \mathbf{j}_0, \mathbf{k}_0\}$ attached to link 0 are employed respectively, shown in Fig. 2. The vector displacement equation of RSPS in three dimensions can be written as

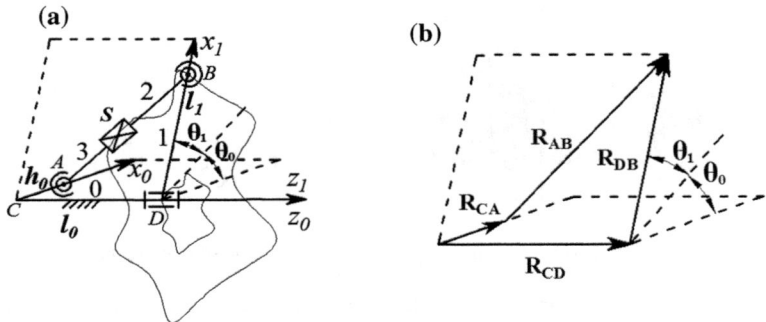

Fig. 2. The parameters of the spatial four-bar linkage *RSPS*

$$\mathbf{R}_{CD} + \mathbf{R}_{DB} = \mathbf{R}_{CA} + \mathbf{R}_{AB} \tag{2}$$

where, $\mathbf{R}_{DB}$ denotes the position of the spherical pair S_B in the moving frame and $\mathbf{R}_{CA}$ does the position of the spherical pair S_A in the fixed frame, $\mathbf{R}_{CD}$ does the position of the moving frame relative to the fixed frame, while $\mathbf{R}_{AB}$ presents the vector from center S_A to S_B. $\mathbf{R}_{DB}$, $\mathbf{R}_{CA}$ and $\mathbf{R}_{CD}$ meet the equation: $\mathbf{R}_{DB} \bullet \mathbf{R}_{CD} = \mathbf{R}_{CD} \bullet \mathbf{R}_{CA} = 0$.

The vector Eq. (2) can be directly solved as a scalar equation by means of the distance between two balls' centers, or a function between the input angle θ_1 around the axis $\mathbf{k}_0$ and the output displacement s, that is,

$$s = \sqrt{l_1^2 + l_0^2 + h_0^2 - 2l_1 h_0 \cos(\theta_1 + \theta_0)} \tag{3}$$

where, l_0, l_1 and h_0 correspond to the values of three vectors $\mathbf{R}_{CD}$, $\mathbf{R}_{DB}$ and $\mathbf{R}_{CA}$ respectively, θ_0 denotes the initial rotary angle between two axes $\mathbf{i}_0$ and $\mathbf{i}_1$, which are the parameters of a spatial four-bar linkage *RSPS*, shown in Fig. 2.

The Eq. (3) set up the relationships among the input angle displacement θ_1, the output displacement s and the parameters (l_0, l_1, h_0 and θ_0) of *RSPS*. In such a situation, the rotator of an R-pair would ideally rotate around the fixed axis $\mathbf{k}_0$ of an R-pair, the output displacement s of *RSPS* can be calculated by the Eq. (3). The line $\mathbf{R}_{AB}$, link 2 and 3, of *RSPS* with parameters (l_0, l_1, h_0 and θ_0) traces a line-trajectory Σ_{AB} in the fixed frame, which is a conical ruled surface but a right cone, while the trajectory of the ball center point S_B is always a circle of the ruled surface Σ_{AB}.

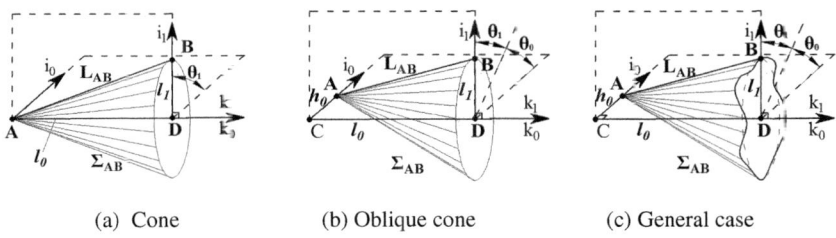

(a) Cone (b) Oblique cone (c) General case

Fig. 3. Three cases of the ruled surface Σ_{AB}

3 The Kinematic Model

Based on the Fig. 2, the output displacement s of *RSPS* can be theoretically calculated by the Eq. (3) in following cases with the parameters (l_0, l_1, h_0 and θ_0), corresponding to the ball-bar instrument assembling cases.

Case 1, two balls of a ball-bar instrument are assembled without any errors, the center point S_A of the fixed ball absolutely locates at the axis of R-pair, or $h_0 = 0$ in Eq. (3) and $s = \sqrt{l_1^2 + l_2^2}$ = constant, and the ruled surface Σ_{AB} degenerates to be a cone, shown in Fig. 3(a), which will not be the case in practice, but it is usually believed to happen in current measuring way.

Case 2, the positions of both the fixed ball's center point S_A and the moving ball' center S_B of a ball-bar instrument are roughly assembled with unknown positions, or the ball-bar instrument has the assemble errors. In the case, the fixed ball's center point S_A locates out of the axis $\mathbf{k}_0$ of an R-pair ($h_0 \neq 0$) and the parameters(l_0, l_1, h_0 and θ_0) need to be determined precisely. Generally, the output displacement s is a theoretical function of the input angle θ_1, other than a constant, described by the Eq. (3), while the ruled surface Σ_{AB} is a conical ruled surface but a right cone, Fig. 3(b).

Case 3, the rotator of an R-pair has an error motion relative to the fixed frame, other than ideally rotates around the fixed axis $\mathbf{k}_0$, the measured data $s^{(i)}$ contains the output values s of *RSPS* theoretically calculated by Eq. (3) and the rotational error of the actual R-pair. In other words, the rotational errors of an actual R-pair can be described by the measured data $s^{(i)}$ deviating from the theoretical output value s of *RSPS* with assemble errors of a ball-bar instrument. In this case, $\mathbf{k}_0$ is regarded as an approximate rotary axis of the actual R-pair and the position of the reference point D follows the rotation of the actual R-pair. As a consequence, the distance between D to B and D to C both vary during a period.

The main goal is to set up a kinematic model to identify the theoretical output s of *RSPS* (theoretical predictions) from the measured result of a ball-bar instrument with assemble errors. Usually, the value of the rotational errors of the rotator of an actual R-pair is greatly less than the output value s of *RSPS* in magnitude, while the latter can always be calculated in terms of the Eq. (3). Thus, the parameters (l_0, l_1, h_0 and θ_0) of *RSPS* are firstly identified from the measured data by the minimax optimization, and the deviation between the measured data and the output value s of *RSPS* is the rotational errors of the rotator of an actual R-pair, whose kinematic model is established as

$$\begin{cases} \Delta_s = \min_{\mathbf{x}} \max_{1 \le i \le n} \left\{ \Delta^{(i)}(\mathbf{x}) \right\} = \min_{\mathbf{x}} \max_{1 \le i \le n} \left\{ \left| s^{(i)*} - s^{(i)} \right| \right\} \\ \text{s.t.} l_1, l_0, h_0 \in (-\infty, +\infty); \theta_0 \in (0, 2\pi) \\ \mathbf{x} = (l_1, l_0, h_0, \theta_0)^T \end{cases} \quad (4)$$

where, $s^{(i)*}$ denotes the measured data of the ball-bar instrument and $s^{(i)}$ does the theoretical output value of *RSPS* calculated by the Eq. (3); Δ_S represents the deviations between the measured data and the output value of *RSPS*; *s.t* are the boundaries of optimization variables; the optimization variables $\mathbf{x} = (l_1, l_0, h_0, \theta_0)^T$ are the design variable vector, while (i) is designated as the number of the discrete position of the rotator and the value $s^{(i)*} - s^{(i)}$ is the rotational error of the rotator along the line $\mathbf{R}_{AB}$ of the ball-bar instrument.

4 Case Study and Experiments

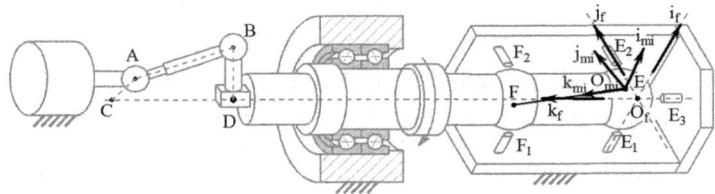

Fig. 4. The actual R-pair and two measuring instruments

In order to illustrate the proposed kinematical model, an actual R-pair, a revolute joint comprised of two bearings, are taken as an example to test and analyze its rotational accuracy by using two measuring instruments, shown in Fig. 4.

The error motion of the rotator of an actual R-pair can be measured by using the double master-ball instrument in six DOFs, fixed to the right end of the rotator, recommended in the ASME standard [1], which means that the error motion of the rotator is known, while the rotational accuracy is tested by a ball-bar instrument, installed to another end of the rotator. In order to describe the error motion of the rotator, a moving frame $\{O_m; \mathbf{i}_m, \mathbf{j}_m, \mathbf{k}_m\}$ is employed and attached to the double master-ball, with the origin point O_m coincident at the sphere center E and the axis $\mathbf{k}_m$ with the center-line EF, respectively. A fixed frame $\{O_f; \mathbf{i}_f, \mathbf{j}_f, \mathbf{k}_f\}$ is established at the holder of the displacement sensors, and the axes $\mathbf{i}_f, \mathbf{j}_f$ and $\mathbf{k}_f$ are coincident with the axes of five sensors E_1, E_2, E_3 and F_1, F_2 shown in Fig. 4.

The measured data of the double master-ball (roundness less than 50 nm) are the distances between the two master balls and the ends of five displacement sensors (error less than 0.25 μm), shown in Fig. 5a. Then, the error motions of the rotator can be calculated [10], and described by three translations $\mathbf{r}_{om} = [x_{om}, y_{om}, z_{om}]^T$ and three

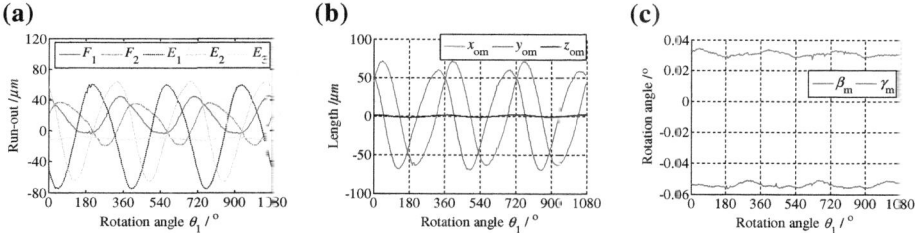

Fig. 5. The measured data by the double master-ball

Eular angles (α_m, β_m, γ_m) of the moving frame $\{O_m; \mathbf{i}_m, \mathbf{j}_m, \mathbf{k}_m\}$ relative to the fixed frame, shown in Fig. 5b and c respectively.

For revealing the kinematic model, a point B (-63198.1, 0, 310757.3) (measured by tool) of the rotator is designated as the moving ball center in the moving frame $\{O_m; \mathbf{i}_m, \mathbf{j}_m, \mathbf{k}_m\}$ and traces a point-trajectory, a spatial curve Γ_B in fixed frame $\{O_f; \mathbf{i}_f, \mathbf{j}_f, \mathbf{k}_f\}$.

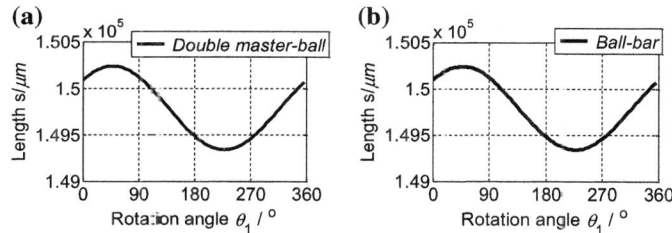

Fig. 6. The measured data and the calculated data

And then, a fixed point A(1028.2, -636.5, 446555.9) (calculated by least square method) is appointed to be the fixed ball center, the distance between them can be calculated and shown in blue curve in Fig. 6(a), which is equivalent to the measured

Table 1. The parameters of *RSPR* corresponding to the ball-bar testing

$l_1/\mu m$	$l_0/\mu m$	$h_0/\mu m$	θ_0/rad	$\Delta/\mu m$
63198.6	135798.9	1063.8	2.3114	6.7

data by a ball-bar instrument. On other hand, the error motion of the rotator of an R-pair is measured by a ball-bar instrument and by the double master-ball simultaneously, and the measured data is expressed in black curve in Fig. 6(b).

Based on the measured data of a ball-bar instrument, a spatial four-bar linkage *RSPS* with parameters (l_0, l_1, h_0 and θ_0) can be optimally calculated by the kinematic model, the Eq. (4), shown in Table 1.

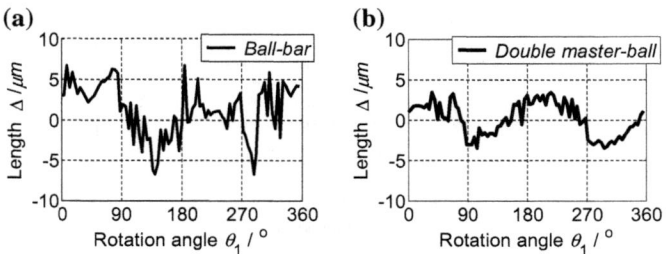

Fig. 7. The deviations corresponding to ball-bar and the double master-ball

By the kinematic model in Eq. (4), the measured data in Fig. 6 is minus the output values of RSPS with the parameters in Table 1, the rotational errors of the rotator of the actual R-pair measured can be described by the deviations, shown in Fig. 7(a).

On other hand, the distances between the moving point B and the fixed point A calculated by means of the measured data of double master ball, shown in Fig. 6(b), are treated in same way as that of a ball-bar, which can describe the rotational errors of the rotator of the actual R-pair, or the deviations shown in Fig. 7(b).

As mentioned above, by comparing the deviations between the outputs of RPSP corresponding to the measured data of the ball-bar and those to the calculated data of the double master-ball, we may find these deviations have much more kinematic geometry meanings. Based on the kinematic invariants of an rotational error motion of a rotator [10], both the parameters of RSPS and the deviations have not been completely revealed yet in kinematic geometry, which needs to be intensively studied further.

5 Conclusions

(1) The kinematic model, a spatial four-bar linkage RSPS with parameters, is presented for a ball-bar instrument to test the rotational accuracy of an actual R-pair, which can easily eliminate the assemble error of ball-bar instrument.
(2) There are three cases for a ball-bar to test the rotational accuracy of an actual R-pair which could not be controlled when test. Most researchers only consider one of them in general.

Acknowledgments. The authors would gratefully acknowledge the support received from the national natural science foundation of China (Grant No. 51275067).

References

1. ASME B89.3.4-2010 (2010) Axes of rotation: methods for specifying and testing. American National Standards Institute
2. ISO 230-7:2006 (2006) Test code for machine tools-part 7: geometric accuracy of axes of rotation. ISO
3. Grejda R, Marsh E, Vallance R (2005) Techniques for calibrating spindles with nanometer error motion. Precis Eng 29(1):113–123
4. Huang N, Bi Q, Wang Y (2015) Identification of two different geometric error definitions for the rotary axis of the 5-axis machine tools. Int J Mach Tools Manuf 91:109–114
5. Jiang X, Cripps RJ (2015) A method of testing position independent geometric errors in rotary axes of a five-axis machine tool using a double ball bar. Int J Mach Tools Manuf 89:151–158
6. Tsutsumi M, Tone S, Kato N et al (2013) Enhancement of geometric accuracy of five-axis machining centers based on identification and compensation of geometric deviations. Int J Mach Tools Manuf 68:11–20
7. Zargarbashi SHH, Mayer JRR (2005) A model based method for centering double ball bar test results preventing fictitious ovalization effects. Int J Mach Tools Manuf 45(10): 1132–1139
8. Lu H, Yang J, Xiang S (2015) Effects of double ball bar setup errors on accuracy of circular tests of machine tools. J Shanghai Jiao Tong Univ
9. Hunt KH (1978) Kinematic geometry of mechanism. The Oxford engineering science series. The Clarendon Press, Oxford University Press, New York
10. Wang DL, Wang Z, Wu Y, et al (2016) Discrete kinematic geometry in testing axes of rotation of spindles. In: ASME 2016 International Design Engineering Technical Conferences and Computers and Information in Engineering Conference

Leg Mechanisms Motion Characteristics

Adriana Comanescu^(✉), Elisabeta Banica, and Dinu Comanescu

University Politehnica of Bucharest, Bucharest, Romania
adrianacomanescu@yahoo.com,
banicaelisabeta29@yahoo.com, dinucomanescu@yahoo.com,
adrianacomanescu@gmail.com

Abstract. The paper deals with the method that achieves the structural synthesis of mono-mobile leg mechanisms. All structural solutions derived from the Watt and Stephenson linkages with two loops are presented. The method may be also applied in the same purpose to the forty linkages with three independent loops and four degrees of freedom. There are also presented some synthesis aspects for bi-mobile and three-mobile systems.

Keywords: Structural synthesis · Leg mechanism · Coupler curve · Active pair · Active modular group

1 Introduction

The leg mechanisms for mobile robots or platforms may have various degrees of mobility [1–4, 6] in function of their applicability.

The simplest mono-mobile 4R mechanism (Fig. 1), which may be used for the leg of different mobile system describes with the T point belonging to the coupler – the opposite element to the basis an algebraic curve of the sixth degree [5].

Its implicit form is usually given by (1)

$$U^2 + V^2 = W^2 \qquad (1)$$

where

$$
\begin{aligned}
U &= b(x^2 + y^2 + a^2 - r^2)[-(x-k)\sin\gamma + (y-h)\cos\gamma] \\
&\quad - ay\Big[(x-k)^2 + (y-h)^2 + b^2 + R^2\Big] \\
V &= ax\Big[(x-k)^2 + (y-h)^2 + b^2 - R^2\Big] \\
&\quad - b(x^2 + y^2 + a^2 - r^2)[(x-k)\cos\gamma + (y-h)\sin\gamma] \\
W &= 2ab\big[(ky - hx)\cos\gamma - (x^2 + y^2 - kx - hy)\sin\gamma\big]
\end{aligned}
\qquad (2)
$$

By selecting some particular geometrical constant parameters (Fig. 1) the T point may have in a part of its curve an approximate straight line. In this situation the Chebyshev mechanism (Fig. 2) is used for leg mechanisms. The Chebyshev mechanism [3] is structurally obtained from the four bar linkage having a single loop (Fig. 3).

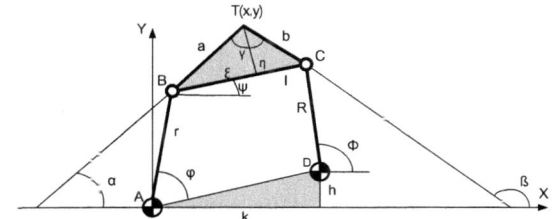

Fig. 1. The 4R mechanism

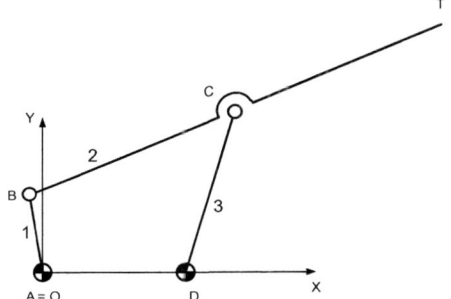

Fig. 2. Chebyshev mechanism

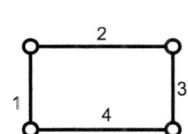

Fig. 3. The four bar linkage

In order to obtain new solutions one may find in the literature [3] two such mechanisms (Fig. 4) in the same construction.

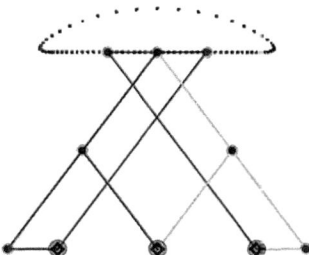

Fig. 4. Two Chebyshev mechanisms in the same solution

The mono-mobile leg mechanisms are structural solutions obtained from planar linkages with four degrees of freedom and different number of contours.

2 Mono-mobile and Multi-looped Leg Mechanisms

The Chebyshev's plantigrade machine (Fig. 5) also named the Lambada mechanism has in view the previously mentioned mechanism [1], which also has in its structure a parallelogram mechanism (1-2″-1′-2‴) characterized by the 2‴ passive link and its two passive adjacent kinematic pairs (Fig. 6). The T1, T4, T3 and T2 extreme points are placed on the 2, 2′, 4 and 4′ elements having a planar motion.

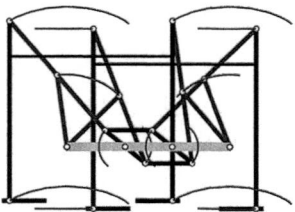

Fig. 5. Chebyshev's plantigrade machine

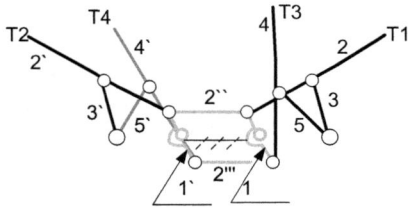

Fig. 6. The Lambada mechanism structural model

The results of its kinematic modeling put into evidence the trajectory (Fig. 7) of each T point placed on the couplers and justifies the Lambada mechanism construction.

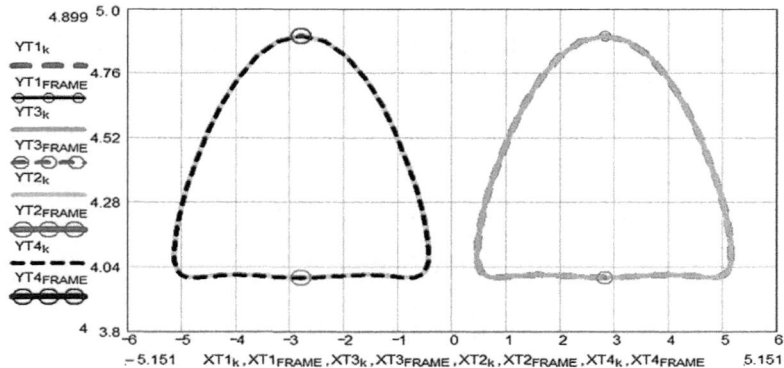

Fig. 7. The trajectory of each T point placed on the mechanism couplers

Other leg mechanisms [3] have their structure based on bi-loop linkages with four degrees of freedom. Some of their structural characteristics are presented in the next chapter.

Other mono-mobile leg mechanisms [3] are designed by means of three loops linkages, which have sixteen distinct solutions [1].

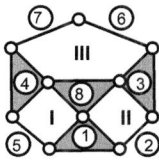

Fig. 8. Three-loop linkage selected from sixteen ones of the same type

The three loops linkage from Fig. 8 is used for the structural synthesis of the leg mechanisms presented in the Figs. 9 and 10.

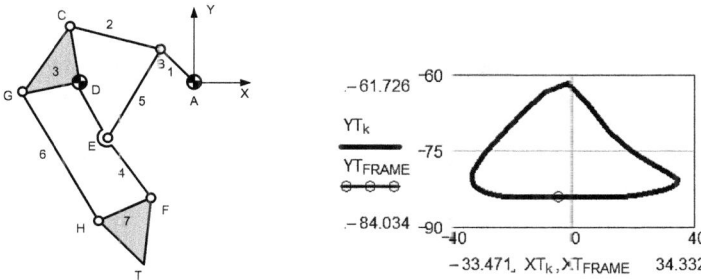

Fig. 9. The leg mechanism with its coupler curve T extreme point

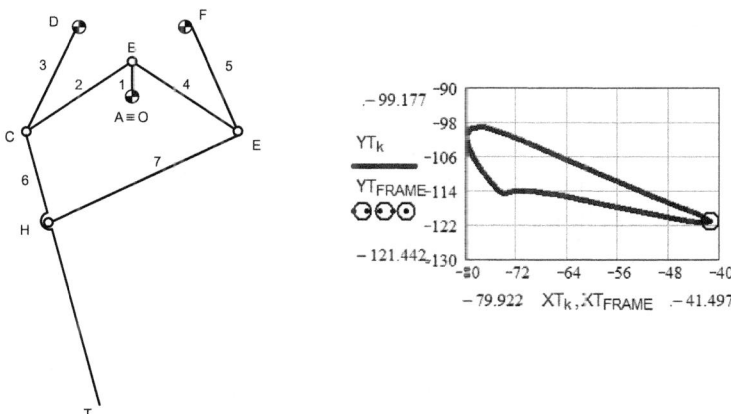

Fig. 10. Leg mechanism with 8 links and with its coupler curve T extreme point

Each mechanism (Figs. 9 and 10) is in fact a structural connection of a single initial modular active group placed at the basis and three passive modular groups of RRR type. The T point trajectory is obtained by their modeling and is given next to the mechanism.

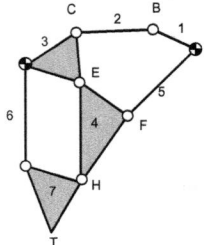

Fig. 11. Leg mechanism with 8 links

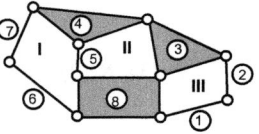

Fig. 12. Three-loop linkage of the leg mechanism

A new mono-mobile leg mechanism (Fig. 11) is based on other three loops linkage (Fig. 12). In the literature [3, 4] the multi-loop planar mono-mobile mechanism (Fig. 13) is also presented.

The mechanism has 15 mobile links (m) and 22 kinematic pairs (c), so that the number of the independent contours is N = c-m, N = 7, which may be identified in its mechanism structure. The S and V points trajectories are also shown in Fig. 13.

3   Bi-mobile Leg Mechanisms

It is also important to mention the planar leg mechanisms with two degrees of mobility [1–3]. Their main characteristic is the possibility to describe with its extreme point belonging to the output link any curve in the mechanism evolution domain [1, 2].

Generally their structural solutions may be obtained by using linkages with 5 degrees of freedom, respectively two degrees of mobility. Such linkages have also different number of independent contours.

In the literature [1] one may find a single mono-loop linkage with five links, four solutions with two loops and seven links and forty ones with three independent contours having nine elements.

The simplest bi-mobile mechanism [3] from Fig. 14 is obtained form the five links linkage (Fig. 15).

The bi-mobile leg mechanism (Fig. 16) selected from literature [3] has in view the bi-loop linkage (Fig. 17) one of the four linkages previously mentioned.

By using the linkage (Fig. 19) with 9 links and three independent loops, one of the forty similar linkages [1, 2] a similar leg mechanism (Fig. 18) is synthesized.

In order to describe any curve by the T point of the output link it is necessary to select an adequate basis and the effector from the bi-mobile linkage. By using its inverse structural models characterized by an instantaneous degree of mobility equal to zero and a connection exclusively made by passive modular groups it is possible to verify the possible basis and its effectors [1, 2].

The leg mechanism with three degree of mobility (Fig. 20) is used for mobile robots or platform, when the output link must follow the ground micro-irregularities. It

Leg Mechanisms Motion Characteristics 51

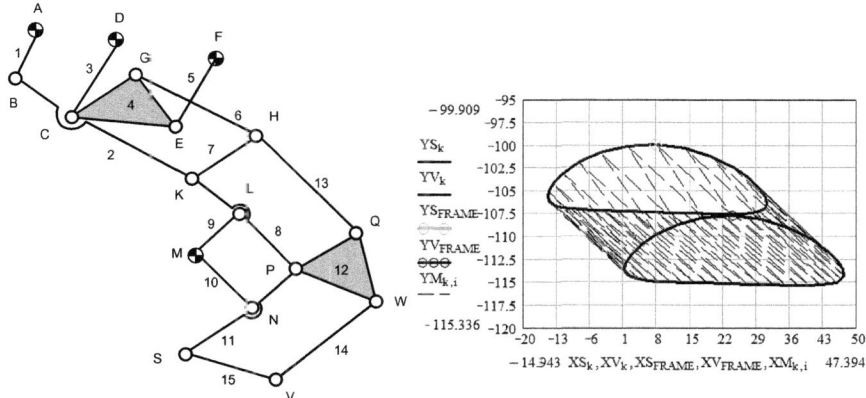

Fig. 13. Multi-loop leg mechanism with 15 links and one degree of mobility

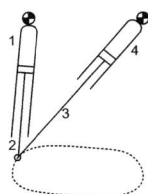

Fig. 14. The simplest bi-mobile leg mechanism

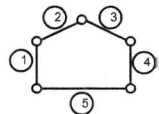

Fig. 15. The five bar linkage

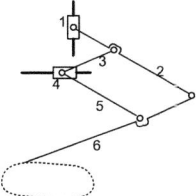

Fig. 16. The bi-mobile leg mechanism

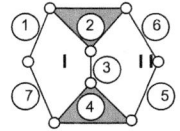

Fig. 17. The seven bar bi-mobile linkage

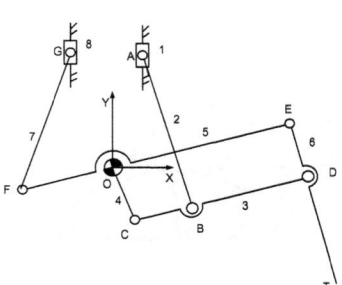 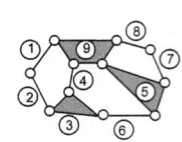 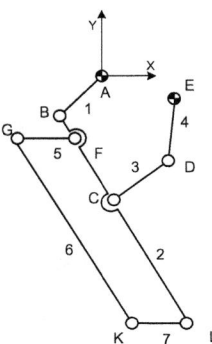

Fig. 18. The three loops leg mechanism

Fig. 19. The bi-mobile linkage

Fig. 20. The three-mobile leg mechanism

may be used for the platform with the start-stop or continuous displacement. The motion is achieved by three active kinematic pairs placed between the following element (0, 1), (0, 4) and (2, 5). The same active pairs determine either the 7 output link or the AE platform motion with displacement laws chosen by the designer [1].

4 Bi-looped and Mono-mobile Leg Mechanisms

The mono-mobile leg mechanisms based on bi-looped linkages with four degree of mobility are generally known from the literature [1, 3, 5]. These linkages presented in Figs. 21 and 22 gives several solutions for mechanisms having an output link with a planar motion and a point with an approximate straight line coupler curve.

Any mechanism derived from the mentioned linkages may have the following PMG - passive modular groups (Fig. 23) and a single AMG - active modular group from those presented below (Fig. 24).

When the output link is selected in the linkage it is necessary to satisfy the following demands: in a four elements contour the respective link is opposite to the basis and consequently it can not be adjacent to the basis.

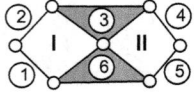 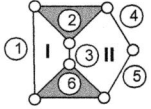

Fig. 21. Watt bi-loop linkage

Fig. 22. Stephenson bi-loop linkage

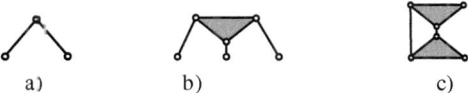

Fig. 23. Passive modular groups (PMG)

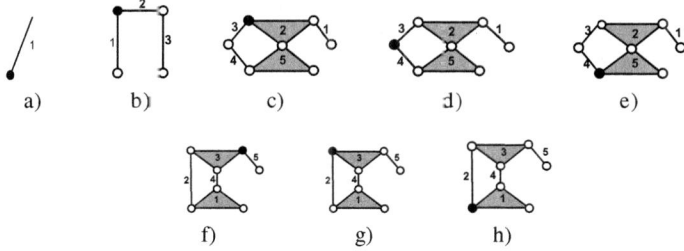

Fig. 24. Active modular groups (AMG)

So the selection of the i basis and the j output link may be synthesized in a symmetrical matrix A(i, j) with null elements of the main diagonal.

This matrix A(i, j) written for the Watt linkage (Fig. 21) is given by (3) with 16 solutions for basis and output link.

$$A(i,j) = \begin{bmatrix} 0 & 0 & 1 & 1 & 1 & 0 \\ 0 & 0 & 0 & 1 & 1 & 1 \\ 1 & 0 & 0 & 0 & 1 & 0 \\ 1 & 1 & 0 & 0 & 0 & 1 \\ 1 & 1 & 1 & 0 & 0 & 0 \\ 0 & 1 & 0 & 1 & 0 & 0 \end{bmatrix} \tag{3}$$

Due to the symmetry of the linkage only two of them are distinct and presented in Fig. 25.

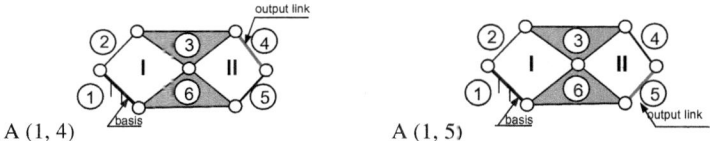

Fig. 25. The distinct solutions for mechanisms derived from Watt linkage

Having in view the AMG active modular groups (Fig. 24) and the PMG passive modular groups (Fig. 23) one may obtain the following variant for mechanisms (Fig. 25):

$$\underline{A\,(1,4)} \quad \begin{array}{l} \text{AMG}[(1,4),2] + \text{PMG}\,(2,3) + \text{PMG}\,(4,5) \\ \text{AMG}[(1,6),6] + \text{PMG}\,(2,3) + \text{PMG}\,(4,5) \\ \text{AMG}[(2,3),3,6] + \text{PMG}\,(4,5) \\ \text{AMG}[2,(3,6),6] + \text{PMG}\,(4,5) \\ \text{AMG}[2,3,4,5,(5,6),6] \end{array} \qquad (4)$$

where the 1 link is the basis and 4 is the effector which point describes the coupler curve. In order to avoid dysfunction the active pair can not belong to the effector.

In the second variant (Fig. 25) the basis is the 1 link and the effector 5. In this case the mechanisms with their active pairs may be those given by (5) the placement of the active pair next to the effector being inacceptable

$$\underline{A\,(1,5)} \quad \begin{array}{l} \text{AMG}[(1,2),2] + \text{PMG}\,(3,6) + \text{PMG}\,(4,5) \\ \text{AMG}[(1,6),6] + \text{PMG}\,(2,3) + \text{PMG}\,(4,5) \\ \text{AMG}[2,(2,3),3,6] + \text{PMG}\,(4,5) \\ \text{AMG}[2,3,(3,6),6] + \text{PMG}\,(4,5) \\ \text{AMG}[2,3,(3,4),4,5,6]. \end{array} \qquad (5)$$

The active pair may be a rotation or a prismatic one.

Only some structural models for leg mechanisms found in the literature [3, 4, 6] are met between solutions included in (4) and (5). From the structural point of view the Stephenson linkage (Fig. 22) may be similarly analyzed.

The matrix A (i, j) is also symmetrical one and has the following form where i is the basis and j is the effector

$$A(i,j) = \begin{bmatrix} 0 & 0 & 1 & 1 & 1 & 0 \\ 0 & 0 & 0 & 0 & 1 & 1 \\ 1 & 0 & 0 & 1 & 1 & 0 \\ 1 & 0 & 1 & 0 & 0 & 1 \\ 1 & 1 & 1 & 0 & 0 & 0 \\ 0 & 1 & 0 & 1 & 0 & 0 \end{bmatrix} \qquad (6)$$

The linkage has the following symmetrical links: $1 \equiv 3$, $2 \equiv 6$, $4 \equiv 5$ and structurally identical the 1-2-4-5-6-1 and 3-2-4-5-6-3 contours.

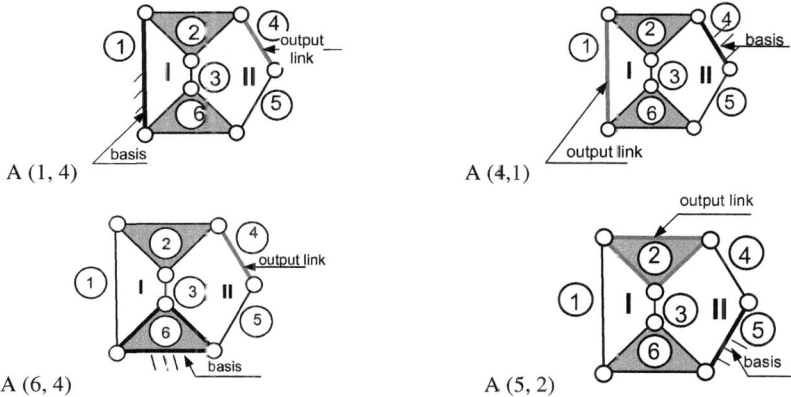

Fig. 26. The distinct solutions for mechanisms derived from Stephenson linkage

In these circumstances the distinct solutions are the following:

For each solution from Fig. 26 the variants have an active pair in their structure and modular groups (Figs. 23 and 24) connected to the structural classical theorems as follows:

$$
\begin{aligned}
\underline{A\ (1,4)} \quad & AMG[(1,2),2] + PMG\ (3,6) + PMG\ (4,5) \\
& AMG[(1,6),6] + PMG\ (2,3) + PMG\ (4,5) \\
& AMG[2,(2,3),3,6] + PMG\ (4,5) \\
& AMG[2,3,(3,6),6] + PMG\ (4,5) \\
& AMG[2,3,6,(6,5),5,4]
\end{aligned}
\tag{7}
$$

$$
\begin{aligned}
\underline{A\ (4.1)} \quad & AMG[(4,2),2] + PMG\ (1,3,6,5) \\
& AMG[(4,5),5] + PMG\ (2,1,3,6) \\
& AMG[2,(2,3),3,6,1,5] \\
& AMG[2,3,(3,4),6,1,5] \\
& AMG[2,3,6,1,(6,5),5]
\end{aligned}
\tag{8}
$$

$$
\begin{aligned}
\underline{A\ (6,4)} \quad & AMG[(6,1),1] + PMG\ (2,3) + PMG\ (4,5) \\
& AMG[(6,3),3] + PMG\ (2,1) + PMG\ (4,5) \\
& AMG[(6,5),5] + PMG\ (1,2,3,4) \\
& AMG[1,2,(2,3),3] + PMG\ (4,5)
\end{aligned}
\tag{9}
$$

$$
\begin{aligned}
\underline{A\ (5.2)} \quad & AMG[(5,4),4] + PMG\ (1,2,3,6) \\
& AMG[(5,6),6] + PMG\ (2,1,3,4) \\
& AMG[6,(6,3),2,1,5] \\
& AMG[2,3,(3,4),6,1\ 4]
\end{aligned}
\tag{10}
$$

It is significant to note that all solutions, which have adjacent active pair to the output link, are eliminated due to the functional purpose of the leg.

Partially some of previously determined structural solutions of mechanism derived from Watt and Stephenson linkages are found in the literature [3, 4, 6]. In [3, 6] there are presented fifteen mechanisms, all obtained from the Stephenson linkages, but only three of them are distinct. These correspond to the following solutions (Fig. 26) for the i basis and the j output link and modular connection given respectively in (10), (7) and (9);

$$\begin{aligned} A\,(5,2) &\rightarrow AMG[(4,5),5] + PMG\,(2,1,3,6) \\ A\,(1,4) &\rightarrow AMG[(1,2),2] + PMG\,(3,6) + PMG\,(4,5) \\ A\,(6,4) &\rightarrow AMG[(6,1),1] + PMG\,(2,3) + PMG\,(4,5) \end{aligned} \quad (11)$$

5 Conclusions

The paper synthesizes and develops the theoretical researches concerning especially the mono-mobile leg mechanisms which partially ensure an approximate straight line for the coupler curves. All structural solutions of such mechanisms with their particular action and obtained from Watt and Stephenson linkages are presented. The same method may be used to achieve new solutions by adopting another linkage from the sixteen ones with three independent contours mentioned in the literature [1, 3].

References

1. Comănescu A, Comănescu D et al (2010) Bazele modelarii mecanismelor. Politehnica Press, Bucharest
2. Comănescu A, Comănescu D, et al (2015) Classical modules applied to the mechanisms structural analysis and synthesis. In: Proceedings of the 14th world congress in mechanism and machine science, Taipei, Taiwan, 25–30 October 2015
3. http://sites.uci.edu/markplecnik/projects/leg_mechanisms/leg_designs/
4. https://en.wikipedia.org/wiki/Leg_mechanism
5. Pelecudi C (1972) Precizia mecanismelor. Editura Academiei
6. Sonawale KH, McCarthy JM (2016) Synthesis of eight-bar linkages by constraining a 6R loop. Mech Mach Theory 105:337–351

Design and Dimensional Optimization of a Novel Walking Mechanism with Firefly Algorithm

Özgün Selvi[✉] and Samet Yavuz

Çankaya University, Ankara, Turkey
{ozgunselvi,syavuz}@cankaya.edu.tr

Abstract. In this paper, a walking mechanism named Atlas is proposed which is a Watt-I type 6 link mechanism. First, the geometry is proposed and then kinematic analysis is done that will be used for the synthesis problem. Furthermore, constraints and an objective function are obtained from kinematic analysis. Then, these constraints are implemented to the Firefly Algorithm and dimensional parameters are obtained for a desired step profile. Finally, these dimensional parameters are tested.

Keywords: Walking mechanisms · Kinematic synthesis · Dimensional optimization · Firefly Algorithm

1 Introduction

Transferring motion from one or more input to one or more output by using attached rigid members requires mechanisms. In dimensional synthesis function, motion and path generation are the solution methods [16]. For the solution of synthesis problems researchers used both analytical, numerical and graphical methods over the years [12]. In path generation synthesis problems with more than 5 points, analytical methods remain incapable. At this point, numerical methods become a part of activity to obtain mechanism dimensions with large number of desired path points. In literature, researchers usually used evolutionary algorithms to overcome synthesis problems for mechanisms. Z. Nariman et al. [12] used hybrid multi-objective genetic algorithms for Pareto optimum synthesis of four bar mechanism with minimizing two objective functions (tracking error and transmission angle error) at the same time. Lin W [20] compounded two different evolutionary algorithm named differential evolution (DE) and real-valued genetic algorithm (RGA) to synthesize four bar mechanism with several design parameters for 6, 10 and 18 points in different cases. Only considered constraints of his work were the Grashof condition, design parameters within specified ranges, rotation range of the crank and relation between input angle and crank. Acharyya S.K. and Mandal M. [1] applied three different type of evolutionary algorithm (GA, PSO and DE) to minimize the error between desired and obtained coupler curve in four bar path generation synthesis. Researcher also compared these methods between each other and selected the best one. H Yu., et al. [5] presented a computer method which uses coupler-angle function curve to synthesize a four bar mechanism.

They practiced a two DoFs additional mechanism to transform coupler-points of the given path to a coupler-angle function curve. They also presented a software which give opportunity to users to define up to 20 points for path. Bulatovic R.R. and Djordjevic S.R. [15] used a direct searching method for four bar synthesis named Hooke-Jeeves's which compares its values at each iteration and changes parameters in order to described objective function. They proposed that the used algorithm in their research does not depend on the preliminary selected variables and showed a four bar example which coupler point draw a straight line. Walking mechanisms are used to simulate the walking motion of human or animals. Researchers presented various mechanisms in this area (Table 1).

Table 1. Walking mechanisms from literature

Theo-Jansen Linkage [6]	Links: 8 Joints: 10	Ghassaei [3]	Links: 7 Joints: 9	Klann Linkage [9]	Links:6 Joints:7
Stephenson [4]	Links:8 Joints: 10	Eight-Bar Leg Mechanism [2]	Links: 8 Joints: 7	Biped Walking Mechanism [14]	Links: 9 Joints: 6

The most popular walking mechanism around the world is Theo Jansen's linkage [6]. Researchers have been studied on this mechanism to optimize its link lengths [3, 8] and to make its dynamic analysis [19]. Researches on Klann mechanism [9] were done in task of optimization [7] as well. For Stephenson's mechanism some researchers improved an algorithm or a method to overcome theirs synthesis problems [4, 11, 18, 22]. Al-Araidah O. et al. [2] presented a single DoFs eight bar mechanism and synthesized it to reproduce a walking mechanism. Erika O. et al. [14] presented and implemented a one DoF biped mechanism for a rickshaw robot which is derived from Chebshev mechanism. After building of the robot they tested the robot in different operating conditions.

In this research, is a Watt-I type 6 link mechanism is proposed for walking. In following section, geometry of the mechanism is presented. In third section, preliminary kinematic analysis of the mechanism for synthesis is done. Forth section lays out the application of the Firefly Algorithm on this problem. Dimensional optimization of the mechanism is done in this section too. Finally, obtained dimensional parameters are tested whether they provide the desired step profile.

2 Geometry of the Walking Mechanism

The selected one DoF mechanism, which is designed to be a walking mechanism, is formed by connecting two four-bar loops (Fig. 1a). This mechanism has been chosen (Watt I type 6-link mechanism) among the mechanisms that will prove the minimum number of links by reviewing the literature. The first four-bar loop is connected to the walking body frame and its link lengths are denoted r_i, and the orientations of these links with respect to the positive X-axis are defined by the θ_i parameters. Here, mechanism is fixed to the walking body frame at joints with parameters θ_2 and θ_4, and θ_2 is selected as the input value. In mechanism, link r_2 can make full rotation. The joints with input and output parameters of the second four-bar loop are placed at the joint ends of the links r_4 and r_5. The second four-bar loop has no fixed links. The link lengths of the second four-bar loop are represented by p_i parameters and the positive X-axis orientations of these links are represent by ψ_i parameters. Stepping direction is selected as positive X-axis and gravitational acceleration is acting in the negative Y-axis direction. In mechanism, the parameters α_1, α_2, α_3 and α_4 have constant values and indicate the angle between the links r_3-p_2, r_4-p_1, r_4-r_5 and p_4-p_5 respectively. The contact point of the walking mechanism to the surface (point P) is free end of the link p_5. The reference coordinate system attached at the joint D. The presented walking mechanism is named as Atlas which has less links and joints compared to the walking mechanisms given in Table 1. Having fewer links and joints gives some advantages

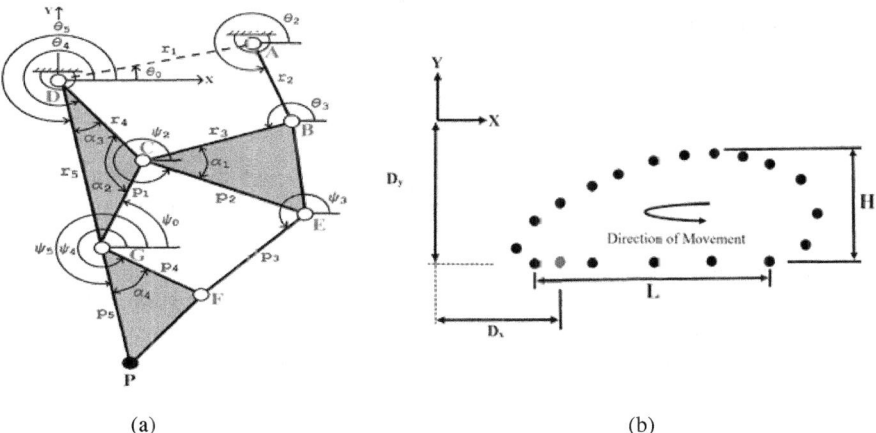

Fig. 1. (a) Presented walking mechanism, (b) Desired step profile

such as: ease of production, less energy loss during walking, less joint failures and more robustness. According to the geometry of the mechanism, this step profile can have different geometries. But a step profile always has the same components. The desired step profile for this mechanism can be seen in Fig. 1b. Here, H is the step height and L is the contact distance with the surface. D_x and D_y are the positions of the point with respect to the X-Y coordinate system attached to the walking body. Numbers of point for desired step profile are selected as 40 and these points range -70, 100 and -320, -290 respectively in X and Y coordinates.

3 Kinematic Analysis and Synthesis of the Mechanism

In this section, kinematic analysis of Atlas is done for both to obtain kinematic equations and to describe constraints for optimization. Before analyzing the mechanism, the relation between two four bar loops should be determined which will transfer the motion from input to output links. In the mechanism, we have a triangle link with parameters r4, r5 and p1. From cosine and sine theorem Eq. (1) is obtained.

$$p_1 = \sqrt{r_5^2 + r_4^2 - 2r_5r_4 Cos(\alpha_3)}, \quad \alpha_2 = ArcSin\left(\frac{r_5}{p_1} Sin(\alpha_3)\right) \quad (1)$$

Second four bar loop joint angle relations with respect to first four bar loop joint expression are described in Eq. (2).

$$\theta_5 = \theta_4 - \alpha_3, \quad \psi_2 = \theta_3 - \alpha_1 + \pi, \quad \psi_0 = \theta_2 + \alpha_2 - 2\pi, \quad \psi_5 = \psi_4 - \alpha_4 \quad (2)$$

For the synthesis of walking mechanism the desired (unknown) parameters are link parameters and appropriate positions of θ_2 and known parameters are coordinate of trajectory point P as Px and Py. After relation between two four bar loops are obtained, a relation is needed to be checked whether dimensional parameters provide the desired step profile. Thus X and Y coordinates of points of the step profile are obtained in Eq. (3) by analyzing a 2-degrees-of-freedom serial kinematic chain that accept the r_5 and p_5 as link lengths and the joint at θ_4 variable as ground.

$$P_x = r_5 Cos(\theta_5) + p_5 Cos(\psi_5), \quad P_y = r_5 Sin(\theta_5) + p_5 Sin(\psi_5) \quad (3)$$

Now, we need to obtain equations for θ_5 and ψ_5 to use them later both as constraints for optimization algorithm and to determine the error. After that two function generation problems will be defined for θ_4 and ψ_4 from θ_5 and ψ_5. Here θ_2 and θ_4 are our input and output values for first, ψ_2 and ψ_4 for second loop respectively. Let's consider we have a 2 degrees of freedom serial manipulator with parameters r_5 and p_5 and joint variables θ_5 and ψ_5 can be computed from Eq. (3). When, loop closure equation is written for the first and second four bar loops, we have the following equations.

$$\vec{r}_1 + \vec{r}_2 + \vec{r}_3 = \vec{r}_4, \ \vec{p}_1 + \vec{p}_2 + \vec{p}_3 = \vec{p}_4 \tag{4}$$

After rewriting Eqs. (4) and (5) explicitly, θ_4 and ψ_4 can be drawn from these equations.

4 Optimization of Dimensional Parameters

To obtain dimensional parameters for walking mechanism Firefly Algorithm is used. First of all constraints and objective function should be determined for optimization. Constraints for Firefly Algorithm are defined as follows. The Grashof conditions should satisfy full rotation for input link of the first four bar loop. For second four bar loop, there is no need of full rotation. Link lengths relations for Grashof condition are given in Eq. (6). Link lengths are defined within the specified ranges. All link lengths parameters are expected to be between 10 mm and 400 mm. Also, situations which would make the $\theta_4, \theta_5, \psi_4$ and ψ_5 equations undefined or imaginary are added as constraints. Objective function is chosen as a maximization problem. To minimize the error of path the dimensional parameters of the designed walking mechanism are aimed to be synthesized by using Firefly Algorithm. Firefly algorithm is developed by Yang [21] and is a nature inspired metaheuristic algorithm [13]. In firefly algorithm, each agents (fireflies) propagate lights and brighter ones pull other fireflies in close [10]. Few researchers in robotics used firefly algorithm to overcome optimization problem for their mechanisms. Nedic N. et al. [17] proposed a cascade load force control design for a parallel robot platform. They used Firefly algorithm for parameter searching. Also they indicate that firefly algorithm is very effective in nonlinear optimization tasks and performs better than other metaheuristic algorithms. They performed four different nature inspired algorithms (GA, PSO, CS and FA) on five different tasks. Researchers have procured that Firefly Algorithm is independent from the complexity of problems, has a better rate of convergence and gives values faster than other tested algorithms. One major reason for choosing Firefly Algorithm is that it can solve highly nonlinear, multimodal problems with high efficiency. Saputra V.B. [10] used firefly algorithm for determination of parallel manipulator's workspaces. From outline derived by Yang X. S. [13], firefly algorithm was applied to our optimization problem as shown in Table 2.

$$r_1 + r_3 - r_2 - r_4 < 0 \ V r_4 + r_1 - r_2 - r_3 < 0 \ V r_4 + r_3 - r_2 - r_1 < 0 \tag{5}$$

Parameters used in the algorithm were selected by taking into consideration of Lukasik S. and Zak S.'s research about firefly algorithm [10]. In algorithm which was given in Table 2, n_f denotes the population of fireflies and is selected 25 and n_c is the number of coordinates and signs the number of variables which are expected to be optimized (We have 16 variables to be optimized). γ is called approach speed or absorption coefficient and it states multifariousness with escalating distance from interacted firefly [10] and is selected as 0.1. β_0 is the attractiveness, it indicates the capability of a firefly to draw in other fireflies and is selected as 0.8. α is defined as randomness and it remarks the how much fireflies move randomly and is selected as 0.1. Light intensity of a firefly is measured by I and it directly impresses the movement

Table 2. Applied Firefly Algorithm for dimensional optimization of Atlas

---
Objective function $f(x_i)$,
$x_i = (r_1, r_2, r_3, r_4, r_5, p_1, p_2, p_3, p_4, p_5, \theta_0, \alpha_1, \alpha_2, \alpha_3, \alpha_4, \theta_{2,i})^T$, where
$f(x_i) = 1/(1 + Sum\,[(Error_1+Error_2), (i, 1, n)])$
Generate initial population of fireflies $x_i (i = 1, 2, \ldots, n_f)$.
Light intensity I at x_i is determined by $f(x_i)$
Define light absorption coefficient γ
for $(m_i; 1, MaxGen)$
  for $i = 1 : n_f$
    for $j = 1 : n_f$
if $(I_i < I_j)$,
$$r_{i,j} = \sqrt{Sum\left[(x_{k,i} - x_{k,j})^2, (k, 1, n_c)\right]};$$
$$Do\left[x_{k,i} = x_{k,i} + \frac{\beta_0}{1+\gamma\, r_{i,j}^2}(x_{k,j} - x_{k,i}) + \alpha\,(Random[\,] - 0.5), \{k, 1, n_c\}\right], \quad \text{(move firefly i towards to j)}$$
      else $Do[x_{k,i} = x_{k,i} + \alpha\,(Random[\,] - 0.5), \{k, 1, n_c\}]$, (move firefly random)
end if
Evaluate new solutions of $f(x_i)$ and update light intensity
    end for j
  end for i
Rank the fireflies and find the current global best g*
end for

---

of fireflies. Here $f(x_i)$ is the objective function and x_i is the solution for parameters which are wanted to be optimized at each iteration. Finally $r_{i,j}$ is the monotonically decreasing function of the distance between fireflies. As the error decreases and approaches zero, the objective function value approaches the maximum value one. The straight line which is followed by the end point of the Atlas has major importancy then the arc part of the step profile. Error weight (w_n) for straight line and arc part are taken as 1 and 0.4 respectively in objective function. $Error_1$ and $Error_2$ in objective function are expressed as follow.

$$Error_1 = \sum_{i=1}^{n} w_n |\theta_{4r,i} - \theta_{4c,i}|, \quad Error_2 = \sum_{i=1}^{n} w_n |\psi_{4r,i} - \psi_{4c,i}| \tag{6}$$

Where n is the number of points, θ_{4r} and ψ_{4r} are the desired output values and θ_{4c} and ψ_{4c} are the obtained output values from Firefly algorithm.

5 Testing of Obtained Parameters

Because of error weight for straight line 1 and arc part 0.4 objective function value comes around 0.6 which also one can be seen from Fig. 2. Obtained step profiles are suitable for a walking mechanism. Value of the objective function can be 1 without error. Firefly Algorithm gave the best value between 0.52 and 0.62 with 5000 iterations. Also, Atlas's geometry gives us one more advantage that reaction forces from ground are faced by links r_5 and p_5 nearly in perpendicular direction relative to surface.

Fig. 2. Objective function value for 5000 iterations

Dimensional parameters are obtained after 5000 iterations from Firefly Algorithm One also is given that the model of the walking mechanism with dimensional parameters (Fig. 3). In this section, it was checked whether the obtained optimum dimensional parameters provided the desired step profile. When we substitute these dimensional parameters into first θ_4, ψ_4, then θ_5, ψ_5 and finally into Eq. (3) we get the step profile for the mechanism. The step size of the found mechanism is desired as 170 mm in x direction, mechanism size become 200 mm in x direction (Fig. 4). The ratio of time taken on ground to the time taken above the ground is found around 0.8 even though it is important for the conversation of potential energy in walking mechanisms the ratio is not considered in this study.

All units for link lengths are mm, for angles are radian.

$r_1=$	160.7	$p_1=$	78.4	$\alpha_1=$	1.3	
$r_2=$	41.5	$p_2=$	132.8	$\alpha_2=$	1.5	
$r_3=$	140.9	$p_3=$	94.3	$\alpha_3=$	0.7	
$r_4=$	86.3	$p_4=$	110.3	$\alpha_4=$	6.9	
$r_5=$	121	$p_5=$	209.5	$\theta_0=$	6.3	

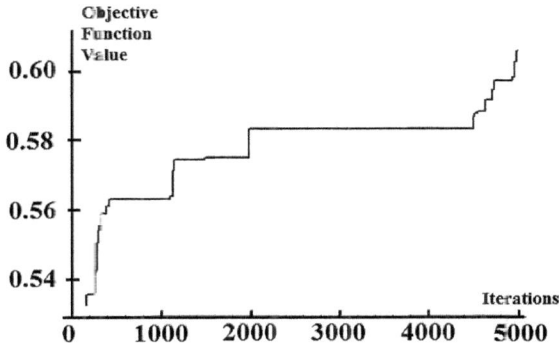

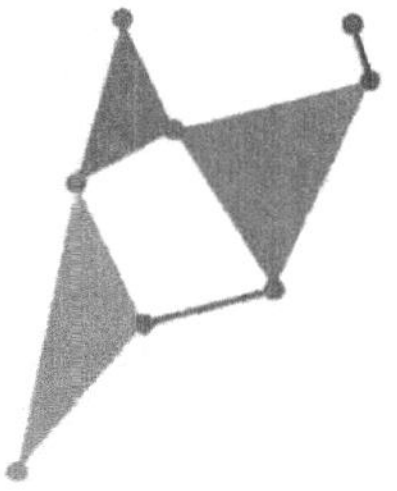

Fig. 3. Presented walking mechanism with optimized dimensional parameters

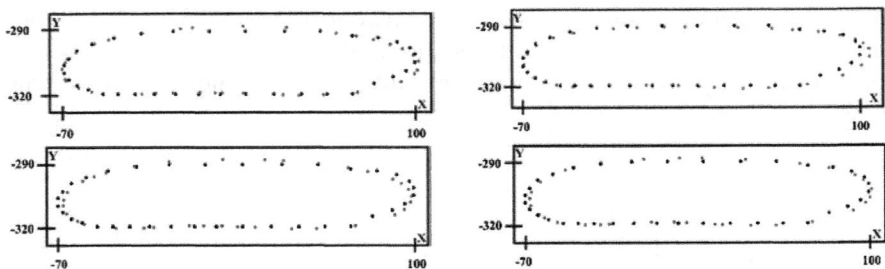

Fig. 4. Desired and obtained step profiles (units are mm) (Red and blue dots indicate the desired and obtained step profiles respectively.) (Color figure online)

6 Conclusion

In this research, a walking mechanism with one DoF named Atlas is presented. Firstly, its dimensional parameters and geometry are shown. A step profile that will be followed by the end point of the walking mechanism is proposed. Inverse kinematic equations of the walking mechanism are extracted to be used as constraints for Firefly Algorithm. Algorithm is run to get the optimized dimensional parameters to provide the desired step profile. After 5000 iterations dimensional parameters are obtained. These parameters are checked to test whether mechanism with optimized dimensional parameters provide the desired step profile. Test results show that obtained step profile is coinciding with the desired one with a straight line. For future works, it is aimed to implement this legs into a rescue robot.

References

1. Acharyya K, Mandal M (2009) Performance of EAs for four-bar linkage synthesis. Mech Mach Theory 44(9):1784–1794
2. Al-Araidah O, Batayneh W, Darabseh T, BaniHani M (2011) Conceptual design of a single DoF human-like eight-bar leg mechanism. Jordan J Mech Ind Eng 5–4:285–289
3. Amanda G (2013) Design and optimization of a crank-based leg mechanism. Pomona College
4. Batayneh W, Al-Araidah O, Malkawi S (2013) Biomimetic design of a single DoF Stephenson III leg mechanism. Mech Eng Res 3(2):43–50
5. Hongying Y, Dewei T, Zhixing W (2007) Study on a new computer path synthesis method of a four-bar linkage. Mech Mach Theory 42(4):383–392
6. Jansen T.: The Great Pretender. Uitgeverij. (2007)
7. Kim H, Jung M, Shin J, Seo T (2014) Optimal design of Klann-linkage based walking mechanism for amphibious locomotion on water and ground. J Inst Control Robot Syst 20–9:936–941
8. Kim S, Kim D (2001) Design of leg length for a legged walking robot based on Theo Jansen using PSO. J Korean Inst Intell Syst 21–5:660–666
9. Klann C (2002) Walking device. U.S. Patent No 6,478,314

10. Łukasik S, Żak S (2009) Firefly algorithm for continuous constrained optimization tasks. In: International conference on computational collective intelligence. Springer, Heidelberg, pp 97–106
11. Mirth J, Chase T (1995) Circuit rectification for four precision position synthesis of Stephenson six-bar linkages. J Mech Des 117(4):644–646
12. Nariman-Zadeh N, Felezi M, Jamali A, Ganji M (2009) Pareto optimal synthesis of four-bar mechanisms for path generation. Mech Mach Theory 44(1):180–191
13. Nedic N, Stajonavic V, Djordjavic D (2014) Optimal cascade hydraulic control for a parallel robot platform by PSO. The Int J Adv Manuf Technol 72(5–8):1085–1098
14. Ottaviano E, Grande S, Ceccarelli M (2010) A biped walking mechanism for a rickshaw robot. Mech Based Des Struct Mach 38:227–242
15. Radovan B, Djordjevic R (2004) Optimal synthesis of a four-bar linkage by method of controlled deviation. J Theoret Appl Mech 31(3-4):255–280
16. Sandor GN, Erdman AG (1984) Advanced mechanism design, analysis and synthesis. Prentice-Hall, New Jersey
17. Saputra B, Ong S, Nee A (2015) A swarm optimization approach for solving workspace determination of parallel manipulators. Robotica 33–03:649–668
18. Shiakolas P, Koladiya D, Kebrle J (2015) On the optimum synthesis of six-bar linkages using differential evolution and the geometric centroid of precision positions technique. Mech Mach Theory 40–3:319–335
19. Shunsuke N, Mohan Rajesh E, Masami I (2003) Dynamic analysis and modeling of Jansen mechanism. In: International conference on design and manufacturing, vol 64, pp 1562–1571
20. Wen-Yi L (2010) A GA–DE hybrid evolutionary algorithm for path synthesis of four-bar linkage. Mech Mach Theory 45(8):1096–1107
21. Yang X (2010) Nature-inspired metaheuristic algorithms. Luniver Press, Bristol
22. Zang H, Zhang S, Hapeshi K (2010) A review of nature-inspired algorithms. J Bionic Eng 7:232–237

Cam Mechanisms

Improving the Kinematics of Motion Curves for Cam Mechanisms Using NURBS

Thi Thanh Nga Nguyen[(✉)], Stefan Kurtenbach, Mathias Hüsing, and Burkhard Corves

Department of Mechanism Theory and Dynamics of Machines,
RWTH Aachen University, Aachen, Germany
{Nguyen-Thi,kurtenbach,huesing,corves}@igm.rwth-aachen.de

Abstract. In cam design process, synthesizing the motion curves is very important because of the effect of motion curves to cam size, force, and vibration. Thus, improving the kinematics of cam curves is significant. This paper presents a general synthesis of motion curves of cam mechanisms. A Non Uniform Rational B-Spline (NURBS) is used to improve the kinematics. The linear system of equations is established to determine motion curves for arbitrary boundary conditions of displacement, velocity, acceleration, and jerk. By controlling the weight parameters of NURBS, the peak values of acceleration and jerk can decrease. Several examples are presented to demonstrate this research. The results are also compared with traditional motion curves.

Keywords: Cam synthesis · Kinematics of cam mechanisms · B-spline · NURBS

1 Introduction

Cam mechanisms play an important role in the operation of many machine classes, especially those of the automatic type, such as shoe machinery, cutting machine, screw machine, etc. The main advantage of cam mechanisms is that they can realize any motion curves through an arbitrary design of the transfer function. The motion curves depend on working requirements and application situations of machines.

Selecting the function of motion curves is the first step in cam design. Traditionally, several basic forms of motion curves are used for the follower of cam mechanisms, such as simple harmonic, modified harmonic, cycloidal, trapezoidal, and polynomial functions [1–3]. T. Kiran and S.K. Srivastava [6] have applied 2-3 polynomial, 3-4-5 polynomial, and 4-5-6-7 polynomial functions for the displacement curve of the follower. A piecewise polynomial method was also used for synthesis of cam motions [7,8]. This method is based on describing the displacement of the follower with a series of polynomial expressions. The continuity conditions throughout the cam motion can be expressed as a set of equations that evaluate the polynomial function at the boundary points.

Apart from the traditional functions, B-spline is one of the methods for synthesizing displacement curves. Cam motions using B-spline curve were presented by [9–12]. Tsay and Huey [9] used B-spline to define the displacement curves that the motion curves satisfy the discrete constraints on follower displacements, velocities, and accelerations. Besides, NURBS curve was also used for cam motion curves. Der Min Tsay and Bor Jeng Lin [13] presented a procedure for synthesizing the dwell-rise-dwell motion of cylindrical cam with oscillating roller follower. This method was shown to improve the pressure angle and the radius of curvature. Applying a NURBS approximation to synthesize motion curves is investigated by N. Sateesh [14]. This research used the boundary conditions of the displacement that is expressed by the basic functions such as simple harmonic, cycloidal, etc. Until now, for a cam design, improving the kinematics (acceleration and jerk) is very important to reduce the force and vibration in mechanisms. Therefore, synthesizing the cam motions to achieve the small value of acceleration and jerk is still interesting for designers.

This paper presents a general synthesis of motion curves using NURBS. For this research, the kinematics of cam motions is improved. The content of this paper is structured as follows: Sect. 2 presents the background of B-spline and NURBS curves. A general synthesis of motion curves is shown in Sect. 3. In Sect. 4, several examples illustrate this research. Section 5 concludes this paper.

2 B-spline and NURBS Curves

B-spline and NURBS have been commonly used in modeling of fitting curves and surfaces. They are also convenient in interpolation for given data points. In the following, the background of B-spline and NURBS curves will be presented in this section.

2.1 B-spline and Its Derivatives

A general expression for B-spline curve of degree p is defined by [4]

$$C(u) = \sum_{i=0}^{n} N_{i,p}(u) \, P_i, \quad u \in [a, b], \tag{1}$$

where P_i, (i = 0, ..., n), are called control points, and $N_{i,p}(u)$ are the B-spline basis functions that are defined over the knot vector

$$\mathbf{U} = \{\underbrace{a, ..., a}_{p+1}, u_{p+1}, ..., u_{m-p-1}, \underbrace{b, ..., b}_{p+1}\}, \tag{2}$$

with $m = n + p + 1$ and $m + 1$ knots.

The B-spline basis functions are computed by using the knot vector as

$$N_{i,0}(u) = \begin{cases} 1 & \text{for} \quad u_i \leq u < u_{i+1} \\ 0 & \text{otherwise} \end{cases}$$

$$N_{i,p}(u) = \frac{u - u_i}{u_{i+p} - u_i} N_{i,p-1}(u) + \frac{u_{i+p+1} - u}{u_{i+p+1} - u_{i+1}} N_{i+1,p-1}(u). \quad (3)$$

Furthermore, the kth derivative of B-spline curve can be computed as

$$C^k(u) = \sum_{i=0}^{n} N_{i,p}^k(u) P_i. \quad (4)$$

Here, $N_{i,p}^k$ are the kth derivative of the basis functions and are defined by [4]

$$N_{i,p}^k(u) = \frac{p!}{(p-k)!} \sum_{j=0}^{k} a_{k,j} N_{i+j,p-k}, \quad (5)$$

with

$$a_{0,0} = 1; \quad a_{k,0} = \frac{a_{k-1,0}}{u_{i+p-k+1} - u_i}$$

$$a_{k,j} = \frac{a_{k-1,j} - a_{k-1,j-1}}{u_{i+p+j-k+1} - u_{i+j}}, \quad j = 1, ..., k-1. \quad (6)$$

$$a_{k,k} = \frac{-a_{k-1,k-1}}{u_{i+p+1} - u_{i+k}}$$

2.2 NURBS Curve and Its Derivatives

A Non Uniform Rational B-Spline (NURBS) curve generalizes B-spline curve. It is the rational combination of a set of basis functions with $(n+1)$ control points P_i and associated weights w_i. A NURBS curve with degree p is defined as [4]

$$C(u) = \sum_{i=0}^{n} R_{i,p}(u) P_i, \quad u \in [a, b]. \quad (7)$$

Here, $R_{i,p}(u)$ are the rational basis functions. They are calculated by

$$R_{i,p}(u) = \frac{N_{i,p}(u) w_i}{\sum_{j=0}^{n} N_{j,p}(u) w_j}, \quad (8)$$

for all weights w_i which are positive.
From Eq. (7), the kth derivative of NURBS is computed by

$$C^k(u) = \sum_{i=0}^{n} R_{i,p}^k(u) P_i. \quad (9)$$

The derivative of NURBS curve with degree p is continuous up to $(p-1)$. The curve and its derivatives can be controlled flexibly by several parameters such as the knot vector, the control points, and the weights.

3 Synthesis of Motion Curves for Cam Mechanisms Using NURBS

A general synthesis of cam motions using Nurbs is presented in this section. For a given number of boundary conditions, $n+1$, we denote, u, the angle of camshaft. The boundary conditions of displacement, velocity, acceleration, and jerk are respectively $C(u_j)$, $C^1(u_k)$, $C^2(u_l)$, and $C^3(u_h)$ at u_j, u_k, u_l, and u_h. Therefore, we can write the boundary conditions as

$$\begin{aligned} [u_j \quad C(u_j)] \quad &for \quad j = 1, ..., d \\ [u_k \quad C^1(u_k)] \quad &for \quad k = 1, ..., e \\ [u_l \quad C^2(u_l)] \quad &for \quad l = 1, ..., f \\ [u_h \quad C^3(u_h)] \quad &for \quad h = 1, ..., g \end{aligned} \qquad (10)$$

where $C(u_j)$, $C^1(u_k)$, $C^2(u_l)$, and $C^3(u_h)$ are the values of displacement, velocity, acceleration, and jerk, respectively (see Eqs. (7) and (9)). Note that the number of boundary conditions, $n+1$, is equal to $(d+e+f+g)$.

According Eq. (7), we can write the linear system of equations

$$\underbrace{\begin{bmatrix} C(u_j) \\ C^1(u_k) \\ C^2(u_l) \\ C^3(u_h) \end{bmatrix}}_{\mathbf{C}((n+1)\times 1)} = \underbrace{\begin{bmatrix} R_{i,p}(u_j) \\ R^1_{i,p}(u_k) \\ R^2_{i,p}(u_l) \\ R^3_{i,p}(u_h) \end{bmatrix}}_{\mathbf{R}((n+1)\times(n+1))} \underbrace{\begin{bmatrix} P_0 \\ P_1 \\ \vdots \\ P_n \end{bmatrix}}_{\mathbf{P}((n+1)\times 1)}, \qquad (11)$$

where matrix boundary conditions $\mathbf{C}$ is expressed by

$$\mathbf{C} = \begin{bmatrix} C(u_1), \cdots, C(u_d), C^1(u_1) \cdots, C^1(u_e), C^2(u_1) \cdots, C^2(u_f), C^3(u_1), \cdots, C^3(u_g) \end{bmatrix}^T \qquad (12)$$

and matrix $\mathbf{R}$ can be written as

$$\mathbf{R} = \begin{bmatrix} R_{0,p}(u_1) & R_{1,p}(u_1) & \cdots & R_{n,p}(u_1) \\ \vdots & \vdots & \ddots & \vdots \\ R_{0,p}(u_d) & R_{1,p}(u_d) & \cdots & R_{n,p}(u_d) \\ R_{0,p}^1(u_1) & R_{1,p}^1(u_1) & \cdots & R_{n,p}^1(u_1) \\ \vdots & \vdots & \ddots & \vdots \\ R_{0,p}^1(u_e) & R_{1,p}^1(u_e) & \cdots & R_{n,p}^1(u_e) \\ R_{0,p}^2(u_1) & R_{1,p}^2(u_1) & \cdots & R_{n,p}^2(u_1) \\ \vdots & \vdots & \ddots & \vdots \\ R_{0,p}^2(u_f) & R_{1,p}^2(u_f) & \cdots & R_{n,p}^2(u_f) \\ R_{0,p}^3(u_1) & R_{1,p}^3(u_1) & \cdots & R_{n,p}^3(u_1) \\ \vdots & \vdots & \ddots & \vdots \\ R_{0,p}^3(u_g) & R_{1,p}^3(u_g) & \cdots & R_{n,p}^3(u_g) \end{bmatrix}. \tag{13}$$

From the solution of linear system of equations Eq. (11), the control points are determined so that the motion curve is established.

4 Application Examples

This section presents several examples to compare the motion curves using B-spline, NURBS, and polynomial. The degree, $p = 5$, is used to calculate B-spline and NURBS because their derivatives will be continuous up to the jerk curve. For applying B-spline and NURBS, we have to compute the knot vector $\mathbf{U}$. For this, here we use the uniform parameterization method. Further detail of this method can be found in S.M.H. Shamsudin and M.A. Ahmed [8].

4.1 Comparison of Motion Curves with Traditional Functions

In the first example, six boundary conditions of displacement, velocity, and acceleration at the start and the end points are given. The traditional transfer function is used with the fifth polynomial. The transfer functions using B-spline and NURBS (see Eqs. (1) and (7)) are achieved by solving the linear system of equations Eq. (11). The comparisons are presented in Fig. 1. It is clearly seen that the fifth polynomial and B-spline motion curves do not change for all of the displacements, velocities, accelerations, and jerks.

Using NURBS curve, the weights, $\mathbf{w} = [1.2, 0.8, 0.9, 0.9, 0.8, 1.2]$, are used for calculation. From Fig. 1, the displacement, velocity, and acceleration curves differ

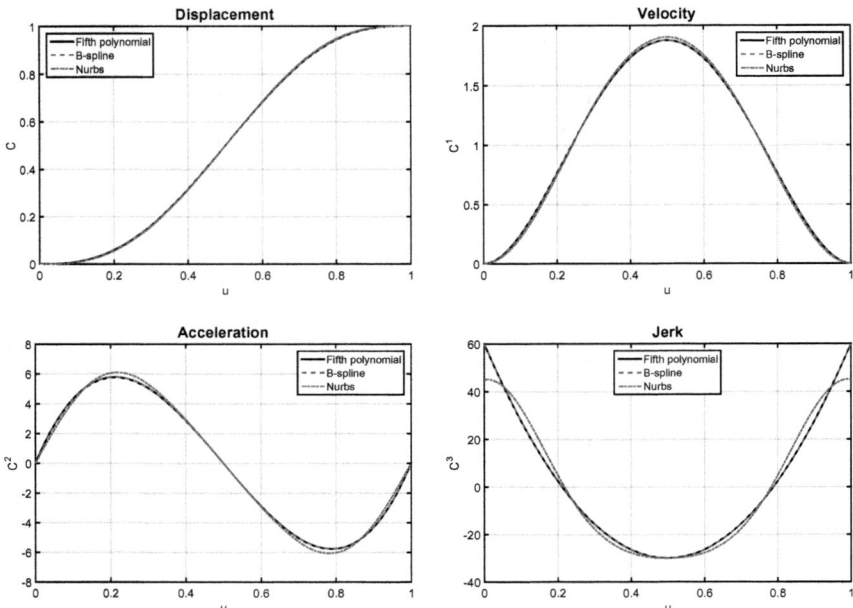

Fig. 1. Comparison of normalized motion curves between fifth polynomial, B-splines and NURBS

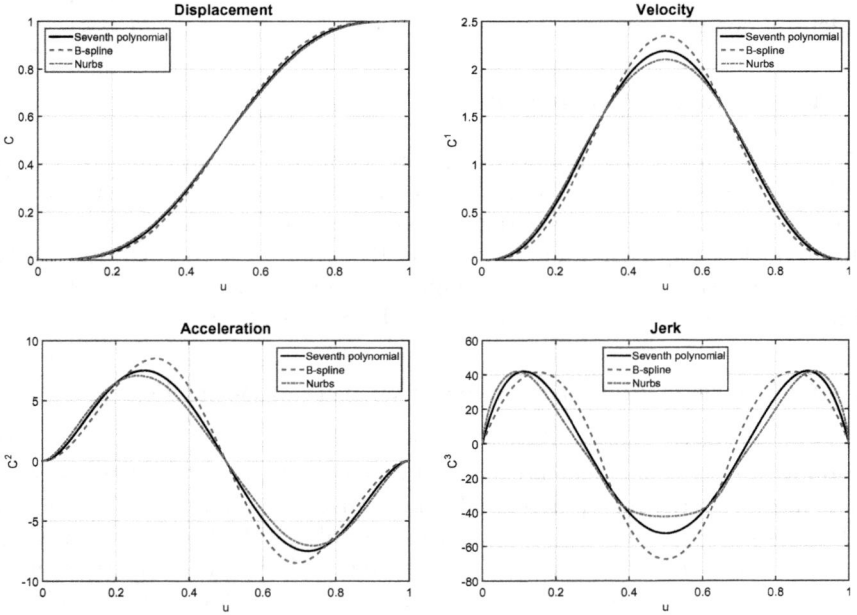

Fig. 2. Comparison of normalized motion curves between seventh polynomial, B-splines and NURBS

slightly between polynomial, B-spline, and NURBS. However, the maximum value of jerk with NURBS is reduced tremendously. This feature is beneficial in cam design because of reduction of vibration.

Next, we consider the second example with eight boundary conditions at the start and the end points of the displacement, velocity, acceleration, and jerk. The weights, $\mathbf{w} = [0.5, 0.8, 1, 1.5, 1.5, 1, 0.8, 0.5]$, are used for this case. These results are shown in Fig. 2. The basic function, seventh polynomial, is used. As shown, the difference of displacements is not much for using polynomial, B-spline and NURBS. Nevertheless, for the acceleration and jerk curves, the maximum values with NURBS (dash-dot line) are much smaller than polynomial (solid line) and B-spline (dashed line).

4.2 Examples for Large Number of Boundary Conditions

Using B-spline and NURBS for synthesis of motion curves is not limited by the number of boundary conditions. To demonstrate this, in the following we consider two more examples to synthesize the cam motion using Nurbs with large number of boundary conditions.

As the third example, we compute the motion curve for cam mechanism in cutting machine shown in Fig. 3 [15]. The position in the rise motion of the blade 4 is listed in Table 1. To determine the profile of cam disk 5, we need to find the displacement function of the follower. This function must satisfy all boundary conditions that are calculated from the given positions of the blade. Due to the discontinuity of the velocity and the acceleration, infinite values of the acceleration and jerk occur, respectively. Thus, to avoid the discontinuity of velocity, acceleration, and jerk, more boundary conditions are added at the start and the end points of those curves, such that their values are equal to zero. The boundary conditions are presented by the star signs in Fig. 4. The displacement is normalized for this calculation. The following weights are used for NURBS, $\mathbf{w} = [0.1, 0.4, 0.7, 1.3, 2.2, 1.1, 0.5, 0.8, 1.1, 1, 0.9, 1.2, 1, 0.9, 0.8, 0.5]$.

Table 1. Position of the blade 4 in rise motion

Angle of camshaft (deg)	0	5	10	15	20	25	30	35	40	45
Position of blade (mm)	0	0.1584	1.0857	3.1001	6.1274	9.8083	13.6077	16.9228	19.1913	20

Figure 4 presents the curves of displacement, velocity, acceleration, and jerk. The displacement using B-spline (dashed line) is slightly oscillating. Thus the velocity, acceleration, and jerk curves are also oscillating. Therefore, the peak values of acceleration and jerk are observed. With the polynomial and NURBS, the difference of displacement and velocity curves does not change much. The maximum value of acceleration with NURBS, compared to polynomial, is not much decreased. However, the maximum value of jerk is decreased enormously from $301.3378(1/rad^3)$ to $266.6077(1/rad^3)$.

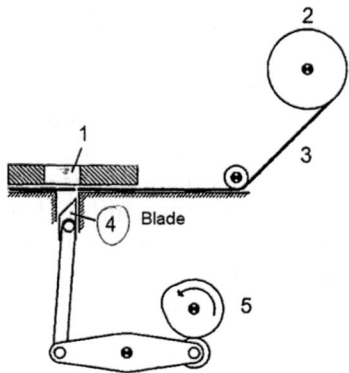

Fig. 3. Cam mechanism using in cutting machine [15]

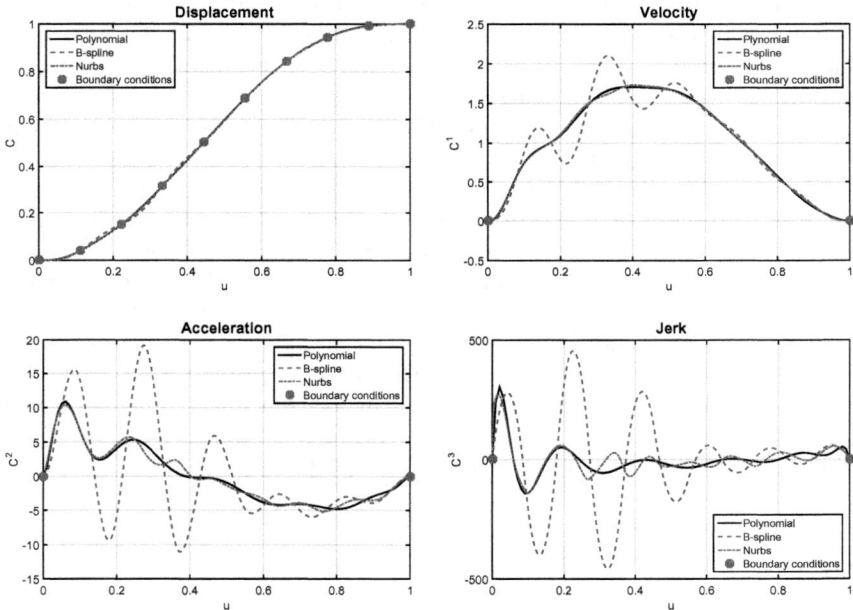

Fig. 4. Comparing normalized motion curves of cam mechanism of cutting machine

As the final example, we consider the cardiovascular mock loop where the motion of the human heart is simulated. The measurement of the displacement follower is shown by star signs in Fig. 5. The boundary conditions are added at the start and the end points of displacement, velocity, acceleration, and jerk to avoid the discontinuity at these points. Figure 5 shows these curves in one cycle of cam mechanism. As seen, the displacement curve does not change much, likewise the velocity curve. For the acceleration with

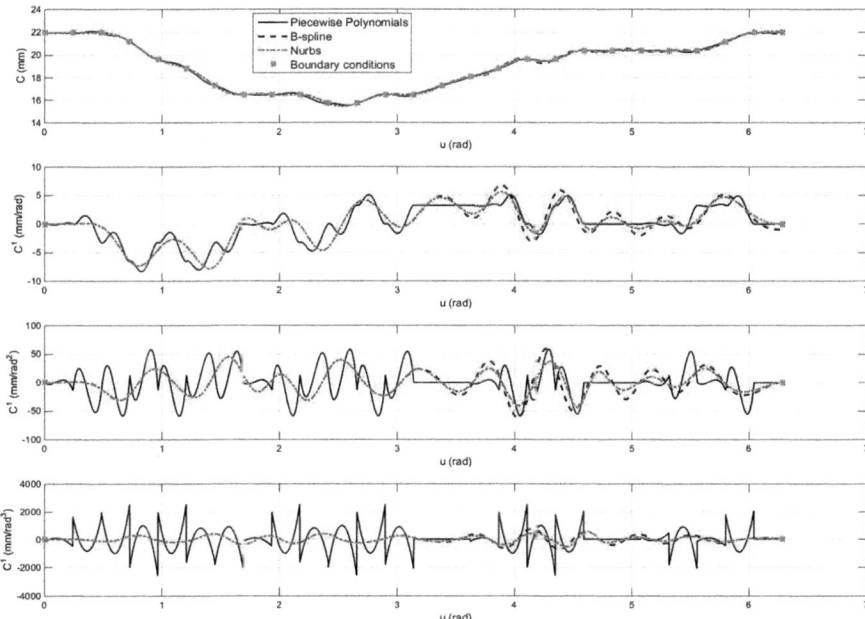

Fig. 5. Comparing SVAJ of piecewise polynomial, B-spline and NURBS

piecewise polynomials of sixth order, we see the break-points. Thus, the large values of jerk occur. Using B-spline gives smooth acceleration and jerk curves, but the maximum value of the acceleration does not decrease (see the Fig. 5). However, controlling the weight factors, the maximum values of the acceleration and jerk curves of NURBS can much decrease. Here, the weights, $\mathbf{w} = [1, 1, 0.5, 0.5, 1, 1, 1, 1, 1, 1, 1, 1, 1, 1, 1, 1, 1, 1, 1, 1, 1, 0.5, 0.5, 1, 1, 1, 0.5, 0.5, 0.2, 0.5, 0.5, 1]$, are used.

5 Conclusions

A general synthesis of motion curves using NURBS is presented in this paper. The examples show that the improvement of kinematics is achieved. The NURBS curve of degree five is very flexible and satisfies the requirements of boundary conditions. As illustrated by several examples above, the motion curves are compared with basic functions of polynomial, B-spline, and NURBS. The acceleration and the jerk values are improved with NURBS. The weights can be controlled to decrease the peak values of acceleration and jerk. In a cam design, this allows for reduction of force and vibration.

References

1. Norton RL (2002) Cam design and manufacturing handbook. Industrial Press, New York
2. Shigley JE, Uicker J (1981) Theory of machines and mechanisms. McGraw-Hill Inc., US
3. Norton RL (1999) Design of machinery: an introduction to the synthesis and analysis of mechanisms and machines, 2nd edn. McGraw-Hill, London
4. Piegl L, Tiller W (1997) The NURBS book, 2nd edn. Springer, New York
5. Shamsuddin SMH, Ahmed MA (2004) A hybrid parameterization method for NURBS. In: Proceedings of the international conference on computer graphics, imaging and visualization (CGIV 2004)
6. Kiran T, Srivastava SK (2013) Analysis and simulation of cam follower mechanism using polynomial cam profile. Int J Multi Curr Res 211–215. ISSN: 2321-3124
7. Mermelstein SP, Acar M (2004) Optimising cam motion using piecewise polynomials. Eng Comput 19:241–254 Springer
8. Wang L-CT, Yang Y-T (1996) Computer aided design of cam motion programs. Comput Ind 28:151–161
9. Tsay DM, Huey CO Jr (1988) Cam motion synthesis using spline function. J Mech Transmissions Autom Des 110:161–165
10. Xiao H, Zu JW (2009) Cam profile optimization for a new cam drive. J Mech Sci Technol 23:2592–2602
11. Naskar TK, Mishra R (2012) Introduction of control points in B-splines for synthesis of ping finite optimized cam motion program. J Mech Sci Technol 26:489–494
12. Sahu LK, Gupta OP, Sahu M (2016) Design of cam profile using higher order B-spline. Int J Innocative Sci Eng Technol 13:327–335
13. Tsay DM, Lin BJ (1996) Improving the geometry design of cylindrical cams using nonparametric rational B-splines. Comput Aided Des 28:5–15
14. Sateesh N (2014) Improvement in motion characteristics of cam-follower systems using NURBS. Int J Des Manufact Technol 8:15–21
15. Construction of planar cam mechanisms practical examples. VDI 2142 Part 3 (2014)

Assessment of the Rolling Contact Fatigue

Monika Hejnová(✉)

VÚTS, a.s, Liberec, Czech Republic
monika.hejnova@vuts.cz

Abstract. There are many methods for the service life estimation of the contact surfaces. The life of the cam mechanisms is commonly estimated theoretically. Experimental testing of the rolling contact fatigue is time consuming and this article deals with the description of the most suitable methods for a fast and easy assessment of rolling contact fatigue. This information can lead to understanding and explaining what happens when rolling contact fatigue occurs.

Keywords: Service life · Theoretical methods · Experimental evaluation · Pitting · Cracks

1 Introduction

VÚTS, a.s. has dealt with the design of cam mechanisms and cam production for more than thirty years. This provides us with a lot of experience in design and calculation of both axial and radial cams. In this context, currently we solve the methodology of the life prediction of the cam surface. On this basis, we can determine the life of the cam mechanism, respectively the time when damage occurs.

A life estimation of the contact surfaces is an important part of the cam mechanism design. The aim of this article is to describe a suitable procedure of the rolling contact assessment.

2 Theoretical Estimation of the Contact Surfaces Life

The proposal of the material for cam production is based on the calculation theories today. We usually use two theories. The first theory [4] is based on the service life estimation by means of the surface hardness. The calculation is simple but the surface hardness does not describe all the mechanical properties of the examined material. The second theory [7] is based on two material parameters. These parameters better describe the mechanical properties of the material but the database of these parameters is limited. It is very expensive and time consuming to obtain the parameters for materials outside the database.

The comparison of these theories results shows that the service life determined by each theory can be very different [1]. Therefore we need to find a suitable experimental procedure which can help verify the accuracy of the theories. Also, we should understand all the processes which lead to rolling contact fatigue (RCF).

3 Experimental Estimation of the Contact Surfaces Life

We have to perceive that there are wide number of facts at the proposal of the suitable experimental testing of the contact surface life. First is substantial preparation of the experiment in the field of the suitable material selecting. This choice is necessary to make in according to the results from the calculations (FEM, analytic methods). It is also necessary to control the supplied material, its chemical composition and micropurity.

In the next step we should determine exactly the parameters of heat, respectively chemical heat treatment. The parameters of these technological processes are very important for ensuring of required properties of the surface layers of the examined component. These are parameters of carburizing, heating etc.

It is necessary to take into account not only the material properties, but the influence of the working conditions on the RCF forming in the experiment too. The most significant is lubricant and its interaction with the contact surfaces and the shape of the contact surfaces too. Information about the interaction between oil and the contact surfaces is important because the behavior can change depending on change of the contact surface properties (change of surface tension, chemical interaction etc.). The properties of the contact surfaces can affect the RCF in two ways. The first one says that if there are cracks on the surface, it can lead to pitting [8]. The second problem can be, when the contact surfaces are too smooth, then can be insufficient space there for oil film [9].

After verification of the properties of the relevant steel, we can produce the test specimen. We produce two types of specimens, the first one is for testing the service life and the second type is for testing the material properties. By these specimens we can define quality and depth of the heat treatment (respectively carburizing, coating, weld deposit etc.). The depth of the carburizing layer can be determined by removing thin layers and then by using basic chemical analysis to determine the carbon content.

Other parameters which are useful we can obtain by nanoindentation measuring. We can get information about mechanical properties of the examined specimens, hardness, modulus, visco-elastic properties and other in the dependence on depth, very detailed in the area of interest.

After choosing the suitable material, production of specimens and determining the right test conditions the specimens can undergo the service life tests. More about these tests in [2]. The picture of the test rig is shown in Fig. 1.

3.1 Evaluation of Results of the Service Life Testing

There are two specimens damaged at the rolling contact in this paper. On these two examples we will present the procedure of the assessment of RCF.

In Fig. 1 specimen A is shown, produced from steel CPM 3V, it was loaded by a force of 6500 N, this corresponds to Hertzian stress of 4217 MPa, theoretical life 0,08 mil. cycles (determined by theory from [4]), in reality reached number of 8,9 mil. cycles to damage.

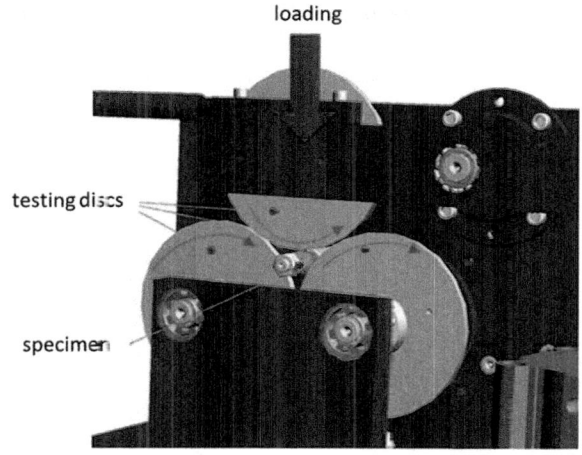

Fig. 1. Test rig

Fig. 2. Specimen A

Fig. 3. Specimen B

In Fig. 3 specimen B is shown, produced from steel 14 220 (carburised), it was loaded by force of 6000 N, this corresponds to Hertzian stress of 2292 MPa, theoretical life 4,4 mil. cycles (determined by theory from [4]), in reality reached number of 61,7 mil. cycles to damage.

After service life testing the tested surface can be observed by stereomicroscope (or other suitable microscope) by low magnification in the first step. The task is to obtain information about surface damage from a macroscopic view (see Figs. 1 and 2).

When we look through the specimens on the stereomicroscope, we can analyze the specimens on the electron microscope (in our case it was a Tescan electron microscope). In the Figs. 4 and 5 is shown pitting on specimens A and B.

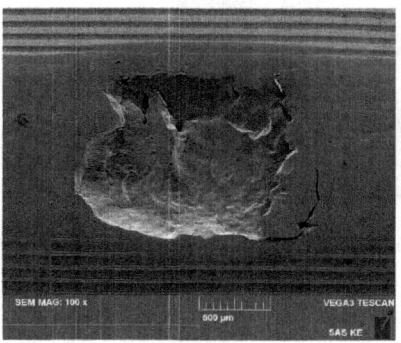
Fig. 4. Damaged surface (specimen A)

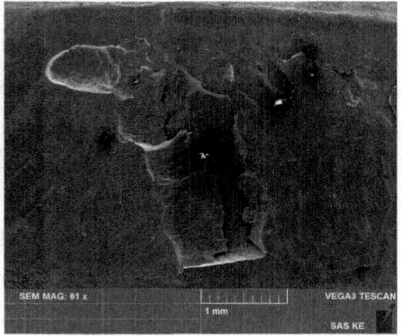
Fig. 5. Damaged surface (specimen B)

The electron microscope allows us to capture the images with a higher magnification and simultaneously performing the chemical composition analysis. The control of the material from the view point of the uniformity of the chemical composition gives more information about the examined system. We can get information about chemical reactions between the material surface and lubricant, presence of inclusions or other required information. For example, uneven steel solidification causes chemical unevenness in the material [5] and it can lead to inhomogeneous properties of the examined specimens.

There is the image shown of the non-damaged surface of specimen A in Fig. 6. In this picture we can see the surface and in this way get more information about the examined material.

Other interesting information we can obtain from Fig. 7. This surface was not loaded, this view is after machining. We can see the structure typical for sintered steels. This structure can lead to cracks forming at the rolling contact.

The next step is to take specimens for cuts in the point of damage by electrospark cutting. These cuts are analyzed on an optic microscope e.g. Olympus GX71 (see Figs. 8 and 9). From these images we can evaluate the integrity of the material and detect defects. A lot of factors can disturb the integrity of the material. There can be pores, cavities, non-metallic inclusions, cracks formed during cooling as a result of internal stress, cracks formed during the material forming, heat treatment, operation overload etc. [5].

After analysis on the optic microscope the cuts can be analyzed on the electron microscope (e.g. Jeol JSM 7000F). This microscope works with high magnification and

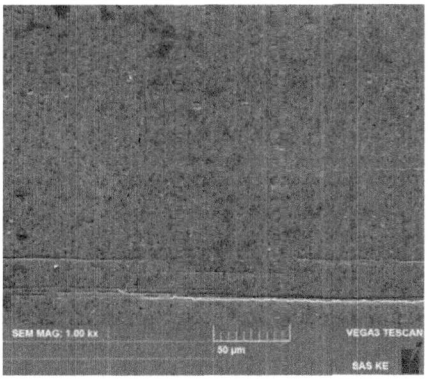

Fig. 6. Non-damaged contact surface (specimen A)

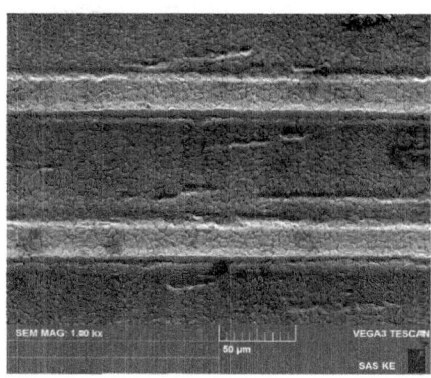

Fig. 7. A conic surface, after machining, not loaded (specimen A)

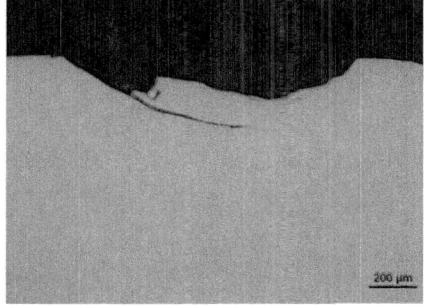

Fig. 8. The general view on the damaged part (specimen A)

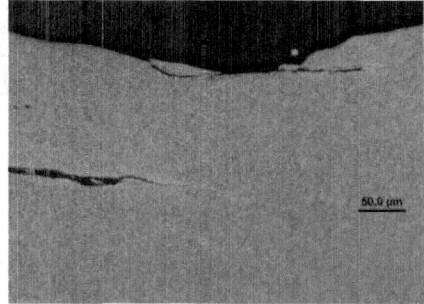

Fig. 9. Detail of cracks (specimen A)

this allows detailed monitoring of the cracks. Except standard display allows display in regimes COMPO and TOPO (see Figs. 10 and 11). Thanks to these regimes we are able to find out more information about the examined specimen. The regime COMPO allows us to identify inhomogeneity on the basis of the differences in chemical composition. It shows grayscale in the dependence of atomic numbers. The regime TOPO shows topographic view at the examined surface.

There is marked a maximal depth of cracks presence in Fig 12. From this image we obtained a value of 0,33 mm. The depth of maximal shear stress is 0,77 mm. This value was determined by analytical calculation [6].

From microscopic analyses we can obtain information about observing the technological process, determining of depth and uniformity of the heat treatment [5]. In Fig. 13 is shown image with a marked depth of the carburised layer.

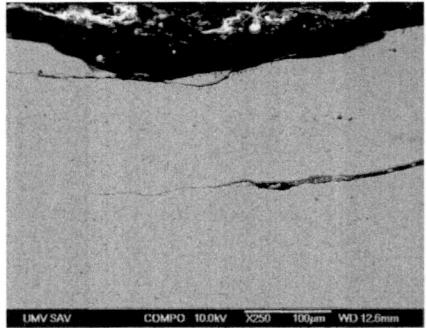
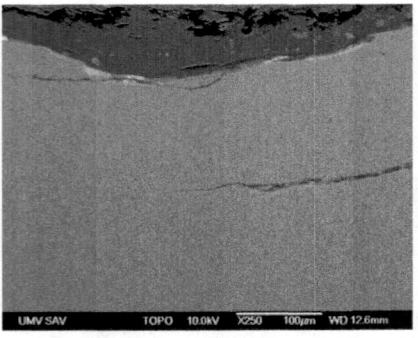

Fig. 10. Detail of the damaged part shown in regime COMPO (specimen A)

Fig. 11. Detail of the damaged part shown in regime TOPO (specimen A)

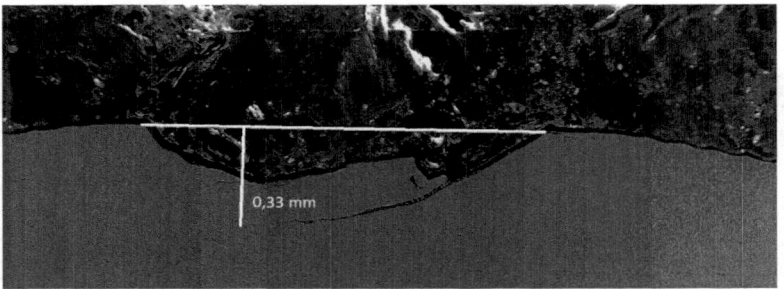

Fig. 12. Depth of damage determining (specimen A)

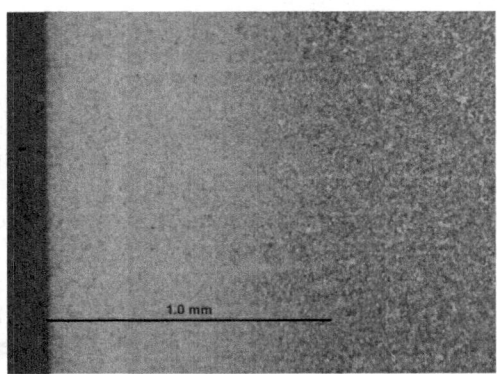

Fig. 13. Depth of the carburized layer is about 0,8 mm (specimen B)

Assessment of the Rolling Contact Fatigue 95

In Fig. 14 shows a detail of the crack root. From this image it is clear that the crack does not form on the grain boundaries. There is not the increase presence of impurities, inclusions, cavities in the examined steel. In the image globular carbides are clearly visible, their chemical analysis was performed.

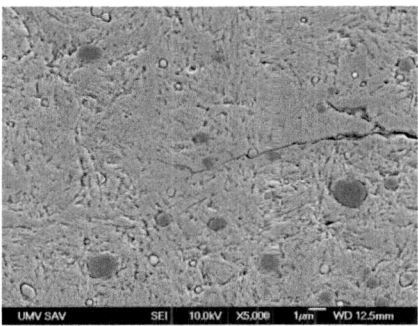

Fig. 14. Crack root shown with carbides (specimen A)

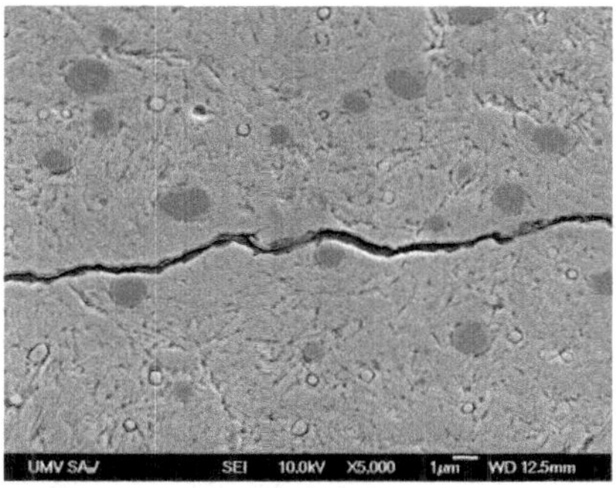

Fig. 15. Detail of the crack (specimen A)

The specimen under the load does not show plastic deformation, evidence is shown in Fig. 15, where it can be clearly observe the follow up structures.

A cut from specimen B is shown in Fig. 16. We can see cracks from the surface. These cracks will meet under the surface and a small part of the material will separate and form the damage called pitting [2].

Fig. 16. Cracks (specimen B)

There is a detail of the cracks from Figs. 17, 18, 19, 20, 21 and 22. We can see here a crack root. From this images we can form results about crack forming. We can say that the crack did not form from any inclusion or impurity or cavity. It was formed only by the exhaustion of mechanical properties of the examined steel.

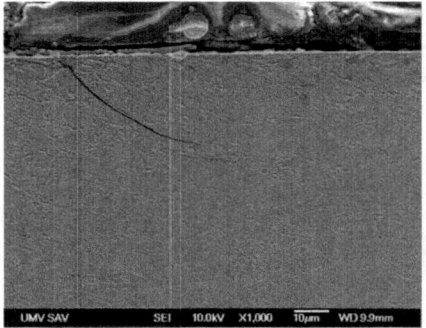

Fig. 17. Crack (specimen B)

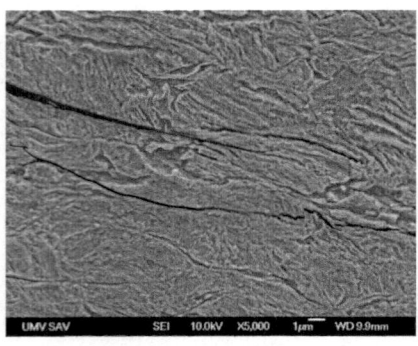

Fig. 18. Detail of the crack from the previous image (specimen B)

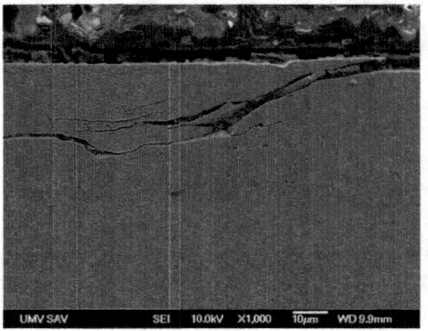

Fig. 19. Detail of the crack 1/3 (specimen B)

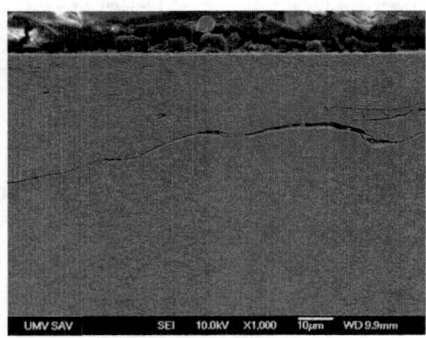

Fig. 20. Detail of the crack 2/3 (specimen B)

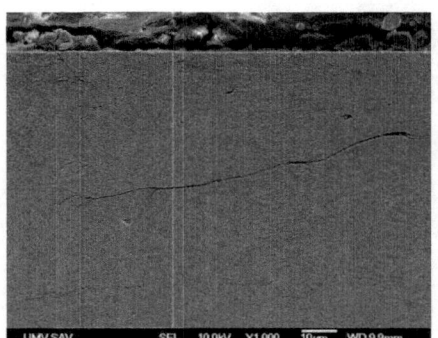

Fig. 21. Detail of the crack 3/3 (specimen B)

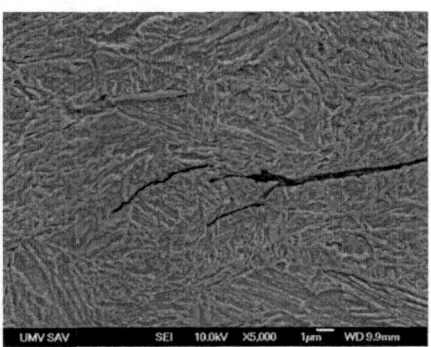

Fig. 22. Detail of the crack root at magnification of 5000x (specimen B)

A part of the cut between the contact surface and non-loaded cone surface is shown in Fig. 23. This area is deformed due to the testing after the first damage formed. We can see the cut, that the surface layer is decarburized. The decarburized layer is formed at steel heating in the environment containing oxygen, carbon dioxide and hydrogen. Decarburization grows with the cooling of the ardent steel in the air [3].

The structure of the examined steel (specimen B) is in Fig. 24. We can see, that here are not many inclusions, impurities or cavities.

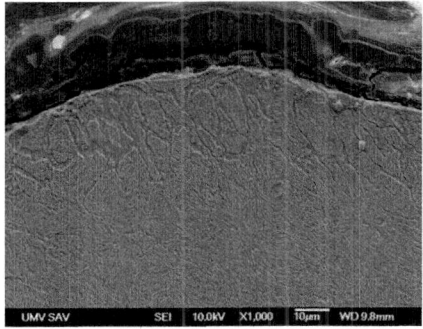

Fig. 23. Transition between loaded the contact surface and the conic surface (specimen B)

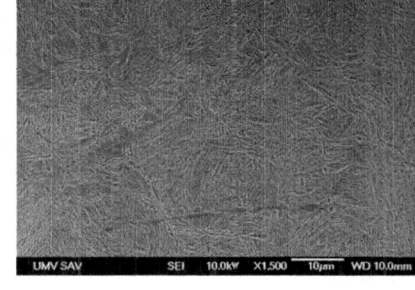

Fig. 24. Structure of the material (specimen B)

3.2 Summary of Results

From analysis can we say whether the structure is uniform, micropurity of the examined material is convenient and to determine the content of the inclusions and cavities. In both the above mentioned cases the damage occurred in the material and the cracks formed on the basis of exhaustion of the mechanical properties of the examined steels. The heat treatment was made in accordance to the requirements, specimen A made from steel CPM 3V is very hardened, low tempered. Specimen B is carburized enough, but the surface layer is decarburized. The plastic deformation did not occur at the load. The evidence is that there was not a displacement of the mutual crack edges.

To have complete information about the examined material, it is useful to measure microhardness and specimens continually (at service life testing) undergo the defectoscopic tests (detection of surface and subsurface defects).

4 Conclusions

The experiment realization can be divided into two parts. The first one solves the accuracy of the mathematical models. On the basis of calculations we can determine the depth of the maximal shear stress below the contact surface. The testing will verify these calculations. The depth of the damage can be determined either on the cut under the microscope or by other ways e.g. by measuring at the 3D CNC system etc.

The second part of the experiments has a task to determine the mechanisms of the crack forming. In the first step we determine the depth of the carburized layer on the basis of calculation results. After the damage occurs we can divide the number of gained cycles into ½ and the next test of the new specimen has to stop at this value. The specimen is analyzed and the cracks identified. If there are cracks, the number of cycles is again divided into ½ and test the new specimen. By this procedure we can determine the process of the cracks forming at the rolling contact.

After the evaluation of these experimental parts an adjustment will be suggested of the mathematical model and possibly technologic parameters of the heat treatment (mainly depth of the carburized layer and carbon concentration).

The most suitable solution of the material problematics of the rolling contact fatigue is to choose the material which will be loaded in its elastic area. This is possible to achieve either material with high yield strength, with sufficient heat or chemical heat treatment surface layer or with weld deposit.

Acknowledgments. The research work reported here was made possible by FR-TI4/801- project supported by the Ministry of Industry and trade.

References

1. Hejnová M, Ondrášek J (2016) Life estimation of the contact surfaces. In: Advances in mechanism design II. Proceedings of the XII international conference on the theory of machines and mechanisms. Springer, pp 43–50. ISBN: 978-3-319-44086-6
2. Hejnová M (2015) Problematika životnosti vačkových mechanismů. In: Mechatronika pohonů pracovních členů mechanismů. VÚTS, Liberec, Chap 21, pp 205–208
3. Jech J (1977) Tepelné zpracování oceli: Metalografická příručka, 3rd edn. STNL, Praha, 400 p
4. Koloc Z, Václavík M (1988) Vačkové mechanismy. SNTL—Nakladatelství technické literatury, Praha, 379 p
5. Koutský J, Šmíd J (1964) Makroskopické a mikroskopické zkoušení oceli. SNTL, Praha, 126 p
6. Mesys: calculation of contact stress. https://www.mesys.ag/?page_id=1220
7. Norton RL (2009) Cam design and manufacturing handbook, 2nd edn. Industrial Press, New York, 592 p. ISBN: 978-0-8311-3367-2
8. Požár R (2009) Rozvoj materiálového poškození při záběru ozubených kol a jeho vizualizace. Brno, 76 p. Diplomová práce. Mendelova zemědělská univerzita v Brně, Agronomická fakulta. http://is.mendelu.cz/zp/portal_zp.pl?prehled=vyhledavani;podrobnosti=30442;download_prace=1
9. Vrbka M et al (2011) Effect of surface texturing on lubrication film formation and rolling contact fatigue within mixed lubricated non-conformal contacts. Meccanica 46:491–498. doi:10.10007/s11012-010-9288-x. http://link.springer.com/article/10.1007/s11012-010-9288-x

The Stress Distribution in the Contact Region of a Cam Mechanism General Kinematic Pair

Jiri Ondrášek[✉]

VÚTS, a.s., Liberec, Czech Republic
jiri.ondrasek@vuts.cz

Abstract. The paper deals with the stress distribution in the contact region of a general kinematic pair. In the case of a general cam mechanism, the general kinematic pair is formed from the contact of the working surfaces of a cam and a follower. In the operation of cam mechanisms, fatigue damage of the contact surfaces of the cam and the follower may appear. This damage is in the form of pitting that develops from cracks on the surface of the active area. This type of damage is due to contact stress, described by the Hertz theory. For the cam mechanism general kinematic pair, the relations for calculating the stress distribution in the contact region were defined. The knowledge of the contact stress is useful for a life estimation of the contact surfaces.

Keywords: Stress distribution · Contact region · General kinematic pair · Cam mechanism

1 Introduction

Cam mechanisms can produce a required working motion within a very precisely prescribed path. They are characterized by a small number of bodies housed inside a relatively small space. Thus, they are still used in the design of various machines and equipment. The cam is the basic member driving the follower by the action of the working surface by means of a general kinematic pair. In technical practice, the contact of the cam contour and the follower may be substituted for an interaction of two cylinders with the parallel axes or a cylindrical body and a non-conforming body. The cylinder represents a cam or a cylindrical cam roller and the non-conforming body represents a crowned cam roller.

Increasing requirements on the performance of combined cam mechanisms cause larger power and inertia effects of moving bodies which raise wear and reduce fatigue life of the contact surfaces. Loading conditions in the general kinematic pair are the cause of the contact stress. If a certain limit of this stress is exceeded, then the fatigue damage on both cam and follower contact surfaces can occur in the course of the operation cycle of cam mechanisms. This damage has the form of pitting developed from cracks on the contact surface. The distribution of the reduced stress at the contact areas of the general kinematic pair and below these points is an important criterion of the service life estimation of the general kinematic pair contact surfaces. The state of deformation and stress existing between the two bodies in contact under load can be established on the basis of the Hertzian contact stress theory for either of the other, on the basis of the use of the finite element method (FEM). In this case, FEM is ineffective.

2 Hertzian Contact Theory Applied to a Cam Mechanism General Kinematic Pair

This section gives only the basic information on how to determine the contact stress distribution of two solids in contact. The Hertzian contact stress theory deals with this issue which is described in detail in [1] and the basic information is available in [3–5]. On the basis of a theory of contact, the shape of the contact area is predicted and the components of deformation and stress in both bodies are calculated in the vicinity of the contact region.

For the Hertzian contact stress theory, the following fundamental assumptions were accepted, see [1, 3, 5]. Both contacting surfaces are smooth and frictionless. When two three-dimensional body are brought into contact they touch initially at a single point or along a line. Under the action of a slightest load F, they will deform and contact is made over a finite area which is small, compared with the dimensions of both bodies. Remote parts T_1 and T_2 of the bodies are approached each other by a distance δ. Points S_1 and S_2 on the approaching contact surfaces are elastically displaced by amount u_{z1} and u_{z2}, as shown in Fig. 1. The shape of each surface in the contact region can be described by a homogeneous quadratic polynomial in two variables [1]:

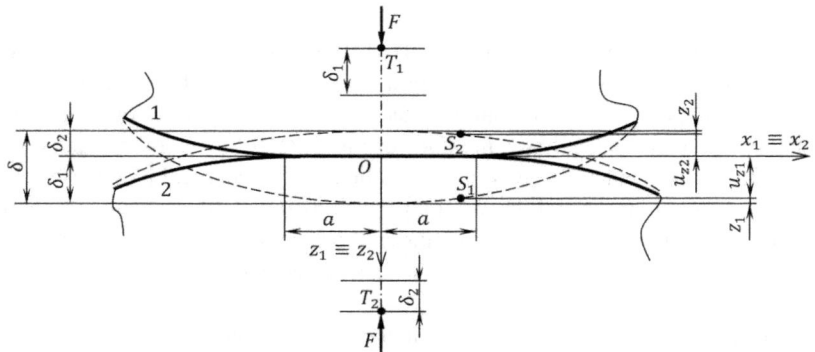

Fig. 1. Contact of two non-conforming bodies after elastic deformation.

$$z_i = \mp \frac{1}{2}\left(\frac{1}{\rho_{xi}}x^2 + \frac{1}{\rho_{yi}}y^2\right), \qquad i = 1, 2 \tag{1}$$

where ρ_{xi} and ρ_{yi} are the principal radii of curvature of the surface at the rectangular coordinate system origin, see Fig. 2. Contact stresses and deformations satisfy the

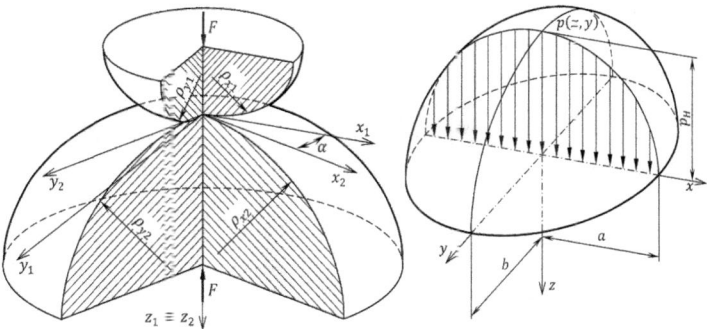

Fig. 2. Elliptical contact.

differential equations for the stress and strain of homogeneous, isotropic, and elastic bodies in equilibrium. The pressure distribution on the contact area is given by the equation [1]:

$$p(x,y) = p_H \sqrt{1 - \left(\frac{x}{a}\right)^2 - \left(\frac{y}{b}\right)^2} \tag{2}$$

where a and b are respective major and minor semi-axes of the elliptical contact area and a maximum value p_H is called Hertzian pressure.

The shape of two unloaded bodies in contact can be described locally with their orthogonal radii of curvature ρ_{xi} and ρ_{yi}, see Fig. 2. The orthogonal coordinate system of one body may be rotated relative to the other by an arbitrary angle α. Based on the individual radii of curvature, an equivalent radius of curvature R_e is introduced by the equation [1]:

$$R_e = \sqrt{R_x R_y} = 1 / \sqrt{4AB} \tag{3}$$

where A and B are positive constants and R_x and R_y are the principal relative radii of curvature along the x and y axes. They are defined as:

$$R_x = \frac{1}{(A+B) - (B-A)}, \quad R_y = \frac{1}{(A+B) + (B-A)} \tag{4}$$

$$(A+B) = \frac{1}{2}\left(\frac{1}{R_x} + \frac{1}{R_y}\right) = \frac{1}{2}\left(\frac{1}{\rho_{x1}} + \frac{1}{\rho_{y1}} + \frac{1}{\rho_{x2}} + \frac{1}{\rho_{y2}}\right),$$

$$|B-A| = \frac{1}{2}\left[\left(\frac{1}{\rho_{x1}} - \frac{1}{\rho_{y1}}\right)^2 + \left(\frac{1}{\rho_{x2}} - \frac{1}{\rho_{y2}}\right)^2 + 2\left(\frac{1}{\rho_{x1}} - \frac{1}{\rho_{y1}}\right)\left(\frac{1}{\rho_{x2}} - \frac{1}{\rho_{y2}}\right)\cos 2\alpha\right]^{1/2}$$

Any radius can be positive (convex) or negative (concave) so long as the principal radii R_x and R_y are positive in their product, see Eq. (3).

Let us now consider the two cylinders in contact with the parallel axes for either of the other, the cylinder and the three-dimensional body, which is described locally in contact with the orthogonal radii of curvature. The elasticity characteristics of the bodies in contact are introduced with the effective modulus of elasticity:

$$E^* = (1 - v_1^2)/E_1 + (1 - v_2^2)/E_2 \tag{5}$$

where E_i and v_i are the respective Young's modulus of elasticity and Poisson's ratio of the individual solids.

2.1 Elliptical Contact

When a cam and a crowned cam roller are in contact, their geometry is described with the crown radius $\rho_{x1} = R$, the cam roller radius $\rho_{y1} = r$, $\rho_{x2} \to \infty$ and the cam radius of curvature ρ_{y2}. The contact area is an ellipse and the pressure is a semi-ellipsoid, see Fig. 2.

The pressure distribution on the contact area is given by the Eq. (2). The maximum contact pressure p_H is at the center of the contact patch. It is given by [1]:

$$p_H = 3F/(2\pi c^2), \qquad c = \sqrt{ab} \tag{6}$$

in which variable c expresses equivalent contact radius and it can be solved as [1]:

$$c^3 = \left(\frac{3FR_e}{4E^*}\right)\left[\frac{4}{\pi e^2}(b/a)^{3/2}\right]\left\{\left[(a/b)^2 E(e) - K(e)\right][K(e) - E(e)]\right\}^{1/2} \tag{7}$$

where $K(e)$ and $E(e)$ are complete elliptic integrals of argument:

$$e = \sqrt{1 - (b/a)^2}, \qquad b < a \tag{8}$$

The argument e denotes an eccentricity of the contact ellipse and major and minor semi-axes of the elliptical contact area are expressed as:

$$a = c(1 - e^2)^{-1/4}, \qquad b = c(1 - e^2)^{1/4} \tag{9}$$

The shape and size of the ellipse of contact are defined by the equation [1]:

$$\frac{R_x}{R_y} = \frac{(a/b)^2 E(e) - K(e)}{K(e) - E(e)} \tag{10}$$

The stress distribution along the z-axis is characterized by the main stress components σ_x, σ_y, σ_z. These quantities are assumed as compressive stresses [1]:

$$\frac{\sigma_x}{p_H} = \frac{2b}{e^2 a}(\Omega_x + v\Omega'_x), \quad \frac{\sigma_y}{p_H} = \frac{2b}{e^2 a}(\Omega_y + v\Omega'_y), \quad \frac{\sigma_z}{p_H} = -\frac{b}{e^2 a}\left(\frac{1-T^2}{T}\right) \quad (11)$$

$$\Omega_x = \frac{T-1}{2} + \xi[F(\Phi,e) - E(\Phi,e)], \quad \Omega'_x = 1 - T(a/b)^2 + \xi\left[(a/b)^2 E(\Phi,e) - F(\Phi,e)\right]$$

$$\Omega_y = \frac{T+1}{2T} - T(a/b)^2 + \xi\left[(a/b)^2 E(\Phi,e) - F(\Phi,e)\right],$$

$$\Omega'_y = T - 1 + \xi[F(\Phi,e) - E(\Phi,e)], \quad = \left[(b^2 + z^2)/(a^2 + z^2)\right]^{1/2}$$

The absolute value of stresses decreases in proportion to a distance from the elliptical area:

$$\xi = |z|/a = \cot \Phi \quad (12)$$

The complete elliptic integrals of the first and second kind $K(e)$ and $E(e)$ and the incomplete elliptic integrals of the first and second kind $F(\Phi,e)$ and $E(\Phi,e)$ are tabulated or have to be computed by the numerical integration, see [6].

2.2 Line Contact

In the case of the contact of two cylindrical bodies, the variable y is constant and Eqs. (1) and (2) are functions of the variable x only. The geometry of the two contacting cylinders is described with the equivalent radius of curvature determined by the expression:

$$\frac{1}{R_e} = \frac{sign(\rho_{x2})}{r} + \frac{1}{\rho_{x2}} \quad (13)$$

where $r \equiv \rho_{x1}$ denotes the radius of a cylindrical cam roller and ρ_{x2} is the cam radius of curvature. When both bodies are pressed in contact by a load f in proportion to the length unit [2]:

$$f = F/(2b) \quad (14)$$

they make contact over an area of the half-width a and half-length b with coordinates located $Oxyz$ in its center, see Fig. 3.

Contact load distribution $p(x)$ along the elliptic cylinder is [1]:

$$p(x) = p_H\sqrt{1 - (x/a)^2} \quad (15)$$

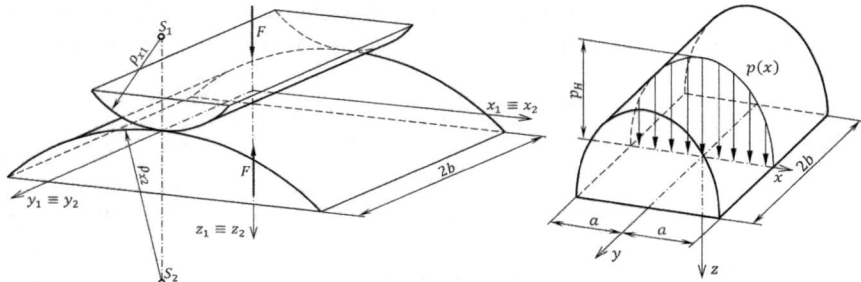

Fig. 3. Line contact.

Hertzian pressure is a maximum compressive stress defined as:

$$p_H = 2f/(\pi a) = \sqrt{fE^*/(\pi R_e)} \tag{16}$$

Rearranging relation (16), the half-width of the contact area is determined as:

$$a = \sqrt{4fR_e/(\pi E^*)} \tag{17}$$

The state of stress existing in the symmetry plane yz is determined by the actual main stress components σ_x, σ_y, σ_z, see Fig. 4. These quantities are assumed as compressive stresses and their absolute value decreases in proportion to a distance z from the contact area. When the proportional independent variable $\xi = |z|/a$ is introduced, the stress components are given by the following expressions [1]:

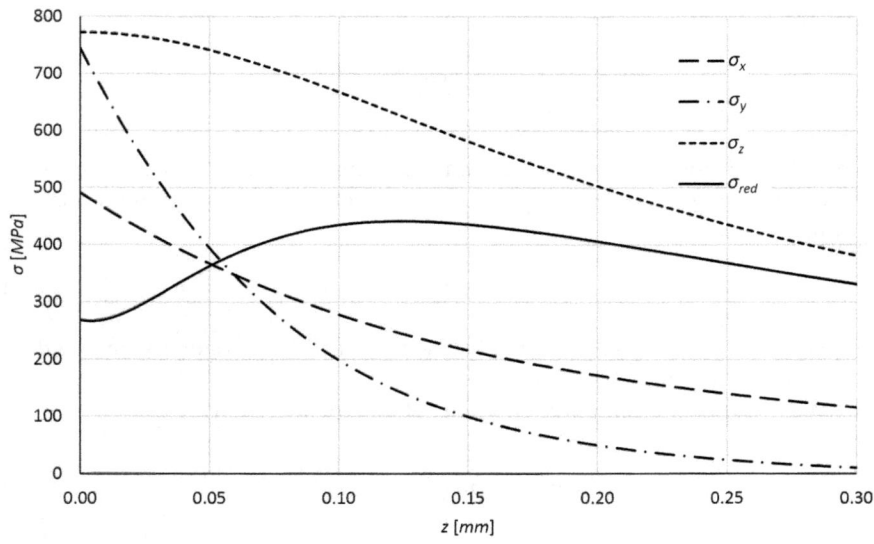

Fig. 4. The state of stress between the crowned roller and the cam in contact.

$$\frac{\sigma_x}{p_H} = -\frac{1+2\xi^2}{\sqrt{1+\xi^2}} + 2\xi, \quad \frac{\sigma_y}{p_H} = -2\nu\left(\sqrt{1+\xi^2} - \xi\right), \quad \frac{\sigma_z}{p_H} = -\frac{1}{\sqrt{1+\xi^2}} \quad (18)$$

2.3 Reduced Stress

At the contact areas of the general kinematic pair and below these points, contact stress becomes a periodical magnitude related to the angular cam displacement ψ. These transitory stresses are characterized by pulses with a periodicity 2π. Strength conditions under a variable load are given by the stress limit σ_h which is in agreement with the disturbance caused by the transitory stress. The reduced stresses $\sigma_{red}(\psi,\xi)$ are limited by the actual strength condition, written as [2]:

$$\max(\sigma_{red}(\psi,\xi)) < \sigma_h, \quad \psi \in \langle 0, 2\pi \rangle, \quad \xi \geq 0 \quad (19)$$

The Tresca yield criterion can be used to find the reduced stress $\sigma_{red}(\psi,\xi)$, according to which this theory can be considered in terms of the maximum sliding stress, proportional to the difference of the main stresses, see [7]:

$$\sigma_{red}(\psi,\xi) = \max\{|\sigma_x - \sigma_y|, |\sigma_z - \sigma_x|, |\sigma_z - \sigma_y|\}, \quad \psi \in \langle 0, 2\pi \rangle, \quad \xi \geq 0 \quad (20)$$

The main stress components σ_x, σ_y, σ_z in the condition (20) are computed by the Eqs. (11) and (18) dependent on the kind of contact of the general kinematic pair. These components must be computed over the whole operation cycle of a cam mechanism to determine their maximum values. From the presented Fig. 4, it is evident that the maximum value of the reduced stress σ_{redMax} is at a distance below the contact patch. In the case of contacting cylindrical bodies, the extreme value reaches the size $\sigma_{redMax} = 0.6 p_H$, see [2].

For steel the usual values are $\sigma_c \approx 0.33 R_m$, $R_{p0.2} \approx (0.55 \div 0.8) R_m$. However, since the transitory stress limit is $\sigma_h \approx 2\sigma_c \approx 0.66 R_m$ the previous relation (19) may be replaced by the condition:

$$\sigma_{red}(\psi,\xi) < R_{p0.2}, \quad \psi \in \langle 0, 2\pi \rangle, \quad \xi \geq 0 \quad (21)$$

The inequality (21) states that no destructive action of elastic deformation is produced in the cam mechanism general kinematic pair under operation, see [2]. The magnitude used in the foregoing text represent, R_m – strength, $R_{p0.2}$ – sliding limit and σ_c – tension-loaded material fatigue limit. In the course of the operation cycle of the cam mechanism, no damage caused by the formation of pits is acceptable on a contact surface under load. Referring to [2], such fatigue damage will not occur if Hertzian pressure is given by the Niemann empirical relation in the form:

$$p_{HMax}(\psi) \leq \frac{K}{N^{1/6}} f(H), \quad K = 4777 MPa \quad (22)$$

where N denotes surface life in million cycles and $f(H)$ expresses the function of surface hardness H as:

$$f(H) = \frac{HB}{1000}, \quad f(H) = 0.251 + \frac{HRC}{100}\left[\frac{HRC}{100}\left(2.74 - 1.22\frac{HRC}{100}\right) - 0.6\right] \quad (23)$$

The surface hardness H related to the maximum Hertzian pressure limit $p_{H\text{Max}}$ in the course of the operation cycle is denoted as HB for the Brinell and HRC for the Rockwell scales. The factor K is the constant, experimentally determined. This criterion is very simple because it is dependent on the only material parameter H.

2.4 Example

Mentioned equations are used to calculate the contact pressure and the components of stress in the contact region of a crowned cam roller and a cam under load $F = 500\ N$. The contact is elliptical due to the shape of both bodies. Therefore the Eqs. (6) to (11) are applied to the calculation. Geometric and material specifications of both bodies in contact are given in Table 1. The stress state is presented in Fig. 4 and its numerical representation is shown in Table 2.

Table 1. Geometric and material specifications.

Description	Symbol	Crowned roller	Cam
Crown radius	$\rho_x = R$ [mm]	500	∞
Radius of curvature	$\rho_y = r$ [mm]	17.5	50
Material: Steel	–	100Cr6	16MnCr5
Modulus of longitudinal elasticity	E [GPa]	210	206
Poisson's ratio	ν [–]	0.3	0.303797468

Another approach to determine the Hertzian pressure and the contact stress, is the FEM use. The FEM-model of the crowned roller and the cam in contact was defined to verify the validity of the Hertzian theory. Solid parabolic tetrahedrons were used to define an elements mesh. The solution of this task was based on the numerical solution of the set of 474219 equations. In this case, the stress state computation converged in 33 contact iterations. The stress state determined by FEM is presented in Fig. 5 and its numerical representation is shown in Table 2.

From the Table 2, the differences between computed values of the contact pressure and reduced stress are evident. In this case, the values differ by approximately 15 percent. In the case of the FEM use, the results of the contact analysis are depended on the element size in the contact region, the used element type, the boundary conditions definition.

Table 2. Hertzian pressure and reduced stress.

Description	Symbol	Hertzian theory	FEM
Maximum contact pressure	$p_{H\text{Max}}$ [MPa]	770	640
Maximum reduced stress	$\sigma_{red\text{Max}}$ [MPa]	440	380

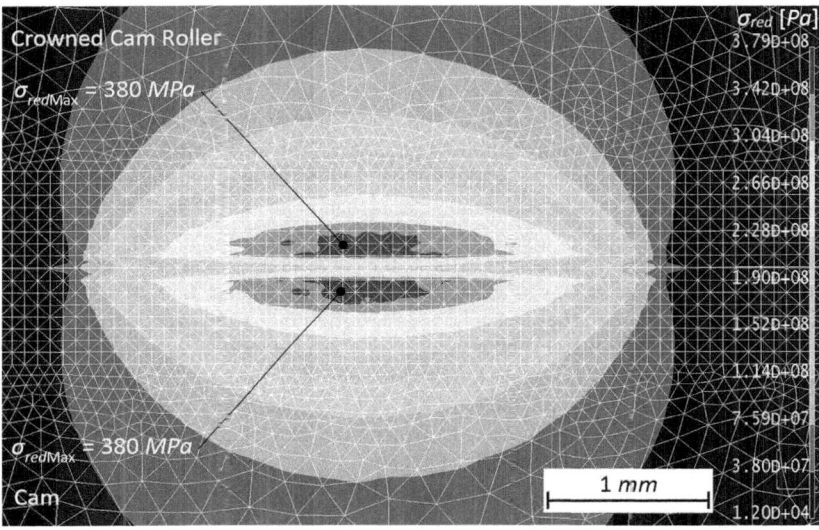

Fig. 5. The stress state between the crowned roller and the cam in contact determined by FEM.

3 Conclusions

The Hertzian contact theory is an important tool for analysis in cam mechanism design. It allows the prediction of the shape and size of a contact region in a general kinematic pair under a load. On the basis of this theory, the components of deformation and stress in a cam and a follower in contact are calculated in the vicinity of the contact region. The results determined on the basis of this theory can also be used to the lifetime estimation of the contact areas of a cam and a follower. The presented relations are quite well adequate for that purpose.

Another approach to determine the Hertzian pressure and the contact stress, is the FEM use, but this approach is time-consuming and therefore inefficient. The very fine mesh of elements must be defined in the contact region, which leads to a large number of solved equations. The computational algorithm of bodies in contact, is based on the numerical iterations of finding of the elements in contact. The FEM use for the computation of the contact stress and deformation components is not quite simple. It requires some experience, while the application of the Hertzian contact stress theory is sufficient for engineering calculations.

Acknowledgments. This paper was created within the work on the project NPU-L012 – Project supported by the Ministry of Education, Youth and Sports of the Czech Republic.

References

1. Johnson KL (1985) Contact mechanics. Cambridge University Press, Cambridge. ISBN 0-521-34796-3
2. Koloc Z, Václavík M (1993) Cam mechanisms. Elsevier, Amsterdam. ISBN 0-444-98664-2
3. Hale LC (1999) Contact mechanics. In: Principles and techniques for designing precision machines. Lawrence Livermore National Laboratory, University of California, pp 414–427. https://de.scribd.com/document/37301862/Principles-of-Precision-Machine-Design
4. Norton RL (2009) Cam design and manufacturing handbook. Industrial Press Inc., New York. ISBN 978-0-8311-3367-2
5. Ondrášek J (2016) The elastic compression in the contact region of a cam mechanism general kinematic pair. In: Proceedings of the XII international conference on the theory of machines and mechanisms. Advances in mechanisms design II. Springer, Switzerland, pp 57–63. ISBN 978-3-319-44086-6, ISSN 2211-0984
6. Arfken GB, Weber HJ, Harris FE (2013) Mathematical methods for physicists, 7th edn. Elsevier, Waltham, pp 927–933 Section 18.8, "Elliptic Integrals", ISBN 978-0-12384654-9
7. Vable M, Mechanics of materials, 2nd ed. Section 10.3, "Failure theories", MTU, pp 486–492. http://www.me.mtu.edu/~mavable/MoM2nd.html

Parallel Manipulators

5 DoF Haptic Exoskeleton for Space Telerobotics – Shoulder Module

Dan Margineanu[(✉)], Erwin-Christian Lovasz,
Corina Mihaela Gruescu, Valentin Ciupe, and Santra Tatar

Politehnica University of Timisoara, Timisoara, Romania
{dan.margineanu,erwin.lovasz,valentn.ciupe}@upt.ro,
corina.gruescu@pt.ro

Abstract. Exoskeletons are used in teleoperation in order to control slave systems replacing the human operator, who would encounter various dangers in outer space, by sending driving commands from the human operator through the exoskeleton system and receiving force feedback. Precise handling, positioning and mounting are possible if teleoperation is assisted both visually and haptic. The present paper describes a lightweight 5DoF haptic exoskeleton, based on a structure which allows rotations of the wrist, elbow and shoulder. As the wrist and the elbow were presented in previous works, the paper emphasizes the construction and functioning of the shoulder.

Keywords: Exoskeleton · Space telerobotics · Haptic feedback

1 Introduction

In the beginnings, the exoskeletons were designed to assist the rehabilitation of physically disabled persons [1]. There are more patents [2–5], proposing different structures and control schemes, meant to rehabilitate the human joints (shoulder, elbow, wrist, knee, ankle or neck) of patients sitting in wheelchairs.

Recently, the exoskeletons' area of use widened. An important domain which implements exoskeletons is telerobotics [6–9]. Outer space implies multiple dangers for the human operator. In order to spare him/her, robotic arms can operate instead. The control of remote robotic arms is best provided by an operator wearing an exoskeleton with haptic reaction.

In several previous works, the authors presented the synthesis and control of two modules of the exoskeleton, namely the wrist and the elbow [10–12]. Different variants of lightweight mechanical structures, actuation solutions and haptic schemes were analyzed. Among potentiometers, encoders, CCD cameras, digital magnetic compasses and 3-axes gyroscopes, the choice for angular sensors or measuring devices went to potentiometers [13]. DC servo- and vibration motors, brakes, linear stepper motors, pneumatic cylinders and miostimulators were discussed and electromagnetic brakes were chosen as most suitable for lightweight space telerobotic applications [13].

The characteristics from different standpoints (static, kinematic, dynamic and economic) of the exoskeleton were defined in [14, 15] and a solution of testing the structure was given in [16].

The present paper focuses on the shoulder module with 2 DoF, i.e. shoulder adduction/abduction and shoulder flexion/extension. Provided with potentiometers to get the rotations to be sent to the slave robot and electromagnetic brakes to give the haptic response, it completes the 5 DoF structure of the exoskeleton.

2 General Design of the 5 DoF Haptic Exoskeleton

The exoskeleton worn by a master human operator is designed to control a slave robotic arm. The structure of the exoskeleton follows he human architecture, and therefore, it contains modules corresponding to the shoulder, the elbow and the wrist.

Considering the requirements regarding the mobility, dexterity, kinematics, ergonomics, weight and economical efficiency, the structure chosen for the exoskeleton is a kinematic chain with 5 degrees of freedom, achieved by means of revolute joints.

The shoulder and the wrist have both two degrees of freedom. Figure 1 presents a general view of the exoskeleton and emphasizes the five modules, each of them being based on a revolute joint. The weight of the elbow, forearm and wrist modules is around 2.5 kg, only half of it being supported by the operator, the rest is taken by the revolute joint on the shoulder flexion/extension module. The weight of the shoulder modules is irrelevant for the study since they are fixed.

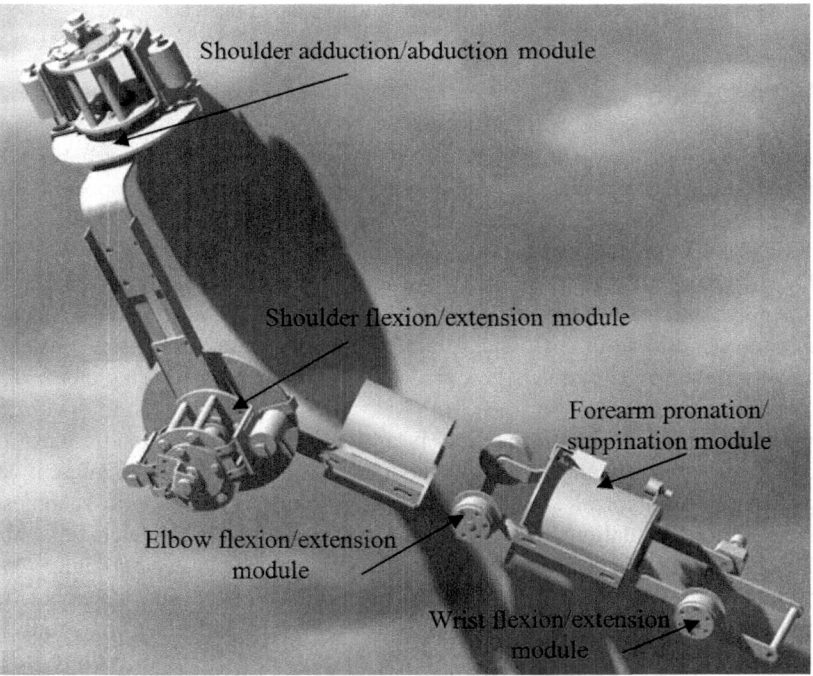

Fig. 1. General view of the 5 DoF exoskeleton

The structural and control design is developed considering the maximal haptic reaction felt by the operator, equivalent to the one created by a 25 N load on his/her palm. The lever arms and the maximal reactive moments taken into calculus are given in Table 1 for each revolute joint. Even though this haptic reaction is just a fraction from the human hand's force capability and cannot block the operator's action at overloading by contact with stiff obstacles, it was accepted as suitable for tests. Larger reaction forces would lead to larger and heavier mechanical structures for the exoskeleton.

Table 1. Lever arms and maximal reactive torque in the exoskeleton's modules

Module	Wrist (flexion/extension)	Forearm (pronation/suppination)	Elbow (flexion/extension)	Shoulder
Lever arm [mm]	100	100	400	800
Maximal reactive torque [Nm]	2.5	2.5	10	20

3 Design of the Shoulder Module

The structure of the exoskeleton corresponding to the human wrist, forearm, arm and shoulder contains five modules:

- The shoulder adduction/abduction module;
- The shoulder flexion/extension module;
- The elbow module;
- The forearm pronation/suppination module;
- The wrist flexion/extension module;

In Fig. 2, a kinematic scheme of the exoskeleton's structure is presented. All modules are based on revolute joints. The CAD model and the practical assembly of the basic joint are presented in Fig. 3. The shoulder adduction/abduction module contains a revolute joint (horizontal/vertical axis, according to the robot's configuration) between the frame and a shoulder plate 1 with adjustable length a_1, used to fit the module to the user's dimensions.

The shoulder flexion/extension module consists of a revolute joint with horizontal axis between the shoulder plate 1 and the arm-shoulder stem 2, also with adjustable length a_2. The two joints are identical and are installed at an angle α_2 of 90° to each other, given by the bending angle of the shoulder plate 1.

The mobile plates 1 and 2 take the rotational movements θ_1 and θ_2 performed by the operator. The movement is transmitted to the disk and the shaft. The potentiometers get the rotation signals from the shafts and send the command to the slave robot to preform the desired motion.

The haptic reaction on the 2 DoF shoulder module's axes is given by two pair of electromagnetic driven brake calipers eb clamping on the brake disks. It produces a

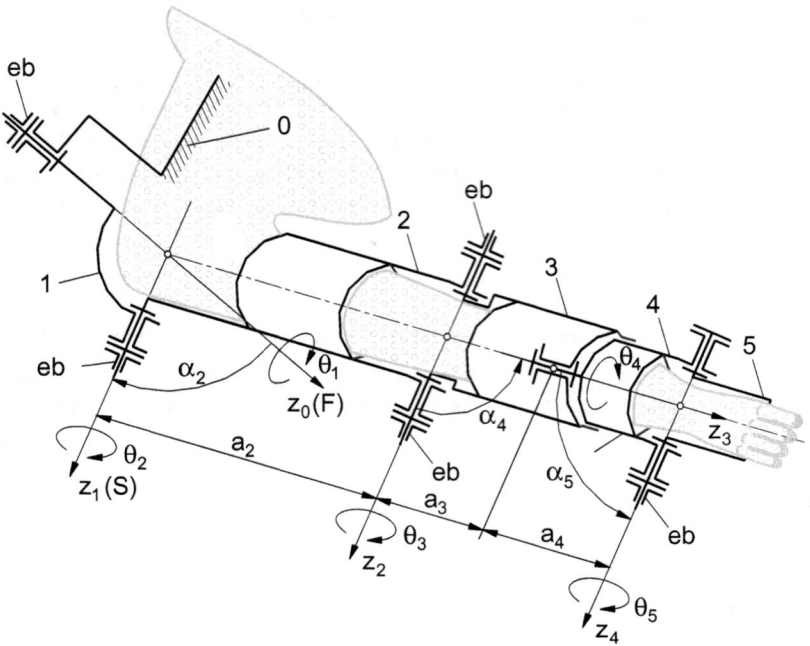

Fig. 2. Kinematic scheme of the exoskeleton's structure

proportional feedback force F_{ps} on the operator's palm according to the lever arm $a_2 + a_3 + a_4$:

$$F_{ps} = \frac{T_{ps}}{a_2 + a_3 + a_4} \qquad (1)$$

Self-adjusting in axial direction due to guides acting as prismatic joints and compensating the mounting errors and the pads' 24 wear, the calipers (23 in Fig. 4) are ensuring an even distribution of the clamping pressure on the friction surfaces.

The elbow module provides the command angle θ_3 for the corresponding robot joint and gives its haptic reaction torque T_{pe} by the action of two concentric compact electromagnetic brakes ep mounted on both sides. A reaction force F_{pe} of:

$$F_{pe} = \frac{T_{pe}}{a_3 + a_4} \qquad (2)$$

will be felt on operator's palm.

Similarly, the forearm pronation/suppination module and the wrist flexion/extension module provide the command angles θ_4 and $\theta 5$ for their corresponding robot joints and give theirs haptic reaction torques. In [12], the linear dependency between joint rotation angle and potentiometer value is described, with a resolution of 0.17°.

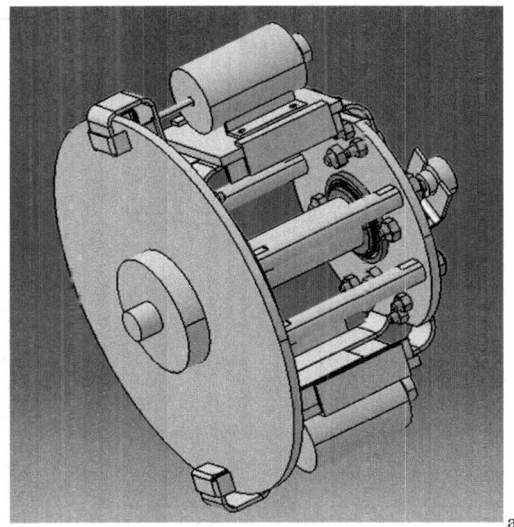

Fig. 3. CAD model (a) and practical assembly (b) of the revolute joint

The section view in Fig. 4 illustrates the construction and the functioning of the shoulder module.

The friction between the disk and the brake pads generates the haptic moment T_{ps} up to 20 Nm. For a friction coefficient μ_{ps} of 1.25 for 2 brake pads on clean and dry d_{ps} = 200 mm aluminum disk, the necessary clamping force F_{clamp}:

$$\left(F_{clamp}\right)_{nec} = \frac{\left(T_{ps}\right)_{nec}}{2 \cdot \mu_{ps} \cdot d_{ps}} \tag{3}$$

is 40 N, as given by a tubular solenoid EMA 3257S.

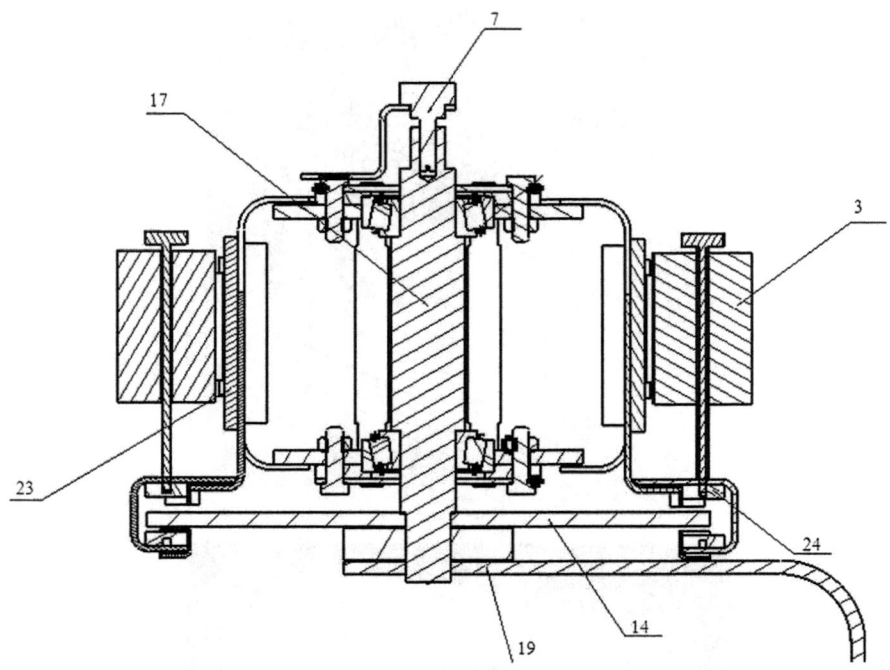

Fig. 4. Construction of the joint (3 – tubular solenoid; 7 – potentiometer; 14 – brake disk; 17 – shaft; 19 – shoulder plate; 23 – caliper; 24 – brake pad)

The clamping pressure p_{clamp} on the brake pads l_p = 10 mm long by w_p = 10 mm wide

$$p_{clamp} = \frac{(F_{clamp})_{nec}}{l_p \cdot w_p} \quad (4)$$

is 0.4 MPa.

4 Exoskeleton Teleoperation Simulation

In order to test the abilities of the completed exoskeleton some teleoperation tests must be conducted. One way of achieving this goal is to develop a slave robot manipulation scene in a virtual environment like V-Rep [17]. The scene consists of a 6-axis serial robot (Fig. 5) which has axes 1, 2, 3, 5 and 6 controlled by the exoskeleton via a custom built electronic interface and a custom script associated with the virtual robot.

The interface is based on an Arduino board and converts the potentiometers' signals into a serial data string [11, 12] for controlling the virtual robot joints (Fig. 6). It also contains the necessary ancillary power electronics for controlling the proportional force feedback provided to the brakes, alongside an axes position information display

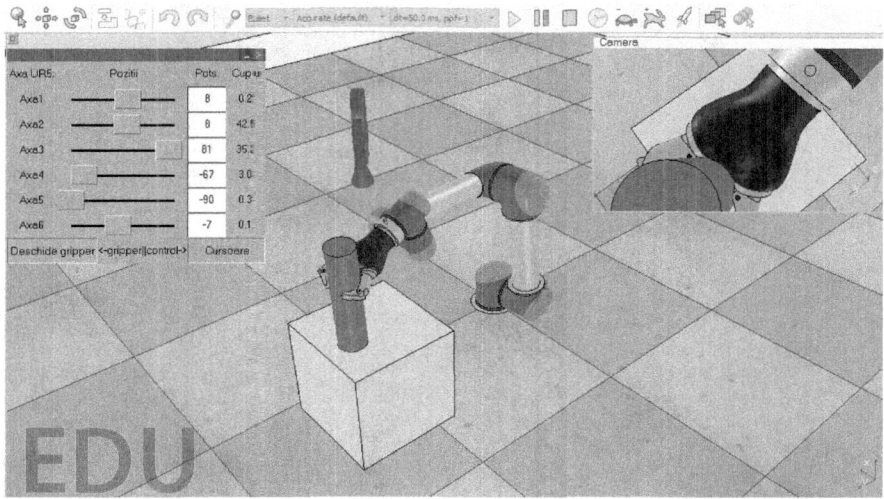

Fig. 5. Virtual robot scene. Axes movement inputs provided by the exoskeleton or via sliders.

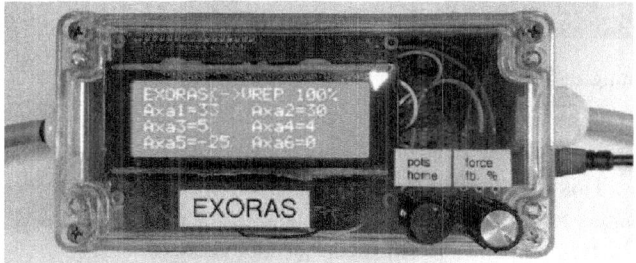

Fig. 6. Exoskeleton electronic interface; schematic and displayed values.

and two user controls for the force feedback gain and for potentiometer offset compensation.

With this setup completed and the exoskeleton connected to the virtual scene via the electronic interface, a human operator can conduct motion/positioning tests and experience the robot provided force feedback (Fig. 7).

For the force reaction to be relevant during tests, the gravity component in the simulation is active and also robot joints reaction torques are active on collision with various surrounding objects introduced in the simulation scene, e.g. a cube, a box or a cylinder, with stiffness properties generated automatically by the simulation environment.

The control characteristics of the brakes, i.e. the dependency between the applied voltage and the reactive torque, presented in [12, 14, 15] were experimentally determined for the desired joints of the exoskeleton. Even though they are non linear, the

Fig. 7. Operating the exoskeleton; virtual robot positioning task with force feedback active.

haptic response on operator's palm is more intense as the reaction on the slave robot joint is greater. This disadvantage may be compensated by a proper special software.

Data exchange between the virtual scene and the real exoskeleton is conducted at a frequency of 20 Hz and it has been noted that this value offers good response time for the operator and delivers a suitable testing environment.

Tests are to be continued in the project and the test results will be published in a future work.

5 Conclusions

The developed exoskeleton may be used in space teleoperation to ensure both safety for the operator against dangers in outer space and fine control by haptic reaction for accurate handling. The lightweight 5DoF haptic exoskeleton is based on a modular structure allowing wrist, elbow and shoulder rotations.

The paper emphasizes the construction and functioning of the 2DoF shoulder module. The two revolute joints are connected through a shoulder plate with adjustable length to fit the user's dimensions. The slave robot gets the commands from potentiometers in each revolute joint and the haptic reaction is provided by custom made electromagnetic brakes.

For testing the completed exoskeleton, a slave robot manipulation scene in a virtual environment V-Rep was developed, consisting of a 6-axis serial robot controlled by the exoskeleton. A custom built electronic interface based on an Arduino board and a custom script associated with the virtual robot were also developed. They control the virtual robot joints and provide the force feedback on the brakes on exoskeleton's joints.

Acknowledgment. The authors would like to express their gratitude towards the Romanian Space Agency (ROSA) for the support through the project New Haptic Arm Exoskeletons for Robotics and Automation in Space (EXORAS).

References

1. Caldwell GD, Tsagarakis GN (2003) Development and control of a 'soft-actuated' exoskeleton for use in physiotherapy and training. Auton Robots 15(1):21–33
2. Dariush B (2010) Exoskeleton controller for a human-exoskeleton system. US 7,774,177 B2. G06F 17/10
3. Joutras FE, Hruska RJ Jr (1999) Exercise apparatus and technique. US005954621A. 5,954,621. A63B 21/02
4. Patoglu V (2012) Exoskeleton. US 20120330198A1. A61H 1/02
5. Perry J, Rosen J (2008) Exoskeleton. US 2008/0009771 A1. A61B 5/103
6. Sicilianao B, Khatib O (eds) (2016) Springer handbook of robotics, 2nd edn. Springer-Verlag Berlin Heidelberg. ISBN: 978-3-319-32550-7. doi:10.1007/978-3-319-32552-1
7. Krishna RA et al (2012) Design and implementation of a robotic arm based on haptic technology. J Eng Res Appl (IJERA) 2(3):3098–3105
8. Letier P, Motard E, Ilzkovitz M, Preumont A, Verschueren JP (2011) SAM: portable haptic arm exoskeleton upgrade. In: Technologies and new application fields. Proceedings of ASTRA 2011, Noordwijk, Netherlands
9. Mauricio J, Motta, ST. Huat LK (2006) Robot calibration: modelling measurement and applications. Industrial robotics: programming, simulation and applications (edn)
10. Lovasz E-C, Mărgineanu D, Ciupe V, Stan SD, Zăbavă E (2013) Synthesis and design of an elbow module for a haptic device used in space telerobotics. In: International scientific conference on advances in mechanical engineering, Debrecen, Hungary
11. Lovasz E-C, Mărgineanu DT, Ciupe V, Maniu I, Gruescu CM, Stan SD, Zăbavă ES (2014) Design and control solutions for haptic elbow exoskeleton module used in space telerobotics. In: Proceedings of 2014 IFToMM Asian conference on mechanism and machine science, Tianjin, China
12. Mărgineanu DT, Lovasz E-C, Ciupe V, Zăbavă ES (2015) 3DoF haptic exoskeleton for space telerobotic. In: Mechanisms, transmissions and applications. In: Proceedings of the third MeTrApp conference 2015, mechanisms and machine science, vol 31, p 279. Springer
13. Lovasz E-C, Ciupe V, Gruescu CM, Margineanu D (2013) Study regarding the actuation choice for the elbow module of an exoskeleton arm. Robotica Manage 18–2
14. Mateaş MC, Lovasz E-C, Mărgineanu DT, Ciupe V, Maniu I, Mărgineanu EZ (2014) Control characteristics of haptic exoskeleton elbow module used in space robotised applications. In: ACME Iaşi

15. Lovasz E-C, Mateas M, Gruescu CM, Mărgineanu EZ, Carabas I, Stan SD (2014) Development of a quality indicator system for haptic exoskeleton modules. In: Proceedings of ACME 2014, applied mechanics and materials, vol 658, pp 648–653. ISSN: 1662-7482
16. Ciupe V, Lovasz E-C, Gruescu CM, Margineanu D, Maniu I. Testing the haptic exoskeleton actuators in a virtual environment. In: The 14th IFToMM world congress, Taipei, Taiwan
17. Copellia Robotics: Virtual Robot Experimentation Platform, October 2015. http://v-rep.eu

Kinematic Design of a Tripod Parallel Mechanism for Robotic Legs

Matteo Russo[(✉)] and Marco Ceccarelli

LARM, University of Cassino and Southern Latium,
Cassino, Italy
{matteo.russo,ceccarelli}@unicas.it

Abstract. In this paper a parallel manipulator with tripod architecture is proposed with the closed-form formulation of its functioning. The mechanism is 3-Degrees-of-Freedom manipulator that can be used as robotic leg for mobile walking robots with two or more legs. The kinematic design is characterized by direct linear drives in the links of the tripod architecture and by a mechanism providing a spherical motion to the foot end-effector.

Keywords: Robot design · Leg mechanism design · Kinematics · Parallel robots

1 Introduction

In the past decades, the development of leg mechanisms for walking robots has been focused on serial architectures, for example the ones in Honda's Asimo [1] and Nao [2]. Few robot legs with parallel architecture can be found in literature, such as the WL family [3], LARM biped locomotor [4], and even quadruped [5] or hexapod [6] robots, despite they are more accurate and able to move larger payloads than serial architectures [7, 8]. They have indeed several disadvantages, first of all their smaller workspace. The step length of a human being is approximately 94% of the leg's height and robot leg motion is expected to be sized in a similar way [9]. However, parallel robots can rarely reach this value. LARM biped locomotor, for example, has a step length equal to 30% of its height [4]. Furthermore, parallel robot kinematics is usually difficult to solve, leading to complex control algorithms.

This paper describes an evolution of the tripod architecture used for LARM biped locomotor, which is described in [10–12]. While in the original manipulator the end-effector of the robot is a platform that is connected to the legs in three different points, the end-effector of the novel mechanism can be modelled as a single point H as shown in Fig. 1c, where the three legs converge in a triple joint. This change simplifies both kinematics and control, while improving the workspace of the mechanism.

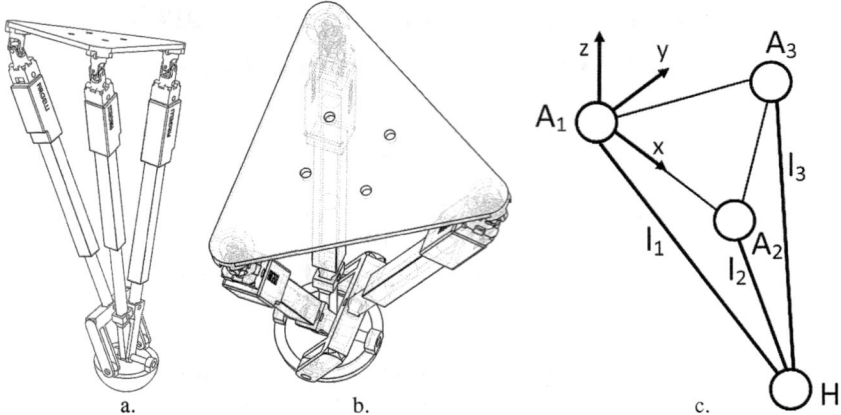

Fig. 1. The proposed mechanism: a. side view; b. upper view; c. kinematic diagram.

2 Designed Structure and Its Kinematics

The proposed architecture is the 3 degrees-of-freedom parallel mechanism that is shown in Fig. 1a and b, whose kinematic diagram is shown in Fig. 1c. The structure is characterised by a fixed, triangular base with vertices in points A_1, A_2 and A_3, three legs with a **UPR** structure, where **U** is a universal joint, **P** is a prismatic coupling and **R** is a revolute joint, and an end-effector body, which is shown in Fig. 2 and acts as a punctiform body in point H. Therefore, the orientation of the end-effector is not taken into account and only its position kinematics is studied. When the coordinate system is the one shown in Fig. 1c, centred in point A_1, the geometry of the mechanism is described by the position of points A_i, which is, respectively, $\mathbf{r_{A1}} = (0; 0; 0)^T$, $\mathbf{r_{A2}} = (a; 0; 0)^T$, and $\mathbf{r_{A3}} = (b; c; 0)^T$, and by the length l_i of the linear actuators, where l_i can be defined as the magnitude of vector $\mathbf{l_i} = \mathbf{r_H} - \mathbf{r_{A1}}$, where $\mathbf{r_H}$ defines the position of the end-effector point H. In the proposed design, the base triangle is an equilateral triangle in which all three sides are equal to a. Therefore, the position of point A_3 can be also written as $(1/2a; \sqrt{3}/2a; 0)^T$.

The end-effector joint, which is shown in Fig. 2a, acts as a universal joint where one of the two revolute joints is connected to two different links. Therefore, two legs are constrained to rotate around its y-axis, shown in Fig. 2b, while the third one can only rotate around the x-axis. All the rotations are centred on point H. The convergence of the three links in a single point simplifies the kinematic formulation of the mechanism and allows for an improved control of its motion. This arrangement also overcomes one of the main disadvantages of parallel manipulators with similar structures, such as the **3UPU** mechanism, that is the high sensitivity to clearance in the joints, which leads to large parasitic motion.

In order to compute the position of the end-effector point $\mathbf{x} = (x; y; z)$ as a function of the joint coordinates $\mathbf{q} = (l_1; l_2; l_3)$, the forward kinematic problem has to be solved. The position of the end-effector is, geometrically, the point of intersection of three

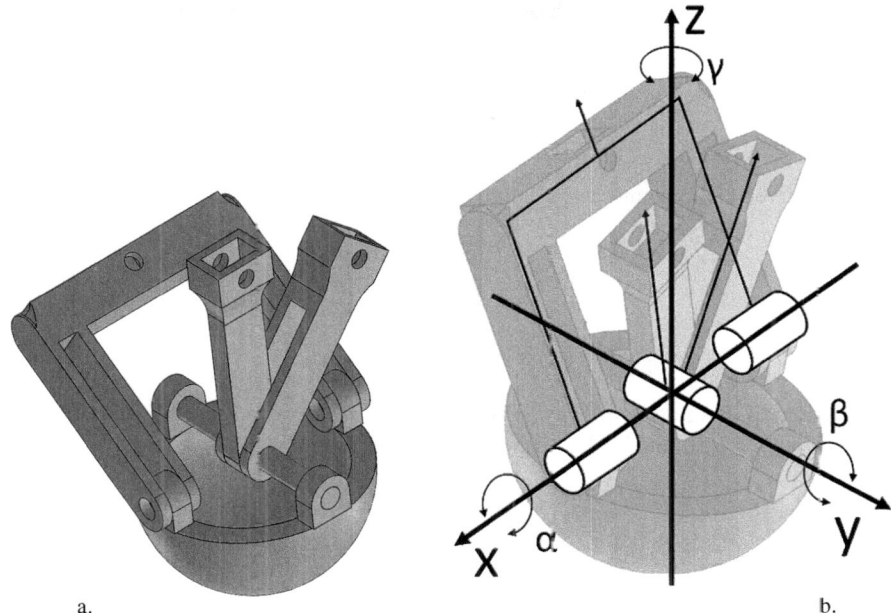

Fig. 2. Structure of the end-effector joint: a. side view; b. kinematic diagram.

spheres centred in A_1, A_2, and A_3, with radiuses equal to respectively l_1, l_2, and l_3. The equations that describe those spheres can be written as

$$l_1^2 = x^2 + y^2 + z^2$$
$$l_2^2 = (x - a)^2 + y^2 + z^2 \quad (1)$$
$$l_3^2 = (x - b)^2 + (y - c)^2 + z^2$$

The inverse kinematic problem can be then solved straightforward from Eq. (1) as

$$l_1 = \sqrt{x^2 + y^2 + z^2}$$
$$l_2 = \sqrt{(x - a)^2 + y^2 + z^2} \quad (2)$$
$$l_3 = \sqrt{(x - b)^2 + (y - c)^2 + z^2}$$

The forward kinematic problem is instead formulated from Eq. (1) as

$$x = \tfrac{1}{2a}\left(l_1^2 - l_2^2 + a^2\right)$$
$$y = \tfrac{1}{2c}\left(l_1^2 - l_3^2 + b^2 + c^2 - 2bx\right) \quad (3)$$
$$z = -\sqrt{l_1^2 - x^2 - y^2}$$

From the results in Eq. (3) it is possible to compute the Jacobian matrix of the manipulator as

$$J = \begin{bmatrix} \frac{l_1}{a} & -\frac{l_2}{a} & 0 \\ \frac{l_1}{c}\left(1-\frac{b}{a}\right) & \frac{bl_2}{ac} & -\frac{l_3}{c} \\ \frac{l_1}{z}\left(1-\frac{x}{a}\right) - \frac{yl_1}{cz}\left(1-\frac{b}{a}\right) & \frac{l_2}{az}\left(x-\frac{by}{c}\right) & \frac{yl_3}{cz} \end{bmatrix} \quad (4)$$

Thus, the inverse Jacobian can be evaluated from Eq. (2) as

$$J^{-1} = \begin{bmatrix} \frac{x}{l_1} & \frac{y}{l_1} & \frac{z}{l_1} \\ \frac{x-a}{l_2} & \frac{y}{l_2} & \frac{z}{l_2} \\ \frac{x-b}{l_3} & \frac{y-c}{l_3} & \frac{z}{l_3} \end{bmatrix} \quad (5)$$

The orientation of the end-effector depends on its position and can be described with the Euler angles α, β and γ shown in Fig. 2b and given by

$$\begin{aligned} \alpha &= \arctan\left(-\frac{y}{z}\right) \\ \beta &= \arctan\frac{x-b}{\sqrt{(y-c)^2 + z^2}} \\ \gamma &= 0 \end{aligned} \quad (6)$$

Equations (1) to (6) are useful both for the analysis of motion performance and for motion planning of the tripod leg mechanism.

3 Workspace Evaluation

The volume of the workspace of the manipulator is computed by discretisation as a point cloud. Since the workspace is the set of points that the robot can reach, it can be formulated as

$$\begin{aligned} x &= \tfrac{1}{2a}\left(l_1^2 - l_2^2 + a^2\right) \\ y &= \tfrac{1}{2c}\left(l_1^2 - l_3^2 + b^2 + c^2 - 2bx\right) \\ z &= -\sqrt{l_1^2 - x^2 - y^2} \\ &\text{for } l_1, l_2, l_3 \in [l_0, l_0 + s] \end{aligned} \quad (7)$$

Figure 3 shows the workspace computed as a cloud of reachable end-effector position, which was obtained by discretising with Eq. (7) the stroke of the actuator in Eq. (5). In order to get the results, the mechanism has been sized with commercial linear actuators, with a = 120 mm and the actuators' length ranging from 232 mm to 332 mm.

Thanks to the particular arrangement of the end-effector joints, the workspace of this mechanism is larger when compared with the one of similar structures of equal size. In particular, the volume of the workspace for the proposed structure obtained with a discrete computation on *Matlab* has been compared with the one of the 3UPU structure proposed in [4], since the geometry of the actuated links and of the base platform is the same for both structures and the only difference is the end-effector structure. As shown

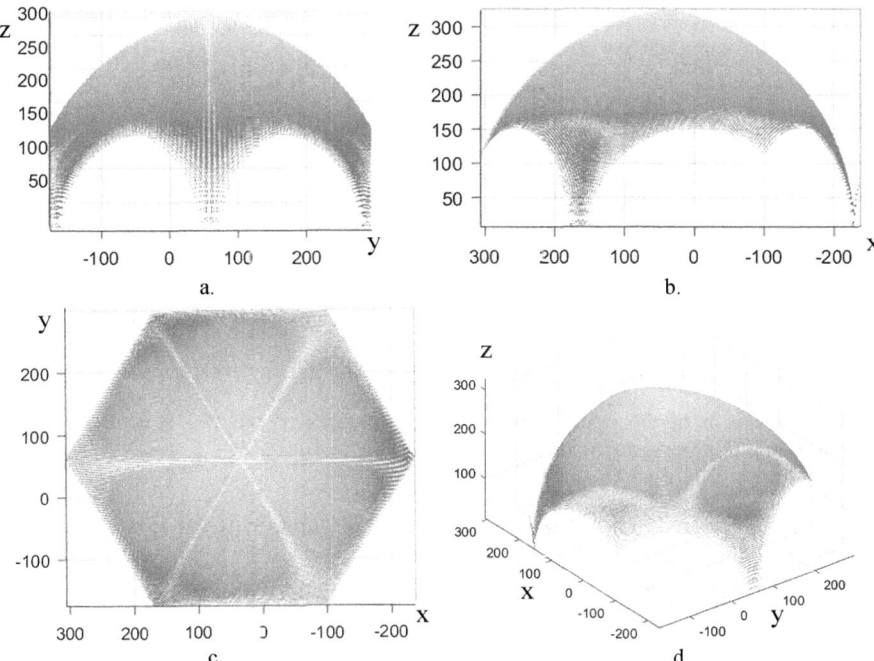

Fig. 3. Workspace computed as a cloud of reachable points, with colour scale representing the z coordinate of each point: a. y-z view; b. x-z view; c. x-y view; d. isometric view.

in Fig. 4b, the proposed structure's workspace is characterized by a shape that is similar to the 3UPU structure's workspace in Fig. 4a, but the surface is larger. In fact, the computed volume for the 3UPU structure's workspace is $5.0986 \cdot 10^6$ mm^3, while the volume of the proposed structure is equal to $7.8431 \cdot 10^6$ mm^3, that is 53,83% larger than the previous one.

Another advantage of this manipulator over the alternative structure is that its workspace and its singularity-free workspace are coincident. In order to compute the singular positions of the proposed structure, its kinematics can be written as

$$\mathbf{A}\dot{\mathbf{q}} + \mathbf{B}\dot{\mathbf{x}} = \mathbf{0} \qquad (8)$$

Matrix **A** and **B** can be expressed as

$$\mathbf{A} = \begin{bmatrix} l_1 & 0 & 0 \\ 0 & l_2 & 0 \\ 0 & 0 & l_3 \end{bmatrix}; \quad \mathbf{B} = \begin{bmatrix} x & y & z \\ x-a & y & z \\ x-\frac{a}{2} & y-\frac{\sqrt{3}a}{2} & z \end{bmatrix} \qquad (9)$$

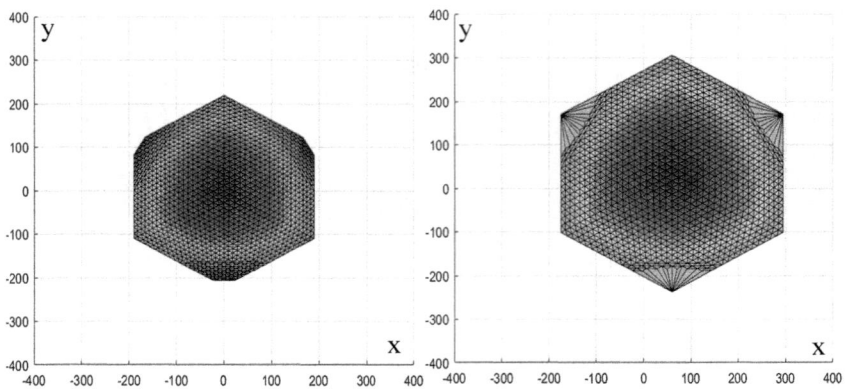

Fig. 4. Comparison of the upper view of the workspaces of the two structures: a. 3UPU structure in [4]; b. proposed structure.

Therefore, the manipulator is in a singular configuration when it is in one of the following situations:

- Redundant input singularity, when a zero twist of the end-effector is obtained for non-zero actuated joint velocities. This singular configuration implies that matrix **A** is singular. This condition is written as

$$\mathbf{x} = (0 \ \ 0 \ \ 0)^T; \ \mathbf{x} = (a \ \ 0 \ \ 0)^T; \ \mathbf{x} = (b \ \ c \ \ 0)^T \tag{10}$$

and therefore defines the singular points as the base joints of the robot. These singular configurations are physically impossible for the end-effector to reach and they are never contained in the workspace of the mechanism.
- Redundant output singularity, when a non-zero twist of the end-effector may be obtained although the actuators are locked. This singular configuration implies that matrix **B** is singular, which is a condition that can be written as

$$z = 0 \tag{11}$$

Therefore, locus of the singularities of the mechanism is the plane which the base platform of the proposed structure lies on. Again, the condition is physically impossible and it never occurs within the workspace of the mechanism. Thus, the manipulator reachable workspace and singularity-free workspace are coincident.

4 A Prototype

In order to check the feasibility of the structure, a prototype has been manufactured and tested. It is characterized by the geometry presented in Sect. 3, which is used for the evaluation of the workspace, and its motion capability. The prototype has been assembled with commercial components and 3D-printed parts. The actuators are L16 linear drives produced by Actuonix, whose datasheet can be found in [13]. Both screws and universal joints are commercial component, while the base platform and the end-effector have been designed in a CAD environment, sliced and printed in PLA plastic with a Flashforge Creator Dual 3D printer [14]. The motors are controlled with an Arduino Control Board [15].

The prototype is shown in Fig. 5. The base side of the manipulator is 120 mm, while its height varies from 250 to 340 mm, depending on the configuration. Its motion capability has been tested by making the end-effector move along several trajectories in the entire workspace. Some pictures of the different poses reached by the end-effector are shown in Fig. 6. Despite materials, universal joints and motors being the same as the prototype in [4], the proposed structure performed better by not showing any parasitic motion due to joint clearance.

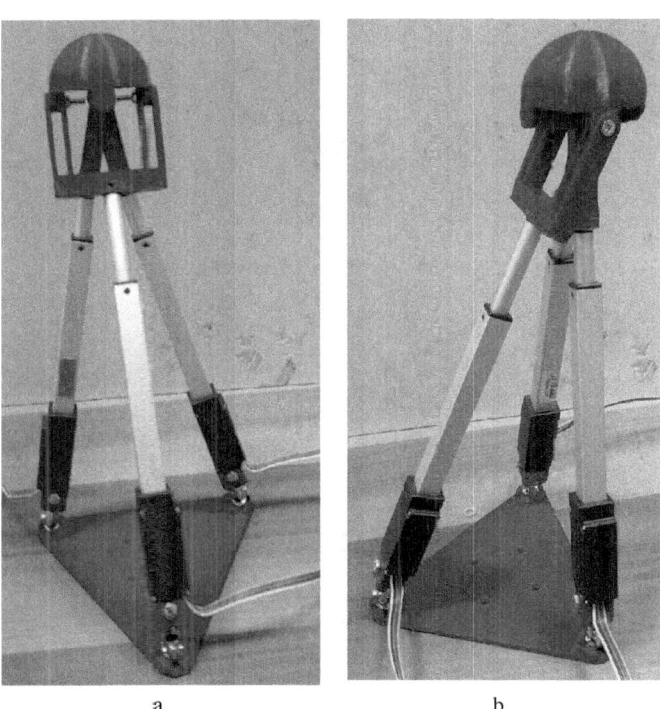

Fig. 5. Prototype of the proposed structure: a. front view; b. lateral view.

Furthermore, the tests proved that the end-effector joint is able to move in the entire workspace without collisions among the links, even if the angular movement that is allowed to them is limited by the particular geometry of the joints. In particular, for the dimensioning used for the prototype, the critical angles δ_1, δ_2 and δ_3 shown in Fig. 7 have to range between 0,1651 rad and 1,4056 rad to avoid interferences.

Fig. 6. Motion test of the proposed mechanism.

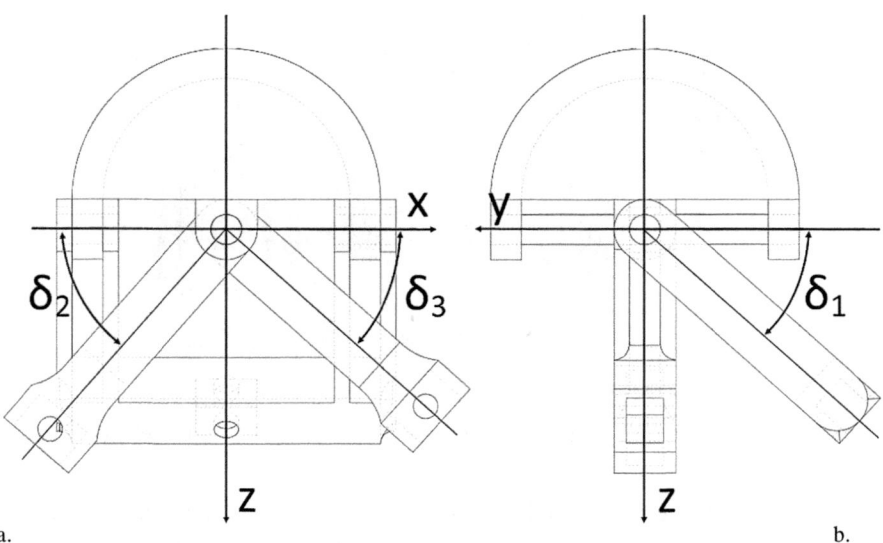

Fig. 7. Critical angles for link interference in the end-effector body.

In fact, during the motion of the manipulator in the whole workspace the three critical angles are in their safe range. As expected, the mechanism prototype did not encounter any singular point during its motion, even when moving at the borders of its workspace. The maximum measured speed of the end-effector point H was equal to 100 mm/s, obtained with a maximum actuation speed equal to 20 mm/s.

Two patents on the proposed structure are pending, one for the foot mechanism [16] and one for the tripod [17].

5 Conclusions

This paper describes the kinematic design for a 3-DoFs robot leg with tripod parallel architecture and with a specific design of the foot mechanism. Forward and inverse kinematic problems are solved for the position of the end-effector, and its orientation has been defined through additional equations. The Jacobian matrix of the mechanism and its inverse have been computed in a closed-loop formulation. The workspace and the singularities of the mechanism have been evaluated, and a first prototype has been manufactured in order to experimentally test and validate the proposed design. The mechanism improves considerably previous designs in workspace volume and performance. Furthermore, it cannot move in singular configurations. Future development on the mechanism design will improve the solution through dynamic and stiffness analysis that will lead to a multi-objective optimization of the geometry of the leg.

References

1. Chestnutt J, Lau M, Cheung G, Kuffner J, Hodgins J, Kanade T (2005) Footstep planning for the honda ASIMO humanoid. In: Proceedings of the 2005 IEEE international conference on robotics and automation. ICRA 2005, April 2005. IEEE, pp 629–634
2. Gouaillier D, Collette C, Kilner C (2010) Omni-directional closed-loop walk for NAO. In: 2010 Proceedings of the 10th IEEE-RAS international conference on humanoid robots (Humanoids), December 2010. IEEE, pp 448–454
3. Lim HO, Takanishi A (2007) Biped walking robots created at Waseda University: WL and WABIAN family. Philos Trans R Soc Lond A Math Phys Eng Sci 365(1850):49–64
4. Wang M, Ceccarelli M (2015) Design and simulation of walking operation of a Cassino biped locomotor. In: New trends in mechanism and machine science. Springer International Publishing, pp 613–621
5. Wang H, Sang L, Zhang X, Kong X, Liang Y, Zhang D (2012) Redundant actuation research of the quadruped walking chair with parallel leg mechanism. In: 2012 IEEE international conference on robotics and biomimetics (ROBIO), December 2012. IEEE, pp 223–228
6. Xin G, Zhong G, Deng H (2015) Dynamic analysis of a hexapod robot with parallel leg mechanisms for high payloads. In: 2015 Proceedings of the 10th Asian control conference (ASCC), May 2015. IEEE, pp 1–6
7. Ceccarelli M (2004) Fundamentals of mechanics of robotic manipulation, vol 27. Springer Science & Business Media, Dordrecht
8. Merlet JP (2012) Parallel robots, vol 74. Springer Science & Business Media, Dordrecht

9. Knudson D (2007) Fundamentals of biomechanics. Springer Science & Business Media, New York
10. Wang M, Ceccarelli M (2015) Topology search of 3-DOF translational parallel manipulators with three identical limbs for leg mechanisms. Chin J Mech Eng 28(4):666–675
11. Wang MF, Ceccarelli M (2013) Design and simulation for kinematic characteristics of a tripod mechanism for biped robots. In: Proceedings of the 22nd international workshop on robotics in Alpe-Adria-Danube region, September 2013, pp 11–13
12. Wang M, Carbone G, Ceccarelli M (2015) Stiffness analysis for a tripod leg mechanism. In: Proceedings of the 14th IFToMM world congress, October 2015, pp 404–410. 國立臺灣大學機械系
13. Actuonix (2016) Firgelli L16 datasheet. http://www.actuonix.com/category_s/1823.htm. Accessed 2 July 2016
14. Flashforge (2016) Flashforge creator dual user manual. http://www.flashforge-usa.com/support/downloads/. Accessed 19 Sep 2016
15. Arduino (2016) Arduino nano control board. https://www.arduino.cc/en/Main/ArduinoBoardNano. Accessed 27 Feb 2016
16. Ceccarelli M, Russo M (2016) Device for the spherical connection of three bodies. IT Patent Application no. 102016000009369, 19 Sep 2016. (in Italian)
17. Russo M, Cafolla D, Ceccarelli M (2016) Device for tripod leg. IT Patent Application no. 102016000097258, 28 Sep 2016. (in Italian)

Parallel Manipulators: Practical Applications and Kinematic Design Criteria. Towards the Modular Reconfigurable Robots

Alfonso Hernández, Mónica Urízar, Erik Macho, and Victor Petuya[✉]

Department of Mechanical Engineering,
University of the Basque Country UPV/EHU, Bilbao, Spain
victor.petuya@ehu.es

Abstract. Modern robotic manipulators play an essential role in industry, developing several tasks in an easy way, enhancing the accuracy of the final product and reducing the executing time. Also they can be found in other fields as aerospace industry, several medical applications, gaming industry, and so on. In particular, the parallel manipulators have acquired a great relevance in the last years. Indeed, many research activities and projects deal with the study and development of this type of robots. Nevertheless, usually, a bilateral communication between industry and research does not exist, even among the different existing research areas. This causes a lack of knowledge regarding works that have been carried out, the ones that are under development and the possible future investigations. Hence, once a specific field of knowledge has acquired a certain level of maturity, it is convenient to reflect its current state of the art. In this sense, the authors of this paper present a review of the different fields in which parallel manipulators have a significant participation, and also the most active research topics in the analysis and design of these robots. Besides, several contributions of the authors to this field are cited.

Keywords: Robots · Parallel manipulators · State of the art · Practical applications · Design criteria · Reconfigurable mechanism

1 Introduction

As known, in a parallel robot the end-effector is linked to the frame by means of independent kinematic chains called legs [1]. This closed-loop architecture has some advantages related to its high load/weight ratio, high stiffness along its workspace, low inertia, high velocities and accelerations and good accuracy. However compared their serial counterparts, they have a limited workspace and a complex kinematics due to the singularities associated to the parallel architecture.

The first modern application of a parallel robot was developed in 1928 by James E. Gwinett who patented [2] an ingenious design for the amusement industry. In 1942, Willard L.V. Pollard designed the first parallel robot for an industrial application (spray painting) [3]. In 1947, the engineer Eric Gough developed the parallel manipulator known as the Hexapod widely used for different applications. The system was originally designed for the Dunlop Tyres Company for tyre quality testing. In 1965, D. Stewart published a paper presenting a parallel robot with very similar characteristics to the one designed by Gough, to be used in flight simulators [4]. Due to their similarities, frequently the Hexapod is referred as the Gough-Stewart platform. Also, the engineer Klaus Cappel, developed and patented independently in 1967 other Hexapod also for flight simulation purposes.

2 Practical Applications of Parallel Robots

In this section some current applications of the parallel robots are presented. Also in [5], other industrial applications of these architectures are discussed.

- *Machine tools.*

Machining operations are very demanding in terms of kinematic and dynamic specifications. Consequently, parallel architectures are a valuable option for this field. Between the different parallel kinematic machines (PKMs) developed in recent years, we highlight the machining centre *Variax Hexacenter* (Giddings and Lewis), and the *Sprint Z^3-head* (Technologie GmbH). The *Octahedral Hexapod VOH-1000* (Ingersoll) with architecture similar to the *Variax* is also a five-axis machine tool based on the Hexapod. *Variax* has an accuracy of 11 µm and the *Octahedral Hexapod* of 20 µm. Other design of Ingersoll is the HOH-600, in which the tool has a horizontal plane. The *Sprint Z^3-head* allows the tool orientation in a range of ± 40° reaching a speed in end-effector of 80°/s.

Between the machining centres presented in [6] we highlight the *Cosmo Centre PM-600* (Okuma), oriented to the production of aluminium dies. The *DECKEL MAHO Pfronten TriCenter DMT100*, has an hybrid architecture combining an 3-dof parallel manipulator and a 2-dof serial chain for milling. The *Vertical Turning Machine Index V100* is used for milling and laser welding.

One of the most successful parallel robots is the *Tricept* (SMT Tricept AB). It has three actuated and one passive kinematic chains obtaining in the end-effector high stiffness and 2R1T motion pattern.

- *Pick and place.*

Parallel robots are very competitive for pick & place applications where low mobility manipulators (less than 6-dof) capable to reach high speeds and accelerations are required. Normally robots with three translational degrees of freedom or four degrees of freedom (Schönflies motion) are required.

The first 3-dof spatial manipulator, the Delta robot [7], was developed in 1990 by Raymond Clavel. This robot has three equal kinematic chains containing a parallelogram. Three rotational motors located in the frame actuate the robot. Also, an actuated telescopic leg connecting the frame with the end-effector provides and additional uncoupled rotational degree of freedom. This concept has been used in successful robots as the *IRB 340 FlexPicker* (ABB) able to perform 150 pick & place cycles in less than one minute with a speed of 10 m/s and a maximum acceleration of 10 g.

There are different variants of the Delta robot that use linear actuators [8] or able to change the orientation of their rotational actuators which is the case of the *NUWAR robot* (Western Australia University) capable of reaching accelerations of 600 m/s². Other parallel robots that generate a 3T motion pattern are: *STAR robot* (Hervé and Sparacino), *3 − RRPaR robot of Maryland University* (Stamper), *3 − UPU robot* (Tsai and Joshi) and the *Cartesian robot 3 − CRR* designed simultaneously in Laval University (Kong and Gosselin) and California Riverside (Kim and Tsai).

Other robots in this field are the *Paraplacer*, which has a hybrid structure, the *Triglide* that is a variant of the Delta robot with linear actuators or the *Hexa* with 6 dof. Recently in the University of Montreal, [9] has been developed the *DEXTAR robot*, based on a kinematic structure 5R, which has the ability to change its configuration between its working modes in order to enlarge the opened workspace [10].

As said the so-called *Shönflies motion*, is quite demanded in pick & place operation. A recent design is the *SMG* (McGill University) [11]. The CompMech research group

Fig. 1. Prototype of Schönflies parallel manipulator. CompMech research group, department of mechanical engineering UPV/EHU/

has also developed a prototype with Schönflies motion (Fig. 1) whose final dimensions were obtained from a multiobjective optimization [12].

- *Automobile industry.*

The automobile industry has been one of the pioneer fields in the use of robots in its production chains. Related to parallel manipulator, the Hexapod is again the most common architecture. Volkswagen in cooperation with Moog has implemented a testing bench in which this architecture is used to test suspension systems and to make fatigue analysis of different components. Also, the parallel robot *F-200iB* (Fanuc) is used for the elevation and positioning of vehicles.

In this sector, we highlight the *MAST (Multi Axis Shaking Tables)*. These machines are able to generate in their end-effector a coupled motion of translations and rotations. They are widely used for the dynamic evaluation of vehicle components. In the last years, different MAST bases on parallel architectures have been developed. Basically they are based on the Hexapod or on a frame locating three or more linear actuators.

- *Medical applications.*

Parallel robots have been also introduced in the medical, rehabilitation and surgical field. In [13] is presented a miniature robot for precision demanding spinal surgical operations. This robot has been the main module of the surgical system *SpineAssist* (Mazor Surgical Technologies).

In neurosurgery we highlight the parallel robot *SurgiScope* (ISIS, Intelligent Surgical Instruments & Systems). This robot is used as a telesurgery tool in neurosurgery.

- *Space applications*

The *Spherical Primary Optimal Telescope, SPOT,* is part of the a research project started in 2003 in the Goddard Space Flight Centre, for developing new concepts for future spatial telescopes. The purpose is to design a robust architecture with three degrees of freedom, tip and tilt rotations and vertical translation, in order to move in a synchronized way the different modules that form the telescope mirror. The final solution is based on the *Tripod* robot with three kinematic chains and linear actuators.

The Aerospace Manufacturing Technology Centre (AMTC) of Canada is exploring the application of the Hexapod for drilling and milling aerospace materials and components.

- *Flight simulators*

As cited in the introduction, one of the first applications of the parallel robots was their implementation in flight simulators. In recent years, we highlight the flight simulator *GRACE - Generic Research Aircraft Cockpit Environment* (National Aerospace Research Lab of Holland). It consists on a hexapod with reconfiguration capacity in order to simulate different aircraft conditions as the Airbus 330 or the Boeing 747-400.

- *Other applications.*

High-speed camera. The *Agile Eye* is a spherical 3 − RRR parallel manipulator designed for locating a camera in its end-effector. It was designed at Laval University [14] obtaining a vision range of 140° with a torsion of ± 30°, reaching velocities and accelerations higher than 1000°/s and 20000°/s^2 respectively.

Haptic devices. They are used to give force and even tactile feedback to the user of a virtual environment. Born for the gaming industry, for example the *Novint Falcon* parallel manipulator, nowadays they are being implemented in other sectors as the medical or industrial fields.

High precision miniaturized robots. In the IWF of Braunschweig University (Germany) some of these robots have been developed. As an example, the *APIS* robot is able to perform a planar motion with three degrees of freedom. Piezoelectric rotational motors actuate it. Other miniaturized parallel robot is *MICABO* with a 3T1R motion pattern. This robot has a 3-PRR architecture where the P joint is actuated via a piezoelectric linear actuator and the R pairs are compliant joints. These robots lie in the field of MEMS (Micro Electro Mechanical Systems) with a great development in recent years.

Cable-driven parallel manipulators. Usually these robots have a hexapod architecture but in this case, cables actuate them. The purpose is to obtain very light structures reaching high accelerations with low energy consumption. In [15] the capabilities of this class of robots are analysed focusing on the project *Tendon-Based Stewart-Platforms in Theory and Application - SEGESTA*. The prototype developed in this project has been implemented in Duisburg University for high-speed manipulability tests. Other potential applications for these robots are: portable rescue cranes [16], aerodynamic tests in wind tunnels [17], medical rehabilitation devices [18], etc.

Three degree of freedom parallel robots. In this category lie the manipulators for orienting a solid in the space and the manipulators with coupled degrees of freedom, as is the case of the 3-RPS and 3-PRS platforms. The most common architecture of this class of robots is an end-effector linked to the frame by three kinematic chains with R joints with axes intersecting in the centre of rotation of the end-effector. This is the case of the Agile Eye previously cited. Other architectures use SPS or UPU kinematic chains. Also, a spherical joint constraints the end-effector's motion to a pure rotation. Some applications of these orientation robots are: solar panels and spatial antennas [19] or component manipulation [20]. Related the robots with coupled degrees of freedom, in [21] is presented a 1T2R robot for telescopic applications that was also implemented in a machine tool and in a vehicle-testing bench. This class of robots has parasitic motions due to the dependency between some output parameters.

3 Kinematic Design Criteria

In this section, the most relevant kinematic criteria for the design of parallel manipulators are reviewed.

- *Workspace assessment*

For the workspace calculation of parallel robots, in literature we can find three types of methods: the discretization methods, the geometrical methods and the analytical methods.

The discretization methods [22] generate a mesh of nodes. Each node corresponds to a location of the end-effector. Then, by solving the inverse kinematic problem it is checked if the node belongs to the manipulator's workspace or not. It has to be taken into account that the inverse kinematic problem is quite straight in parallel robots. These methods are quite easy to implement in a computer but they have a high computational cost and the accuracy depends on the refinement of the mesh. In this category, we can find the Interval Analysis method [23].

The geometrical methods are widely used to obtain the boundaries of the workspace. The first step is the definition of the surfaces generated be the ending point of each kinematic chain. Then, the surfaces are intersected to obtain the complete workspace. These methods are limited to the calculation of 3D workspaces.

Finally, the analytical methods are based on the mathematical resolution of the non-linear constraint equations derived from the manipulator's topology. The complexity of the workspace calculation depends highly on the architecture of the manipulator under study.

In this research field, the CompMech research group has developed the *GIM software* that calculates and analyses the workspace of fully parallel manipulators [24]. It is based on a hybrid analytical-discrete procedure that evaluates a high density of discrete nodes with relatively low computational effort. The key of the method is that it uses the analytical equations of each kinematic chain of the manipulator. GIM allows the visualization and analysis of the workspace but also of other kinematic entities as the jointspace, reduced configuration space and singularity maps.

- *Multiobjective optimisation.*

When a robot is being designed, one of the main criteria is to maximize the workspace. However, other criteria need to be also taken into account as the regularity of this workspace [25] or other additional ones in order to improve the performance of the robot. To take all into account, a multiobjective optimisation procedure has to be done. This optimisation finds a good compromise between all these criteria. To do this, some authors use the Pareto optimality and other authors try to find an optimal solution taking into account the minimum predefined requisites for each function. In [26] are optimized the workspace, the dexterity and the energy consumption of the actuators.

Recently, the CompMech research group has presented a design methodology to increase the workspace (including also a multiobjective optimisation step) based on the assessment of the non-singular transitions [27]. These transitions appear in the cuspidal manipulators joining different solutions of the direct kinematic problem without loss of control of the robot [28].

- *Structural synthesis*

The structural synthesis analyses and selects the kinematic chains suitable to obtain a particular motion pattern in the end-effector desired for the application to be used. The motion pattern defines the number and type of the degrees of freedom in the end-effector as well as the directions of the pure rotations and translations. There are three approaches for the structural synthesis: the screw theory, the theory of groups of displacement and the linear transformation theory.

Most of the designers use the second one as it works with finite displacements. This theory uses the mathematical properties of the Lie Group. J. Hervé proposed it for the structural synthesis of mechanisms. It uses the so-called motion generators that are kinematic chains that generate a particular motion in the end-effector. These generators verify the properties of the Group Algebra. So, in the general category of rigid-body displacements are: translational subgroups (dimension 1, 2 and 3), rotational, cylindrical, Schönflies, etc. For a parallel robot, each leg is a motion generator applied to the end-effector. From these motion generators the final motion pattern of the manipulator is obtained. In any case, in order to select the best architecture for a parallel robot, additionally to the motion pattern some other characteristics have to be taken into account: fabricability, economical cost, maintenance and wear, etc. A brief review in this field can be found in [29].

- *Accuracy*

Accuracy and sensitivity indicators are particularly relevant in order to evaluate the quality of a design. Accuracy tries to quantify the influence of some error sources as the clearances in the joints and manufacturing and assembling tolerances in the real position of the end-effector. Generally, the errors derived from the manufacturing and assembling tolerances are compensated via calibration techniques. Different procedures have been proposed for the analysis of other factors as the clearances in the joints and actuators [30].

Sensitivity quantifies the different influence of the design parameters in the positioning of the end-effector. In [31] a method for sensitivity analysis based on the interval linearization is presented.

- *Multioperacionality*

Currently, designers are focusing on increasing the flexibility and versatility of the manipulators. The goal is to design robots able to adapt to different tasks. This is a promising research line quite active in recent years. In the *PARAGRIP project* (RWTH Aachen) [32], the object to be manipulated becomes the end-effector after different kinematic chains are linked to it. Other strategies are based on varying the dimensions of the links or blocking different actuators or kinematic joints in the manipulator [33]. These new design criteria will be introduced more in detail in the next section.

4 Towards the Modular Reconfigurable Robots

Traditionally, the design of robotic systems has focused on achieving robust and efficient models, optimized for a specific task. However, at the end of the last century, a novel idea started to grow up which consisted in designing machines and mechanisms capable of carrying out several tasks, thus offering higher flexibility and versatility. To sum up, the target is to get multioperational systems in such a way that the robots can change their configuration to execute multiple tasks. Nowadays the technical chance exists to design and install them in several fields such as mounting small series of large components in the aeronautical and eolic field, flexible tasks in packing and folding, deployable antennas and trusses, biomedical and rehabilitation applications, etc.

Bearing this in mind, the authors are leading the *MoMaR research project*. The main objective of this new design approach is to develop a methodology for modular design of reconfigurable parallel manipulators with variable topology. The design of parallel manipulators capable of achieving the highest number of motion patterns in their moving platform is proposed, basing on the reconfigurability of 6 degree of freedom manipulators. This reconfiguring ability is accomplished by blocking different actuators of the manipulator. In this way, a simpler machine is obtained from the operative point of view. Because of working with a lower number of actuators in each operation mode the manipulator will have a simpler kinematics and accordingly a faster and more efficient control.

For the structural synthesis of the reconfigurable parallel robots, the authors propose a modular design approach (Fig. 2). From the joints and links, the kinematic chains are generated. These chains and different typologies of end-effectors will be implemented in a virtual library of GIM software. This library will be used by the designer to define the architecture of the reconfigurable parallel robot. As future work, the analysis capabilities of GIM software will be increased to evaluate the kinematic design criteria presented in

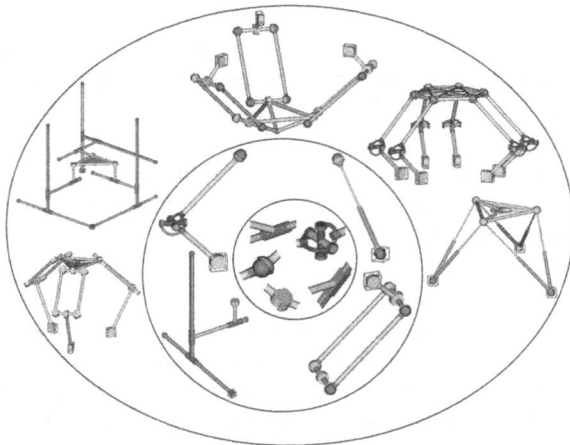

Fig. 2. Modular design approach: joints, kinematic chains and manipulators.

Sect. 3 for this class of parallel robots. This will help the designer to define the dimensional parameters of the links and joints of the parallel robot.

5 Conclusions

In this paper, the authors present a bibliographical review about research lines, new designs and applications of parallel robots in recent years. Also, in this paper the main kinematic design criteria of parallel robots are presented and discussed. They are focused in improving the designs in terms of workspace, singularity avoidance, etc. Some references related the works of the CompMech research group in these fields are included. Finally, the authors highlight the potentiality of reconfigurable parallel robots in order to increase the flexibility and versatility of these parallel architectures.

Acknowledgments. The authors wish to acknowledge the financial support received from the Spanish Government through the "Ministerio de Economía y Competitividad" (Project DPI2015-67626-P (MINECO/FEDER, UE)), the financial support from the University of the Basque Country (UPV/EHU) under the program UFI 11/29 and the support to the research group, through the project with ref. IT949-16, given by the "Departamento de Educación, Política Lingüística y Cultura" of the Regional Government of the Basque Country.

References

1. Merlet JP (2006) Parallel robots. Springer, Heidelberg
2. Gwinett JE (1928) Amusement device. US Patent No. 1,789,680
3. Pollard W (1942) Position controlling apparatus. US Patent No. 2,286,571
4. Stewart D (1965) A platform with six degrees of freedom. Proc IMechE 180(15):371–385

5. Patel YD, George PM (2012) Parallel Manipulators Applications-A Survey. Mod Mech Eng 2:57–64
6. Pandilov Z, Dukovski V (2012) Parallel kinematics machine tools: Overview- from history to the future. Ann Fac Eng Hunedoara-Int J Eng
7. Clavel R (1990) Device for the movement and positioning of an element in space. US Patent No. 4,976,582
8. Stock M, Miller K (2003) Optimal kinematic design of spatial parallel manipulators application to linear delta robot. ASME J Mech Des 125(2):292–301
9. Campos L, Bourbonnais F, Bonev IA, Bigras P (2010) Development of a five-bar parallel robot with large workspace. In: Proceedings of the ASME 2010 international design engineering technical conferences and computers and information in engineering (IDETC/CIE 2010), Montreal, Canada, 15–18 August 2010
10. Macho E, Altuzarra O, Pinto C, Hernández A (2008) Workspace associated to assembly modes of the 5r planar parallel manipulator. Robotica 26:395–403
11. Angeles J, Caro S, Khan W, Morozov A (2006) Kinetostatic design of an innovative schönflies-motion generator. Proc Inst Mech Eng Part C J Mech Eng Sci 220(7):935–943
12. Altuzarra O, Hernandez A, Salgado O, Angeles J (2009) Multiobjective optimum design of a symmetric parallel Schonflies motion generator. ASME J Mech Des 131(3)
13. Shoham M, Burman M, Zehavi E et al (2003) Bone mounted miniature robot for surgical procedures: Concept and clinical applications. IEEE Trans Robot Autom 19(5):893–901
14. Gosselin C, Hamel J (1994) The agile eye: A high-performance three-degree-of-freedom camera-orienting device. In: IEEE international conference on robotics and automation, pp 781–786
15. Hiller M, Fang S, Mielczarek S, Verhoeven R, Franitza D (2005) Design, analysis and realization of tendon-based parallel manipulators. Mech Mach Theory 40:429–445
16. Daney D, Merlet J-P (2010) A portable, modular parallel wire crane for rescue operations. In: IEEE international conference on robotics and automation (ICRA), Anchorage, USA
17. Sturm C, Bruckmann T, Schramm D, Hiller M (2011) Optimization of the wire length for a skid actuated wire based parallel robot. In: 13th world congress in mechanism and machine science, Guanajuato, México, 19–25 June 2011
18. Homma K, Fukuda O, Sugawara J, Nagata Y, Usuba M (2003) A wire-driven leg rehabilitation system: Development of a 4-DOF experimental system. In: Proceedings of the 2003 IEEE/ASME International Conference on Advanced Intelligent Mechatronics (AIM 2003)
19. Robertson J (2006) Application of the Trio-Tri-Star carpal Wrist for use in Solar Array Tracking Mechanism for the Momentum-eXchange. NASA Marshall Space Flight Center
20. Callegari M, Carbonari L, Palmieri G, Palpacelli M-C (2013) Parallel wrists for enhancing grasping performance. In: Carbone, G. (eds.) Grasping in robotics. Springer, Heidelberg, pp 189–219
21. Carretero JA, Podhorodeski R, Nahon MA, Gosselin CM (2000) Kinematic analysis and optimization of a new three degree of freedom parallel manipulator. J Mech Des 122(1):17–24
22. Masory O, Wang J (1995) Workspace evaluation of Stewart platforms. Adv Robot 9(4):443–461
23. Merlet J-P (2009) Interval analysis for certified numerical solution of problems in robotics. Int J Appl Math Comput Sci 19(3):399–412
24. Macho E, Pinto C, Amezua E, Hernández A (2011) Software tool to compute, analyze and visualize workspaces of parallel kinematics robots. Adv Robot 25(6):675–698

25. Li Y, Xu Q, (2007) Design and application of a new 3-DOF translational parallel manipulator. In: IEEE workshop on advanced robotics and its social impacts (ARSO 2007) doi:10.1109/ARSO.2007.4531432
26. Altuzarra O, Pinto C, Sandru B, Hernández A (2011) Optimal dimensioning for parallel manipulators: Workspace, dexterity and energy. ASME J Mech Des 133(4), 041007-7
27. Hernández A, Altuzarra O, Petuya V, Macho E (2009) Defining conditions for nonsingular transitions between assembly modes. IEEE Trans Rob 25(6):1438–1447
28. Urízar M, Petuya V, Amezua E, Hernández A (2014) Characterizing the configuration space of the 3-SPS-S spatial orientation parallel manipulator. Meccanica 49(5):1101–1114
29. Meng X, Gao F, Wu S, Ge QF (2014) Type synthesis of parallel robotic mechanisms: Framework and brief review. Mech Mach Theory 78:177–186
30. Chen G, Wang H, Lin Z (2013) A unified approach to the accuracy analysis of planar parallel manipulators both with input uncertainties and joint clearance. Mech Mach Theory 64:1–17
31. Tannous M, Caro S, Goldsztejn A (2014) Sensitivity analysis of parallel manipulators using an interval linearization method. Mech Mach Theory 71:93–114
32. Müller R, Riedel M, Vette M, Corves B, Esser M, Hüsing M (2010) Reconfigurable self-optimising handling system. In: Ratchev S (eds) Precision assembly technologies and systems. Springer, Heidelberg, pp 255–262
33. Grosch P, Di Gregorio R, López J, Thomas F (2010) Motion planning for a novel reconfigurable parallel manipulator with lockable revolute joints. In: IEEE international conference on robotics and automation (ICRA2010), Alaska, USA

Two-Degree-of-Freedom Special Parallel Manipulator for Laser Interferometry-Based Tracker

Baris Celik, Takaaki Oiwa[✉], Kenji Terabayashi, and Junichi Asama

Shizuoka University, Hamamatsu, Japan
oiwa@shizuoka.ac.jp

Abstract. This study deals with a two-degree-of-freedom special manipulator for tilting a set of a laser light source and a detector around two axes. Six sets of this manipulator is planned for feedback sensors to measure six-degree-of-freedom relative motions between a tool spindle and surface plate of a precision machine tool or a coordinate measuring machine. This report describes an outline of the laser tracking system and 2-degree-of-freedom parallel kinematic manipulator based on a parallelogram linkage. In this laser interferometry-based tracker, the detector head is required not to be rotated around the laser light axis to ensure distance measurement accuracy even when the head is tilted by actuators around x- and y-axes. Thus, such a parasitic angular motion was measured in experiments based on the motion capture device.

Keywords: Parallel manipulator · Laser tracker · Precision machine system

1 Introduction

Recently, to realize submicron-order volumetric accuracy of spatially moving mechanisms for precise machining or coordinate measurement, there has been serious interest in a machine system able to generate accurate relative motion between the tool and work piece, as well as accuracy improvements for each element of the machine [1, 2]. In general orthogonal coordinate machine structure, feedback control is implemented for each feed drive mechanism, which is based on one axis displacement measurement or a one degree-of-freedom (DOF) motion measurement, even though the machine consists of three feed axes. In each feed mechanism, 5-DOF motion errors, including three angular motion errors, are usually not measured to compensate for the Abbe errors caused by the angular motion errors and the Abbe offsets [3]. On the other hand, prediction methods have been investigated to compensate for the thermal deformation of the machine structure based on temperature sensors and thermal deformation analysis. However, thermal deformation is very difficult to predict precisely under rapid temperature change [4, 5]. This study has proposed a feedback control system based on six laser interferometry-based trackers for measuring the 6-DOF relative motions between the tool and workpiece [6]. In this report, we first describe a concept of the machine system and the laser tracker based on a two-degree-of-freedom parallel manipulator consisting of spherical joints and a parallelogram linkage. In experiment with a prototype manipulator, an angular motion error around the laser light axis was measured by a motion capture system employing a digital camera and a grid pattern.

2 Machine System Based on 6-DOF Motion Feedback Control

Figure 1 shows the components of the proposed machine system based on a measurement device for the 6-DOF motions between the tool and the workpiece. This system consists of a six-degree-of-freedom motion measurement device, a conventional

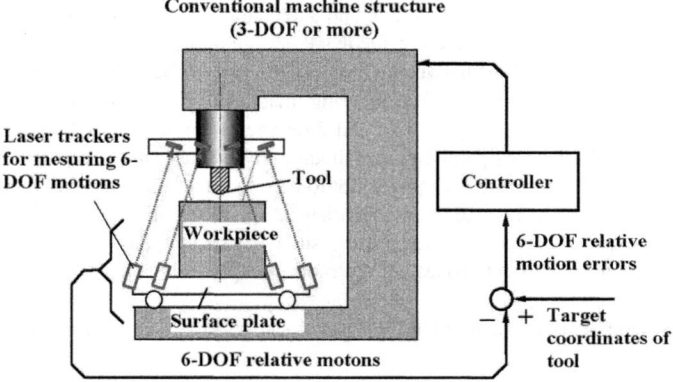

Fig. 1. Fundamentals of proposed machine system based on 6-DOF measurement device between tool and workpiece

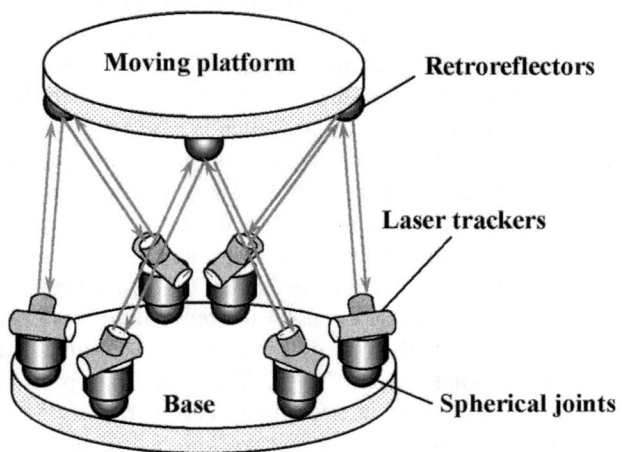

Fig. 2. Six-DOF motion measurement device consisting of six laser interferometry-based trackers

machine structure and a controller. To measure the relative 6-DOF motions between the tool and workpiece, the measurement device is mounted between the machine spindle and the surface plate. Besides, the device consists of six laser trackers and their targets, which is based on the laser interferometric length measurement, as shown in Fig. 2. As each laser tracker mounted on the surface plate measures change in distance between the tracker and a target or a retroreflector mounted on the spindle, the 6-DOF motions can be calculated by the forward kinematics of typical hexapod-type PKMs. Consequently, the controller compensates for the motion errors and accurately actuates the tool of the machine. Because the measurement device measures the 6-DOF motions, including angular motions, no Abbe error is produced even if the orthogonal coordinate machine structure is employed. Moreover, this measurement device is independent of any elastic and thermal effects and motion errors of the machine structure because the device is separated from main structure of the machine [7]. In other words, the system can not only compensate for the systematic motion error caused by geometrical deviations of the machine elements, but non-repetitive motion errors caused by the elastic and thermal deformations if the 6-DOF motion can also be precisely measured. When this measurement device is used for the coordinate measuring machine, the coordinates of the probe tip are directly obtained by the device.

3 Laser Tracker

In this report, a laser interferometer measurement system (Laser Encoder®, Renishaw) is employed for the laser tracker. To measure change in a distance from the tracker to the target, a detector head of the Laser Encoder® emits a laser light and receives the laser light reflected at the target retroreflector. Because the target position relative to the tracker varies during machine operation, the detector head always has to track the target so that the reflected laser light returns to the detector. In general, conventional laser scanners have often employed two sets of a galvano scanner consisting of a plane mirror and a rotational servo motor [8]. Moreover, commercially-available laser tracker often employs two-axis gimbal set which is rotating the laser head around two axes [9, 10]. However, in two types of trackers mentioned above, misalignment or offset of these two rotation axes often causes difficulty calculating the light path length because the trilateration is based on measured path lengths. Figure 3 depicts an outline of the laser tracker in this report. The tracker mechanism is based on a 2-DOF spatial rotational manipulator consisting of a spherical joint, a parallelogram linkage and a simple linkage. Thus, the laser head mounted on a moving platform is simply tilted around a rotational center of the spherical joint without the offset between two rotational axes.

If an elastic deformation of the joint, caused by variation of load, is small, the rotational accuracy can be almost guaranteed by only the sphericity of the ball in the joint. Then, two rotational axes intersect at the center of the ball. Thus, our tracker has a potential of greater accuracy over the galvano scanner and the gimbal set.

Each linkage is driven by a linear VCM (voice coil motor) actuator so that the beam reflected at the retroreflector always comes into a center of a quadrant photodiode. However, during the tilting, the detector head is required not to be rotated around the laser light axis to keep the laser beam parallel to a line between the spherical joint and

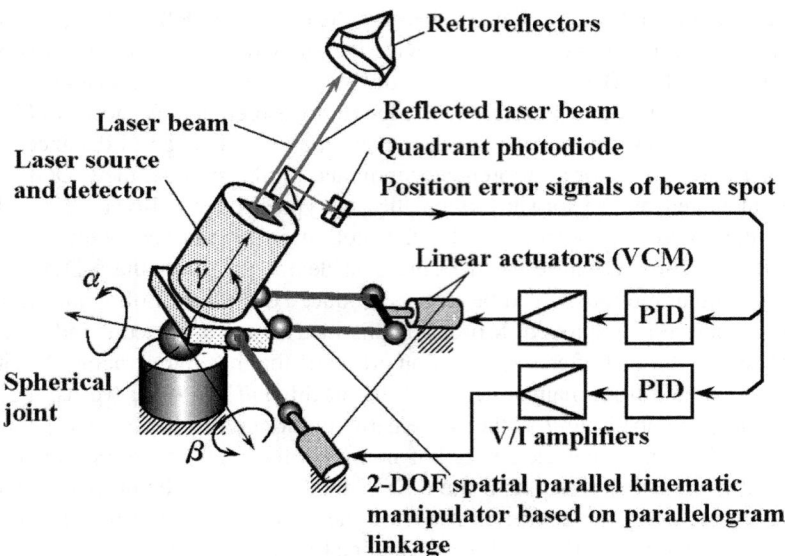

Fig. 3. Laser tracker based on two-DOF spatial parallel kinematic manipulator actuated by two linear voice coil motors

the corner of the retroreflector because such a misalignment has a possibility of producing any length measurement error, such as the cosine error. Because a beam separation or a distance between two beams is d [mm] (Renishaw RLD10: 7 mm), it is expected that angular motion $\Delta\gamma$[rad] around the beam axis produces a misalignment of $d\,\Delta\gamma$. Consequently, the misalignment of $d\,\Delta\gamma$ introduces a length measurement error shown in the following equation,

$$\Delta L \approx \frac{d^2}{L}\Delta\gamma^2, \qquad (1)$$

where, L is a distance between the laser head and the target retroreflector. For instance, when L is 500 mm, an angular motion error of $\pm 5°$ (± 17.5 Mrad) causes a length measurement error of ± 0.75 μm. If a length measurement error within ± 0.5 μm is required, the angular motion error must be smaller than $\pm 4.1°$.

4 Experiment

4.1 Experimental 2-DOF Parallel Mechanism

To demonstrate the 2-DOF motions and investigate the unfavorable parasitic motion around the laser beam axis during the tilting of the moving platform, an experimental setup of the 2-DOF spatial parallel kinematic mechanism was constructed as shown in Fig. 4. First, a moving platform was mounted on a spherical joint consisting of a steel

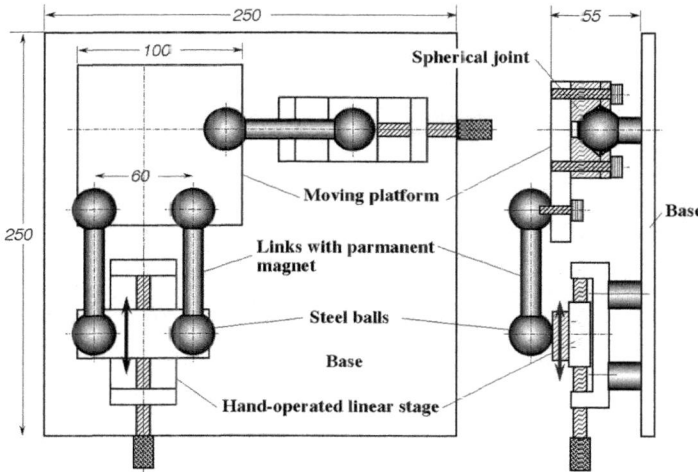

Fig. 4. Two-DOF spatial parallel kinematic mechanism consisting of spherical joint and parallelogram linkage

shank with a 1/2″-diameter steel ball (grade: 28), brass-made holder and a brass-made rid. Three sets of a chain consisting of two steel balls and a permanent magnet link connect the platform to two hand-operated linear stages. Then, two of them set to be parallel to restrict the angular motion around z axis. The stages give the platform two angular motions in α and β directions.

4.2 Measurement Method

In order to measure three-axis angular rotations of the moving platform driven by two linear stages, image process technique was used. In the technique when we shoot a chessboard we can evaluate extrinsic camera parameters that give us the rotation and translation vectors by using Python software®. Figure 5 shows the experimental setup.

We used a 1.3-megapixel USB camera. Due to the low quality of the camera, we mounted an A4 size plate on the moving platform to improve the resolution of the chessboard image. First we find the measurement resolution in order to understand how the measurement system is accurate. Measurement resolution calculated by seven images captured in different pose of the parallel manipulator. Table 1 shows the measurement resolution results of X-, Y- and Z-axis rotations or α, β and γ directions. As a result, it is confirmed that the measurement system has an angular measurement resolution better than 0.02° or 0.35 Mrad.

To compensate the distortion of captured images, we identified the intrinsic camera parameters by using MATLAB®. The intrinsic parameters were used to calculate extrinsic camera parameters. While the moving platform was located at 99 different positions, the images were captured. After that, three angular motions were calculated from these captured images and the extrinsic camera parameters.

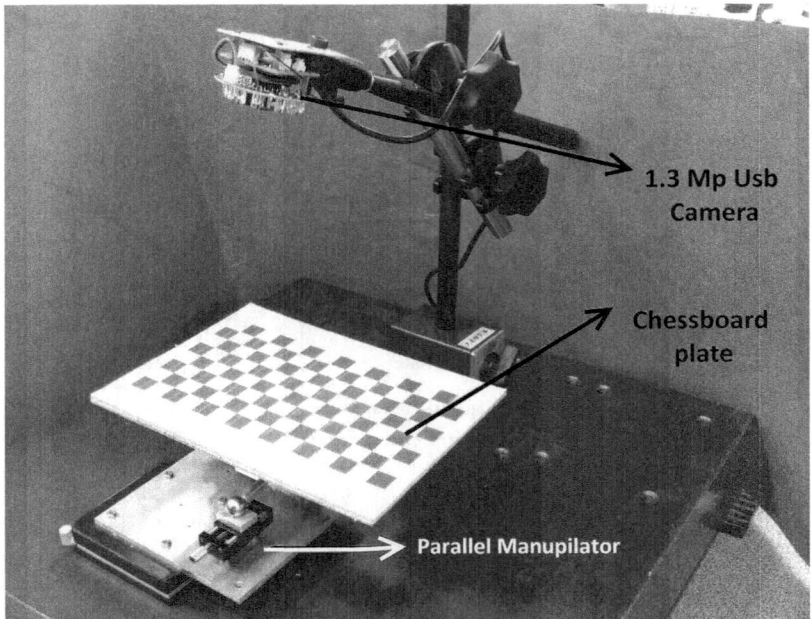

Fig. 5. Experimental setup for measuring three-axis rotations of moving platform

Table 1. Measurement resolution

Pose	α	β	γ
1	0.022°	0.031°	0.010°
2	0.003°	0.005°	0.002°
3	0.004°	0.005°	0.004°
4	0.044°	0.036°	0.026°
5	0.005°	0.008°	0.002°
6	0.006°	0.010°	0.001°
7	0.003°	0.004°	0.001°
Average	0.012°	0.014°	0.006°
RMS	0.019°	0.018°	0.011°

4.3 Measurement Result

Figures 6, 7 and 8 show three-axis rotations measured by the motion capture system. When the linear stages positioned to −2 mm and −3 mm respectively, the moving platform was parallel to the base. Ordinate values in each 3D chart represent the angular rotation corresponding to the positions of the linear stages. The moving platform tilted from approximately −10° to +40° around x axis when the x stage moved to ±5 mm. Besides, the platform tilted from approximately −30° to +7° around Y axis.

Two-Degree-of-Freedom Special Parallel Manipulator   147

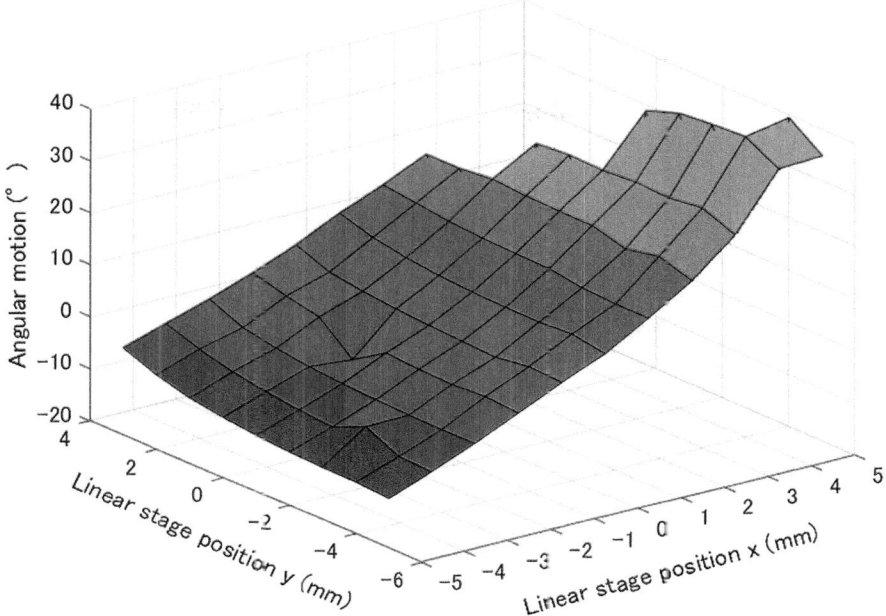

Fig. 6. X-axis rotational change

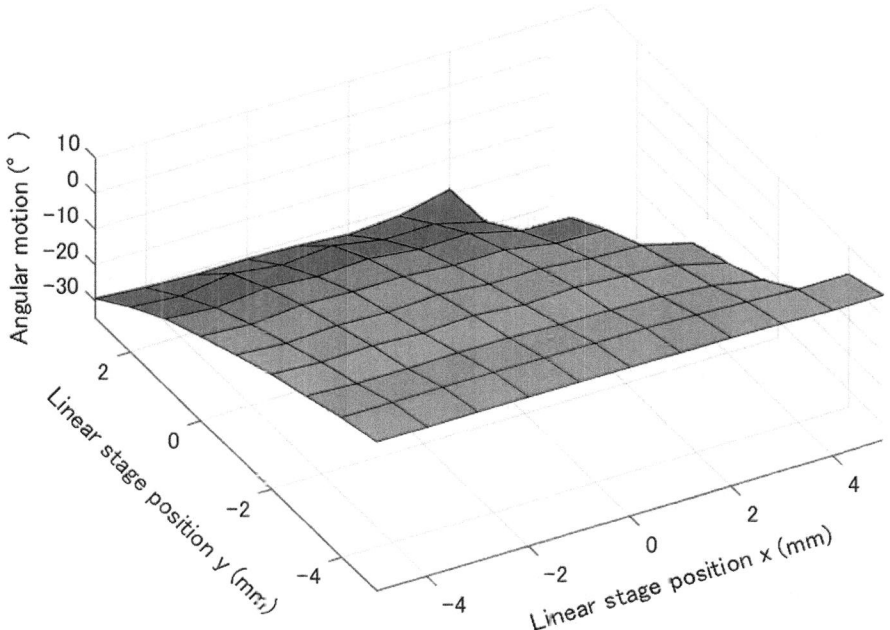

Fig. 7. Y-axis rotational change

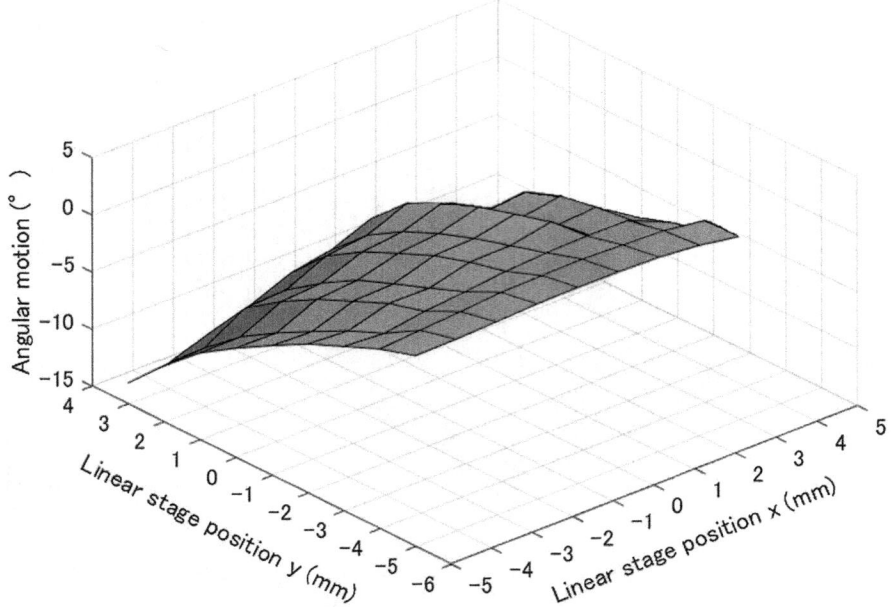

Fig. 8. Z-axis rotational change

On the other hand, the z-axis rotation changed from approximately $-13°$ to $+2°$. This angular motion undesirable for the tracker was larger than expected in Sect. 3, or was almost doubled.

4.4 Discussion

According to the Eq. 1, change in the rotational angle of 15° has a possibility to generate a length measurement error worse than 2 μm. In particular, the moving platform most tilted and rotated when each stage moved in positive direction while the platform less tilted when the stage moved in negative direction. An experimental mechanism demonstrated that the unfavorable parasitic angular motion around the beam axis was smaller than 15° when the mechanism tilted by approximately 50° around two horizontal axes. One of the main reasons for this is the fact that the links was a bit longer than a design value of the distance between the platform and the stages.

5 Conclusions

We described a precision machine system based on six laser trackers measuring for six-degree-of-freedom motion errors between the tool and workpiece. A two-degree-of-freedom spatial parallel kinematic mechanism was proposed for tilting the laser detector head. Based on Z-axis angular results at the present stage, an experimentally-manufactured parallel mechanism cannot be used as in the laser tracker. Because of the error overvalue the mechanism should be improved. With adjusting properly the link length may reduce the angular error. In addition, to further decrease the motion error, the distance and the links should be longer. However, such an arrangement of the manipulator might enlarge the size of the tracker. With the future work, the special parallel manipulator can be using instead a gimbal system.

Acknowledgments. The authors would like to thank Mr. K. Ogasawara of Shizuoka University for his help in carrying out the measurements.

References

1. Ramesh R, Mannan MA, Poo AN (2000) Error compensation in machine tools - a review part I: geometric, cutting-force induced and fixture-dependent errors. Int J Mach Tools Manuf 40:1235–1256
2. Lee GD, Suh DJ, Kim SH, Kim MJ (2004) Design and manufacture of composite high speed machine tool structure. Composites Science and Technology 64:1523–1530
3. Abbe E (1890) Messapparate für Physiker (Measuring Equipment for Physicists). Zeitschrift für Instrumentenkunde 10:446–447 (in German)
4. Ramesh R, Mannan MA, Poo AN (2000) Error compensation in machine tools - a review part II: thermal errors. Int J Mach Tools Manuf 40:1257–1284
5. Lee SD, Choi YJ, Choi D-H (2003) ICA based thermal source extraction and thermal distortion compensation method for a machine tool. Int J Mach Tool Manuf 43:589–597
6. Oiwa T (2007) An ultraprecise machining system with a hexapod device to measure six-degree-of-freedom relative motions between the tool and workpiece. Int J Precis Eng Manuf 8(1):3–8
7. Bryan JB, Carter DL (1978) Design of a new error-corrected co-ordinate measuring machine. Precis Eng 3:125–128
8. Gilby JH, Parker GA (1982) Laser tracking system to measure robot arm performance. Sens Rev 2(4):180–184
9. Lau K, Hocken R, Haynes L (1985) Robot performance measurements using automatic laser tracking techniques. Robot Comput Integr Manuf 2(3/4):227–236
10. Hughes EB, Wilson A, Peggs GN (2000) Design of a high-accuracy CMM based on multi-lateration techniques. CIRP 49(1):391–394

Control Applications

Design of Artificial Neural Network Predictor for Trajectory Planning of an Experimental 6 DOF Robot Manipulator

Sahin Yıldırım[✉] and Burak Ulu

Erciyes University, Kayseri, Turkey
{sahiny,burakulu}@erciyes.edu.tr

Abstract. Nowadays, the use of robots is continuously increasing in industry. Especially, robotic Gas metal arc (GMA) welding is widespread used as manufacturing process. Because of this increase in the use of robots in the industry, there is a need to study on a number of improvements. This paper presents an experimental research on the robot manipulator, using image processing to detect location of welding seam for the planning optimal trajectory. This new study provides the weld seam trajectory to be created without being affected by manufacture faults. Firstly, communication interface between the robot and the computer is developed by using previous related software library. Then, the weld seam trajectory are automatically generated using image processing via Matlab and reference points are determined on the trajectory for tracking of the manipulator. The values of this points are sent to the robots for calculation of the joint angles by the software on the robot side. Furthermore, the related parameters are tested with neural network predictor to predict optimal trajectory on resulting image using image processing. The results show that this approach improved that neural network predictor can increase trajectory accuracy for quality welding process.

Keywords: Robotic arc welding · Neural network · Trajectory planning · Industrial robots

1 Introduction

The importance of studies on trajectory problems encountered in robotic welding process is increasing day by day. At this experimental study, it is worked on positioning and determine optimum weld seam trajectory. Normally, industrial robots allow user to programming using teach and playback methods. However, the methods can take a long time and it can also restrict controlling the robot path.

Some of significant papers have remarked about these problems. Numerous researches have focused on automatic weld seam identification [1–3]. Mitchell Dinham and Gu Fang were present a new weld joint detection method using stereo vision system for automatically generate programme paths. This approach was experimentally tested on an industrial robot manipulator [4]. Neural networks are proposed as an approach for solution of complex equations. Ikbal Eski et al. were presented a methodology to fault detection on robot manipulators based neural network.

The efficiency of the approach was experimentally implemented on a KUKA six-axis industrial robot manipulator [5]. Francisco Rubio et al. introduce an efficient trajectory planning for industrial robots [6].

In this paper, neural network predictors are used to predict of optimal joint angles of the robot manipulator. The proposed methods can track determined trajectory on surface of the workpieces. Experiments are implemented using KUKA KR 6-2 industrial robot manipulator in the laboratory to prove that the recommended method can be used for industrial implementations. The paper is organized as follows: Sect. 2 identifies the theory of robot manipulator. Section 3 gives some details the proposed neural network predictor. The experimental and simulation results are given in Sect. 4. Finally, this paper concluded and discussion in Sect. 5.

2 Description of the Welding Robot Manipulators

In this experimental study, KUKA KR 6-2 is used as industrial robot that it has six degrees of freedom. It is employed to analyze the optimal angles of joints. The robot manipulator are driven by electromechanical AC servo motors. Maximum speed of robot manipulator's end-effector is 6000 mm/s. The positioning repetition accuracy of the robot manipulator is 0.1 mm [5].

2.1 Robot Communication via Computer

Several researches on Kuka industrial robots communication have been presented. Normally, KUKA.Ethernet RSI XML that an add-on technology package of Kuka is used to provide communication between Kuka industrial robots and external devices. The exchanged data are transmitted via the Ethernet TCP/IP or UDP/IP protocol as XML strings [8]. However, this software interface restrict user for controlling the robot. Therefore, Francesco Chinello et al. developed a new interface using Matlab platform [10]. Thanks to this interface, the Kuka industrial robots are controlled via the computer and various applications are implemented. But for this interface to be used, the robot interface packages were needed. So the program couldn't provide the desired flexibility. Finally, F. Sanfilippo et al. succeeded in developing a different communication interface than the previous ones. They have developed program libraries that allow communication with robots in a flexible way without requiring any additional programming using the Java language.

In this study, it is aimed to develop an interface that can compatible work in Matlab by using JopenShowVar library to communicate with KUKA via TCP/IP protocol [7]. The generated interface is being used in the experimental work to be done, and the development is still going on.

2.2 Movement Approach

Industrial arc welding robots permit the user to programming using teach and playback methods. In this method, three different motion types are used. These motion types are

PTP (Point-to-Point), LIN (Linear), CIRC (Circular). The point-to-point (PTP) motion is quickest way of moving the tip of the tool (Tool Center Point: TCP) from the current position to a programmed end position. However, the methods can take a long time and it can also restrict controlling the robot path [8].

Normally, the programmer teaches the determinated points to robot for programming the robot. Because of that points are manually teached by the programmer, both it can take a significant amount of time and the positions of the points are determined relatively. In this study, the points on trajectory are determined using image processing method and position variables of the points are sent via developed communication interface from computer to Kuka industrial robot controller. For this reason, more accurate results are obtained.

3 The Proposed Seam Identification Method

An overview of the trajectory determination method and neural network predictor are presented in this section. (1) The first step is to capture image of the workpiece and get the clean image using image processing. (2) Neural network predictor is generated to remove unwanted faults on the image. Then, the cartesian co-ordinates of the detected points are calculated. (3) Finally, weld seam trajectory is planned and position variables are sent to robot controller.

3.1 Image Processing and Seam Detection

In this experimental study, image processing is implemented using matlab for seam detection. The 1024 × 575 sized colour images captured by the camera is converted to a grayscale image which is shown in Fig. 1(a). The conversion from RGB to grayscale can be achieved using equation:

$$I_{gray} = \begin{bmatrix} 0.29 & 0.58 & 0.11 \end{bmatrix} \begin{bmatrix} I_{rgb}\{R\}(u,v) \\ I_{rgb}\{G\}(u,v) \\ I_{rgb}\{B\}(u,v) \end{bmatrix}. \quad (1)$$

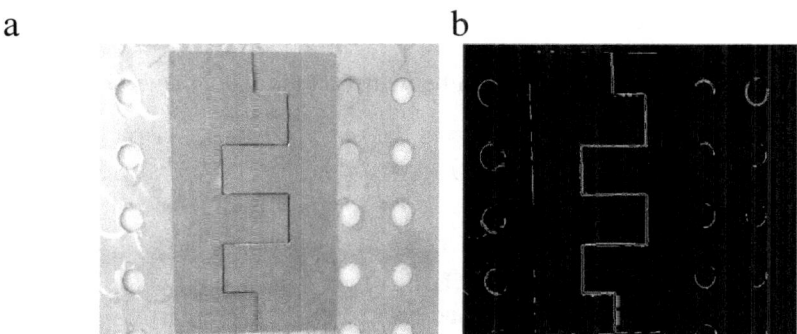

Fig. 1. (a) Grayscale image, (b) edge image with prewitt algorithm.

where I_{gray} is calculated grey image from the camera, (u, v) is the pixel co-ordinate, $I_{rgb}\{R\}(u,v), I_{rgb}\{G\}(u,v), I_{rgb}\{B\}(u,v)$ are the red, green and blue components of the left RGB image respectively [11].

After image was converted to grayscale, edge detection algorithm prewitt is used for identify weld seam on image of the be welded workpiece in Fig. 1(b). But after the algorithm implemented, undesired edges appear with the weld seam on the background. For this reason, bwareaopen (BW, P) operation of the Matlab is implemented to removes all connected object that have fewer than P pixels from the binary image BW, producing another binary image [12]. The resulting image size is cut to 260 × 576 pixels for easier focus on the weld seam, shown in Fig. 2(a). Finally, the resulting image size is minimized to 116 × 53 pixels, shown in Fig. 2(b), for the number of pixels has been reduced to be used as input data of the neural network.

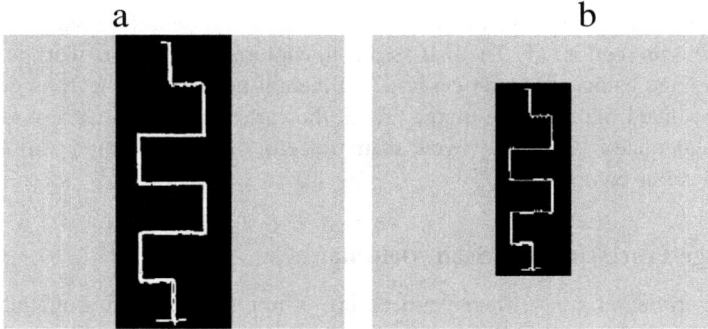

Fig. 2. (a) Dividend image and (b) 116 × 53 pixels resulting image.

3.2 Neural Network for Seam Identification

ANNs are widely accepted in the artificial intelligence research where a non-linear mapping between input and output parameters is required for a function approximation [9]. In this study, feedforward neural network with three layered is used to train images of the weld seam. Schematic representations of a multi-layer neural network are shown in Figs. 3 and 4. Feedforward backprop training algorithm is used to establish the network.

The output signal of network can be written in the following equations:

$$y_i = f(w,x) = f\left(\sum_{i=1}^{n} w_i x_i + b\right). \tag{2}$$

where x_i is the input of the network, w_i is the weight of the connection between the input layer neurons and the hidden layer neurons, b is the bias of the connection between a fixed input of unit value and neuron.

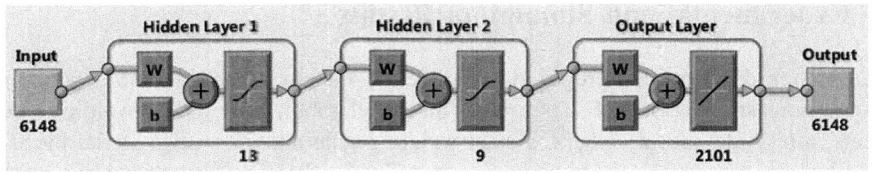

Fig. 3. Neural network architecture for predicting target image.

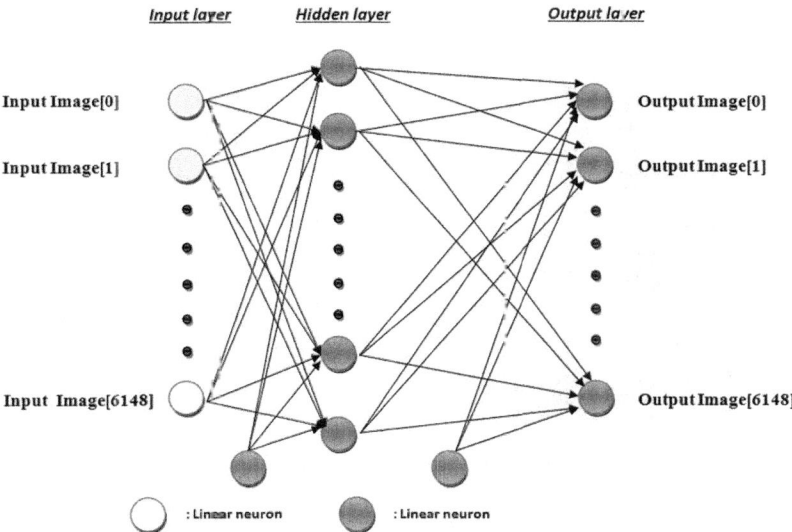

Fig. 4. Schematic representations of the neural network parameters.

In Fig. 4, Input image[] is an array that it has 6148 indexes. Each element of the array contains the sequential pixels of the resulting image. Totally 6148 pixel values from Input Image[0] to Input Image[6147] is used as input parameter for the developed neural network.

3.3 Trajectory Planning and Sending of Position Data

To find which of the weld joint and then to determine the motion points, new neural network predictor is developed in this study. The cartesian co-ordinates of the detected points are sent to the industrial robot controller via computer as axis angle values. This communication is provided by the developed new interface on the computer side. On the Kuka side, KUKAVARPROXY is used for data transfer [7]. The robot act to points that was sent for complete the planning trajectory.

4 Experimental and Simulation Results

This section describes the proposed experimental system apparatus. The experimental system consists of a 6 DOF Kuka KR6 industrial robot, Fronius arc welding equipments, a high definition camera, a steel welding table and 20 steel material pieces as experimental pieces for testing and computer. The workpieces was placed in a random position on the steel material welding table. Then, camera that mounted above the table capture the images. The experiment system is represented in Fig. 5 for good understanding the process.

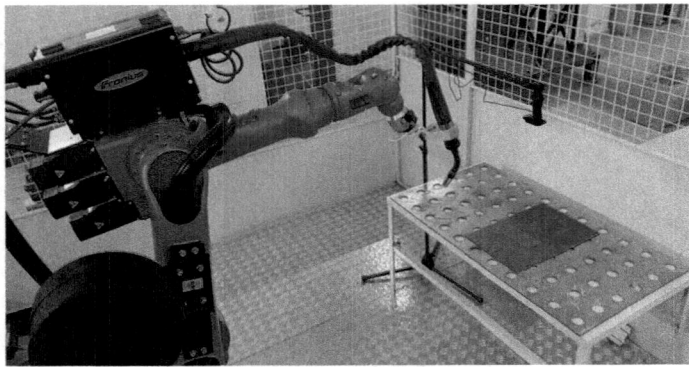

Fig. 5. Experiment system with industrial welding robot and camera.

The captured images are corrected using image processing algorithms. However, weld seam isn't purely detected on the resulting image. For this reason, a proposed feedforward neural network is used to predict prescribed trajectory of the robot manipulator with using backpropagation learning algorithm. In addition to this, Fig. 6 (a), (b) present the differences between standard and the proposed neural network

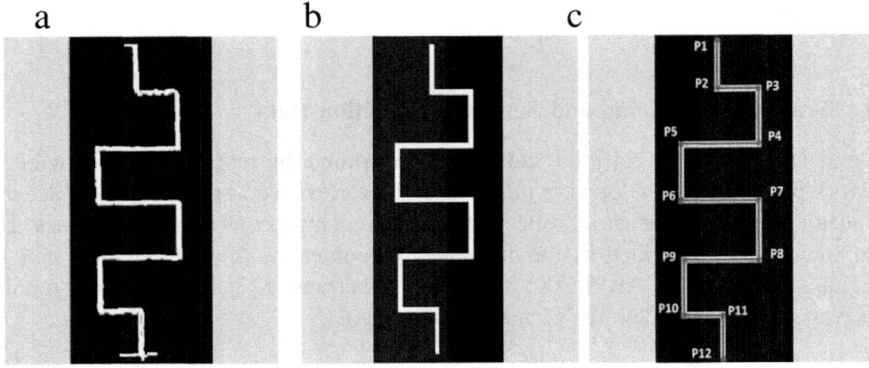

Fig. 6. (a) The resulting image, (b) after NN is implemented, (c) detection points on final image.

predictor approaches. On the other hand, determinated points on the trajectory are described and shown in Fig. 6(c).

Detection points on target trajectory isn't possible using the resulting image from image processing. Therefore, this resulting image is converted binary 116 × 53 sizes image. Then, the 6148 × 1 sizes array is obtained from this binary image. All elements of the array are used as input of the neural network. Similarly, target image that was generated for optimal trajectory is used as output of the neural network. The generated neural network model is trained using feed forward backpropagation algorithm. Later, neural network is tested to predict captured new image of the workpiece. Resulting output image is very clearly for detection target points. Figure 7 shows the best training performance with 9.8454e-06 at 732 iterations.

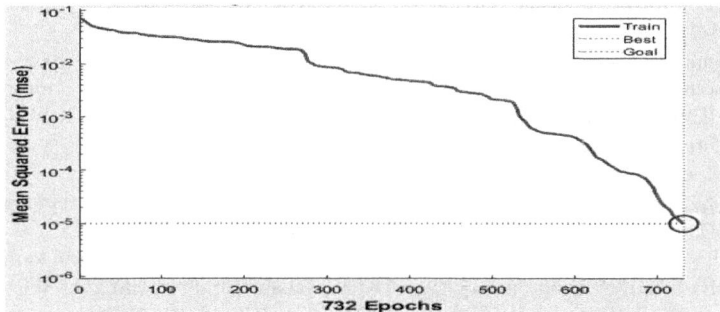

Fig. 7. The curve of the learning performance.

5 Conclusions

In this experimental based work, a proposed process method for autonomous trajectory of the 6 DOF robot manipulator with arc welding process detection is presented. For this, image processing technique as described in Sect. 3.1 is used to determine weld seam on the captured images. There some faults on the resulting image that obtained using image processing. A proposed neural network based predictor is used to achieve successful results to correct these unwanted faults as depicted in Fig. 6(a) and generate optimal trajectory. The proposed method improved that a new approach for identification weld seam using image processing with neural network predictor. Finally, the proposed NN predictor can employed in real time applications such as robotic based welding process.

Acknowledgments. We would like to thank Erciyes University Scientific Research Projects Coordination Unit (ERU/BAP) and TUBİTAK.

This work is supported by the Erciyes University Scientific Research Projects Coordination Unit (ERU/BAP) under project codes FYL-2016-6348 and TUBİTAK 2210-C program under project codes 1649B021507250.

References

1. Micallef K, Fang G, Dinham M (2011) Automatic seam identification and path planning in robotic welding. Robot Weld Intell Autom 88:23–32
2. Chen XZ, Chen SB (2010) The autonomous detection and guiding of start welding position for arc welding robot. Ind Robot Int J 37:70–78
3. Qin J, Ma G, Liu P (2011) Image processing algorithm of weld seam based on crawling robot by binocular vision. In: Proceedings of the 2011 IEEE second international conference on mechanic automation and control engineering MACE2011, Inner Mongolia, pp 337–340
4. Dinham M, Fang G (2014) Detection of fillet weld joints using an adaptive line growing algorithm for robotic arc welding. Robot Comput Integr Manuf 30:229–243
5. Eski I, Erkaya S, Savas S, Yildirim S (2011) Fault detection on robot manipulators using artificial neural networks. Robot Comput Integr Manuf 27:115–123
6. Rubio F, Llopis-Albert C, Valero F, Suner JL (2016) Industrial robot efficient trajectory generation without collision through the evolution of the optimal trajectory. Robot Auto Syst 86:106–112
7. Sanfilippo F, Hatledal LI, Zhang H, Fago M, Pettersen KY (2014) JOpenShowVar: an open-source cross-platform communication interface to kuka robots. In: Proceeding of the 2014 IEEE international conference on information and automation ICIA2014, Hailar, pp 1154–1159
8. KUKA, expert programming (2003) KUKA Robotics Corporation
9. Kim I-S, Son J-S, Lee S-H, Yarlagadda PKDV (2004) Optimal design of neural networks for control in robotic arc welding. Robot Comput Integr Manuf 20:57–63
10. Chinello F, Scheggi S, Morbidi F, Prattichizzo D (2010) The KUKA control toolbox: motion control of KUKA robot manipulators with Matlab. In: Proceeding of the 2010 IEEE international conference on robotics and automation ICRA2010, Anchorage, pp 4603–4608
11. Dinham M, Fang G (2013) Autonomous weld seam identification and localisation using eye-in-hand stereo vision for robotic arc welding. Robot Comput Integr Manuf 29:288–301
12. Matworks bwareopen operation-(web) (2016). https://www.mathworks.com/help/images/ref/bwareaopen.html. Accessed 5 Dec 2016

Fault-Tolerance Experiments with a Kinematically Redundant Holonomic Mobile Robot

Osman Nuri Şahin, Onur Çelik, and Mehmet İsmet Can Dede[(✉)]

Izmir Institute of Technology, Izmir, Turkey
{osmansahin,onurcelik,candede}@iyte.edu.tr

Abstract. Indoor locomotion with mobile robots has found applications in industrial part moving, factory floor investigation and cleaning tasks. Holonomic mobile robots have the advantage of moving in tight and winding passages without the need of steering. In this study, a four omni-directional wheeled mobile robot is considered. The mobile robot is kinematically redundant having four of these wheels and this is used in providing fault-tolerance due to a malfunction in one of the wheels. In this paper, the redundancy resolution for this mobile robot is explained providing a solution to a fault in one of the wheels during operation. A top level controller to compensate for the orientation errors is introduced. Finally experimental set-up is presented along with the result of the fault-tolerance experiments.

Keywords: Holonomic mobile robot · Fault tolerance · Kinematic redundancy · Omni-directional wheels

1 Introduction

Mobile robots find application domains in an increasing trend. They were first employed autonomous ground vehicles (AGV) in factory floors [1] and later mobile robots entered the households as automatic cleaning agents [2]. Among the key components of these robots, localization equipment, battery management and locomotion mechanism are perhaps the most important ones. In terms of locomotion, mobile robots can be classified as wheeled, tracked and legged mobile robots.

A subtype of indoor wheeled mobile robots is the holonomic mobile robots. Holonomic refers to the relationship between controllable and total degrees of freedom of a robot. If the controllable degree of freedom is equal to total degrees of freedom, the robot is holonomic. A robot which has omni-directional wheels is a good example of holonomic mobile robots. Holonomic mobile robots are capable of translational motion in two degrees-of-freedom (DoF) on a plane and a single DoF rotational motion about the normal to the plane independently and simultaneously. As a result of this advantage, holonomic mobile robots are widely used in many applications, especially in tasks carried out in narrow spaces such as transporting of goods in hospitals, public cleaning, factories and sheltered workshops for disabled people [3].

One problem that arises with the use of autonomous vehicles is the need of fault tolerance against the possible malfunction of component during operation or manufacturing faults. In this study, mechanism level fault tolerance is worked out by introducing redundancy to a holonomic mobile robot. The faults considered in this paper are a single wheel malfunction and faults during manufacturing and assembly of the mobile robot. The next section provides a brief overview of holonomic mobile robots and the omni-directional wheels used for these robots. In the following section, the holonomic robot designed and built in this work is presented along with the kinematics equations. The redundancy resolution algorithm that enables the fault tolerance due to a malfunction of a wheel is explained and the top level controller for compensating the orientation errors, which occur due to the malfunctioning wheel and/or manufacturing errors, is introduced. Finally, experimental results are provided and discussed.

2 Overview of Holonomic Mobile Robots

Holonomic motion of ground vehicles can be constructed with ordinary caster wheels [4] or special omnidirectional wheels [5] (mecanum and universal). Holonomic robots, which have caster wheels, firstly changes wheel orientation according to motion direction and then robot moves. This procedure brings delay in operation and robot cannot perform a smooth continuous motion. Omnidirectional wheels allow motion that is perpendicular to the rotating direction of the wheel through smaller free rollers which are placed on the outer diameter of the wheel. In holonomic mobile robots that use caster type of wheels, at least three caster wheels and six actuators among which 3 actuators are required for rotation of wheels and 3 actuators are required for changing the wheels' orientation. In holonomic mobile robots that use omnidirectional wheels, only three universal or four mecanum wheels and an actuator for every wheels are enough. Therefore, omnidirectional wheels are more efficient for holonomic ground robots in terms of minimized use of the actuators. Universal wheels (Fig. 1) are selected for the locomotion of developed holonomic mobile robot in this study.

Fig. 1. The omni-wheel used in this study.

Omnidirectional wheels have some problems despite their advantages. There are not many commercial omnidirectional wheels developed for holonomic robots. Some wheels utilized in these robots are designed as conveyor rollers. Therefore, material of these wheels is not suitable for traction, which causes slippage problems. Another, disadvantage is that universal omnidirectional wheels requires dust-free environments. Most important disadvantage of them, which affects motion capability of holonomic mobile robot, is that outer circle of wheel is completed with radius of each roller and this is not a smooth circle. Although effect of this problem is decreased with two row rollers, the problem causes vibration on mobile robot, decreases road-holding and makes it difficult to carry out odometry from encoders mounted on wheel actuators. While rollers are placed perpendicular to the circular plane of the wheels in universal wheels, rollers are placed with an angle in mecanum wheels. This enhances the road-holding of the robot so that mecanum wheels are commonly preferred for industrial holonomic lifting and carrying robots [6].

3 Design and Governing Equations of the Holonomic Mobile Robot

In this study, a mobile robot is designed to have kinematic redundancy by using 4 universal wheels. It is actually enough to use 3 universal wheels for constructing a holonomic mobile robot. A general type of mobile robot can be employed at distant or dangerous environments for human, such as in space or in nuclear reactors. Tolerating the faults emerged during operation is crucial for operational success. Therefore, main reason of selecting this configuration is that if any wheel is broken down while in operation, robot can tolerate this fault and continue the operation with three wheels without losing its holonomic property.

Figure 2 represents location of mobile robot according to global coordinate frame and wheel velocities of the robot. In this figure, V_i and ω_i (i = 1, 2, 3, 4) are linear and angular velocities of the wheels respectively, θ is the orientation frame attached to the mobile robot (x-y frame) with respect to a fixed frame, V_x and V_y are velocity components of the mobile robot along the X_w and Y_w axes of the fixed frame, and ω_v is angular velocity of mobile robot about the Z_w axis of the fixed frame. All four wheels are placed to have angles between the neighboring wheels to be 90°. Distances between wheels and center of mass are the same (250 mm) and shown with the parameter L in Fig. 2.

Kinematic equations of mobile robot derived by assuming that there is no slippage during operation are given below,

$$V_x = V_1 \cdot \sin\theta - V_3 \cdot \sin\theta - V_2 \cdot \cos\theta + V_4 \cdot \cos\theta \quad (1)$$

$$V_y = -V_1 \cdot \cos\theta + V_3 \cdot \cos\theta - V_2 \cdot \sin\theta + V_4 \cdot \sin\theta \quad (2)$$

$$\omega_v = -(V_1 + V_2 + V_3 + V_4)/L \quad (3)$$

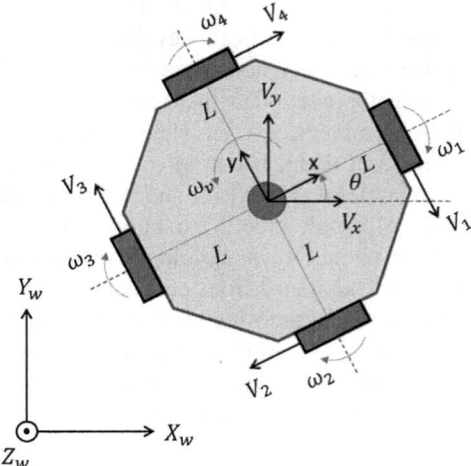

Fig. 2. Top view of four wheeled mobile robot

$$\begin{bmatrix} V_x \\ V_y \\ -\omega_v.L \end{bmatrix} = J . \begin{bmatrix} V_1 \\ V_2 \\ V_3 \\ V_4 \end{bmatrix} \quad (4)$$

$$J = \begin{bmatrix} \sin\theta & -\cos\theta & -\sin\theta & \cos\theta \\ -\cos\theta & -\sin\theta & \cos\theta & \sin\theta \\ 1 & 1 & 1 & 1 \end{bmatrix} \quad (5)$$

Because of the redundancy, Jacobian matrix given in Eq. 5 is not a square matrix. Hence, pseudo inverse method used to find minimum norm of wheel velocities. Inverse pseudo Jacobian matrix, J^+, for robot is derived as below:

$$\begin{bmatrix} V_1 \\ V_2 \\ V_3 \\ V_4 \end{bmatrix} = J^+ \begin{bmatrix} V_x \\ V_y \\ -\omega_v.L \end{bmatrix} \quad (6)$$

$$J^+ = J^T \left(J.J^T \right)^{-1} \quad (7)$$

4 Fault Tolerance Aspects

Fault tolerance studies in literature are achieved in two aspects; fault diagnosis and fault tolerance. Also, faults can be occurred in software or hardware of mobile robots. This study is focused on tolerating hardware faults of the developed holonomic mobile robot.

One of fault tolerance scenario is the case of failure in the actuators coupled to the wheels. The method of weighted pseudo inverse of Jacobian matrix $(\hat{J}_w^+)$ is applied to deal with this type of fault. The direct kinematics equation is modified to Eq. 8 and inverse kinematics making use of $(\hat{J}_w^+)$ is calculated as shown in Eq. 9.

$$\begin{bmatrix} V_1 \\ V_2 \\ V_3 \\ V_4 \end{bmatrix} = \hat{J}_w^+ \begin{bmatrix} V_x \\ V_y \\ -\omega_v L \end{bmatrix} \tag{8}$$

$$\hat{J}_w^+ = \hat{W}^{-1} \cdot \hat{J}^T \cdot \left(\hat{J} \cdot \hat{W}^{-1} \cdot \hat{J}^T \right)^{-1} \tag{9}$$

$\hat{W}$ is a diagonal weighting matrix.

$$\hat{W} = \begin{bmatrix} W_1 & 0 & 0 & 0 \\ 0 & W_2 & 0 & 0 \\ 0 & 0 & W_3 & 0 \\ 0 & 0 & 0 & W_4 \end{bmatrix} \tag{10}$$

$$\begin{aligned} V_1 = &- [0.5/(W_1 + W_2 + W_3 + W_4)][\omega_V.L(W_2 + W_4) \\ &+ V_y(\sin\theta\,(W_2 - W_4) + \cos\theta\,(W_2 + 2W_3 + W_4)) \\ &+ V_x(\cos\theta\,(W_2 - W_4) - \sin\theta\,(W_2 + 2W_3 + W_4))] \end{aligned} \tag{11}$$

W_1, W_2, W_3, W_4 each represents the chosen weighing constant for the designated motor. Increasing any W_i constant with respect to other three, in Eq. 7, will decrease the corresponding motor's contribution in operation. If a weight is to be chosen infinite, the desired motion will be carried out with the other three since the system is already redundant. The calculation of the linear velocity requirement of one of wheels for a desired task space velocity profile is given in Eq. 11.

Fault tolerance is adapted to the redundant mobile robot system to be used in case of losing power in one motor completely or decrement in motor efficiency. In order to evaluate the efficiency of this algorithm, the failure is generated artificially in the control interface by cancelling motor control or decreasing its control input signal.

Additionally, two different sets of omnidirectional wheels with different friction properties are used for further fault tolerance tests. The wheel assembly with the actuation system can rotate in both directions. The rollers of two universal wheels are synthetic rubber coated polypropylene and other two are nylon. This fault causes various road-holding performance of the robot in each direction.

The other faults focused on in this study are manufacturing errors and deficiency of omnidirectional wheels. Even if angles between wheels are set as 90°, because of manufacturing process, these angles can have small deviation, which can result in uneven input from the wheels. Also, because of disadvantages of omnidirectional wheels mentioned in Sect. 2, road-holding of any wheel can be different from others. These problems cause undesirable orientation changes. In order to deal with this

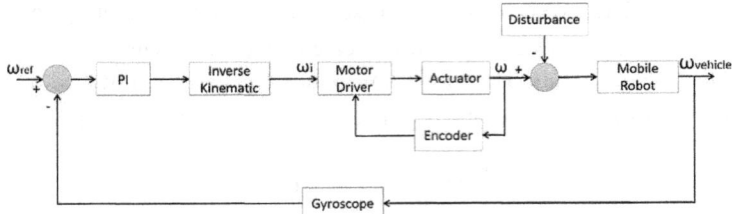

Fig. 3. Top level control algorithm

problem and to control the angular velocity of mobile robot, a top level closed-loop control is developed at velocity level that uses the angular velocity of robot measured through a gyroscope attached on mobile robot as the feedback signal. Figure 3 shows the Top-Level controller algorithm.

In Fig. 3, reference input ω_{ref} is the desired angular velocity of the mobile robot. After the required velocity change is regulated by the PI controller, the task level velocities are decomposed to wheel velocities and low-level wheel actuator controllers regulate the wheel speeds accordingly. The Top-Level controller (PID or PI) parameters are selected experimentally in order to make the controlled system to have a 5 rad/s bandwidth. Red dashed lines show the 3 dB range in Fig. 4. PID controller (blue line) has a resonant frequency at about 3 rad/s. However, with PI controller (orange line), it was possible to have a flat magnitude plot until 5 rad/s. The parameters of PI controller are $K_p = 4$, $K_i = 1, 25$.

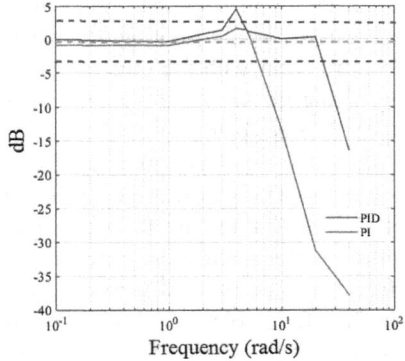

Fig. 4. Bode plot for PID and PI controller

System with the Top-Level controller produced the transient state characteristics for a step input as listed in Table 1. Where, t_r rise time, t_p peak time and t_s settling time.

Table 1. Top-level controller transient response characteristics.

Parameters	
t_r	0.614 s
t_p	0.796 s
t_s (for 2%)	8 s
Percent overshoot	18

5 Experimental Verification of Fault Tolerance Features

The robot manufactured in this study (Fig. 5) is a four-wheeled redundant holonomic mobile robot. The robot is driven with four 24 V DC motor (Dunkermotoren G30.0) and Maxon ADS 50/10 motor drivers. Energy need is supplied with two 12 V lead acid batteries. Energy supply unit has a step-up circuitry to increase voltage from 12 V to 24 V for driving actuators and step-down circuitry to supply energy to sensors and the DAQ system. UEISIM 600-1G GigE Simulink target Cube which have two modules (analog input and output modules) is utilized for data acquisition. Data transfer between host pc to target UEISIM 600-1G is provided by UDP protocol. The wireless communication is used to embed control algorithms, which are generated with Matlab Simulink, into the robot's controller and for monitoring and logging the data from UEISIM DAQ during experimental studies. Wheel angular velocities are measured with magnetic encoders attached at the rear end of the motors. Also developed mobile robot consists of an angular velocity measurement unit (analog gyroscope) and eight infrared sensor (SHARP GP2Y0A02YK0F) to detect obstacles surrounding of the robot.

The controller embedded in the mobile robot utilizes the weighted pseudo inverse of the Jacobian matrix as explained in Sect. 4. In case of the existence of a faulty actuator, all actuator motion demands are updated by changing weighing constant of

Fig. 5. Four omnidirectional-wheeled holonomic mobile robot (left: Perspective view, right: top view; in the top view yellow box is the UEISIM DAQ, blue box is the wireless router, red boxes are the motor drivers, orange boxes are the batteries, orange ellipses are the infrared sensors, green box is the power inverter and dashed yellow box is the I/O cards.) (Color figure online)

each wheel in weighing matrix. During the experiments, faults are generated manually in the algorithm as a performance drop in one of the actuators. The objective of this study does not include fault detection therefore, it is assumed that fault is detected and the time of occurrence is known.

The fault is defined as proportional to the performance loss. During the experiments one actuator is subjected to performance loss up to 70%. After 70% of performance loss, velocity demands of the working wheels rises to levels that they can no longer provide the demanded velocity due to traction problems. According to the test scenario, while the mobile robot is following a linear path, one of its actuators providing this linear motion is subjected to a performance loss at 5th second.

The non-idealities such as, skewness in actuator axes, friction coefficient difference of the wheels and translation in mass center, leads to the need of observing fault tolerance algorithm in different motion directions selected as x- and y-directions. The experiments are run for different levels of performance loss and the performance loss of the actuators is increased by 10% until the actuators are completely disabled. The angular velocities of wheels measured with magnetic encoders are illustrated for the motion along x-direction in Fig. 6(a) and y-direction in Fig. 6(b). In both experiments, motion of the robot is initiated at second 2 and the performance loss for one actuator is issued at second 5. The performance loss level is kept at 50%.

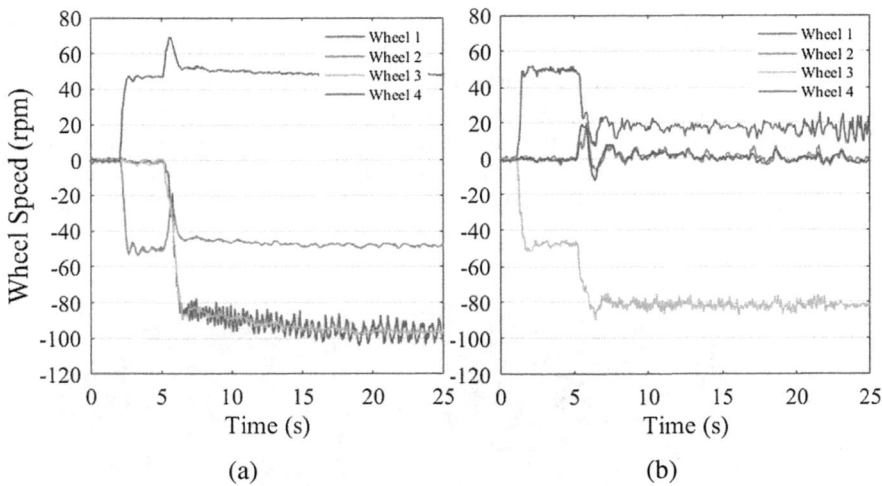

Fig. 6. Velocity distribution of mobile robot for experiment (a)- moving in x direction, (b)- moving in y direction

Wheels 1 and 3 that are responsible for the motion along y-direction have the rollers with relatively smaller friction coefficients. As a consequence of this, it can be observed from Fig. 6(a) that these two wheels start to operate at fast speeds in order to compensate for the faulty wheel 2. The encoder reading from the faulty wheel 2 shows that the performance did not drop to 50% however, this angular velocity measurement is due to the rolling of the wheel with friction as the mobile robot moves along the x-direction.

In Fig. 6(b), wheels 1 and 3 provide the motion along the y-direction. When 50% fault is issued to wheel 1, its angular velocity drops to 50%. Wheels 2 and 4 with relatively better friction coefficients hold the road better and thus provide the necessary traction with minimal effort. Therefore, the angular velocity contributions of wheel 2 and 4 are very limited as it can be observed from Fig. 6(b).

It can also be observed from Fig. 7(a) and (b) that after the transient state the angular velocity of the mobile robot is kept at about zero condition in both of the experiments. This shows that the Top-Level controller is successful in regulating the angular velocity of the mobile robot independent of the difference in road-holding characteristics of the wheels.

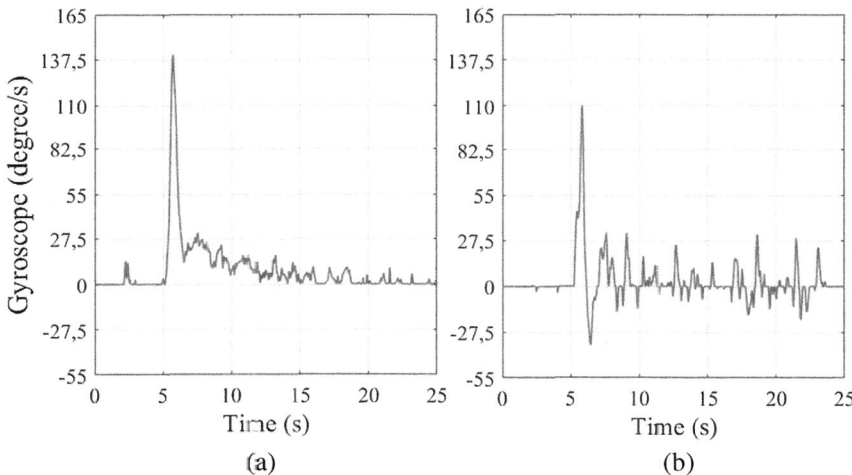

Fig. 7. Gyroscope measurements of mobile robot with Top-level controller (a)- moving along x direction, (b)- moving along y direction

6 Conclusions

In this study, experimental results of fault tolerance algorithm devised for a four omnidirectional wheeled holonomic mobile robot are presented. Faults focused on this study are the performance loss of any wheel of robot and undesired orientation changes caused by manufacturing errors and uneven friction coefficients of the omnidirectional wheel. Because the mobile robot is designed to have kinematic redundancy, performance loss or disability of any wheel during robot in task can be tolerated with other the three wheels and the robot can perform the task without losing its holonomic motion ability. This fault is tolerated by using weighted pseudo inverse of the Jacobian matrix derived for the robot's kinematics.

The fault tolerance method developed in this work is applied in robot controller and fault tolerance tests are carried out. According to test results given in previous section, fault tolerance algorithm tolerates performance loss of any wheel during a motion in any direction. However, the test results for both direction are different from each other because of the differences in wheels' friction characteristics (rollers of two wheels are coated synthetic rubber, other two are nylon).

It should be noted that during fault tolerance tests performed without top-level controller, mobile robot rotates uncontrollably since the wheels, which compensate angular velocity changes of the robot, have different road-holding capabilities. Top-level controller is necessary to eliminate this fault in the system due to uneven wheel characteristics.

References

1. Takashi G, Koh-ichi I, Hirokazu M (1994) The development of a fully autonomous ground vehicle (FAGV). In: Intelligent vehicle 1994 symposium. IEEE, pp 62–67
2. Joseph LJ (2006) Robots at the tipping point: the road to iRobot Roomba. IEEE Robot Autom Mag 13(1):76–78
3. John ME (1994) HelpMate: an autonomous mobile robot courier for hospitals. In: Intelligent robots and systems 1994, pp 1695–1700
4. Robert H, Oussama K (2000) Development and control of a holonomic mobile robot for mobile manipulation tasks. Int J Robot Res 19:1066–1074
5. Martin U, Karl I (2009) Analysis, design, and control of an omnidirectional mobile robot in rough terrain. J Mech Des 131:121002
6. Olaf D, Aparna B, Glen B, Johan P, Sylvester T (2002) Improved mecanum wheel design for omni-directional robots. In: Australasian conference on robotics and automation. pp 117–121

Dual-Loop Motion Control for Geometric Errors and Joint Clearances Compensation of a Planar 2-PRP+1-PPR Manipulator

Santhakumar Mohan[1(✉)], Jayant Kumar Mohanta[1], Mathias Huesing[2], and Burkhard Corves[2]

[1] Indian Institute of Technology Indore, Indore, India
{santhakumar,phd1401103005}@iiti.ac.in
[2] RWTH Aachen University, Aachen, Germany
{huesing,corves}@igm.rwth-aachen.de

Abstract. The existence of geometric errors (misalignment and kinematic errors) and joint clearances is an inherent problem in manipulators. In addition, friction between surfaces and gear backlash errors are unavoidable factors in manipulators using geared motors as their joint actuators. This paper presents a potential solution for the above issues through the application of a dual-loop control scheme. The proposed control scheme uses a redundant feedback strategy, i.e., individual joint displacements (at the joint space level) and, end-effector positions and orientation (at the task space level) are obtained as a feedback signal using appropriate sensors. Using this redundant feedback information, the actual error of the joint displacements are computed and rectified the desired joint space trajectory in joint-space trajectory tracking control to achieve the desired task space trajectory. To demonstrate the effectiveness and show overall performance of the controllers, real-time experiments are performed on an in-house fabricated 2-PRP+1-PPR planar parallel manipulator. The experiment results show that the manipulator tracking performance is significantly improved with the proposed dual-loop control scheme. In addition, the controller parameter sensitivity and robustness analyses are performed.

Keywords: Dual-loop control · Error compensation · Trajectory tracking control · Parallel manipulators

1 Introduction

Robot manipulator is not only a mechanical architecture, it also slaved by a controller that impacts the overall system performance. Robot manipulators similar to many mechanical engineering applications, it is impossible or very difficult to obtain an exact dynamic model of the robot due to many unavoidable reasons. Therefore, control of manipulators is a challenging task. In addition, the real-time systems involves both mechanical hardware and electronic control, where the aim is to achieve accurate tracking of a target trajectory (mostly end-effector trajectory), but the errors incurred in the system can be attributed to the mechanical errors, caused due to imperfect modelling of geometric parameters, inaccuracies in fabrication and assembly, inherent limitations of the mechanical structure and un-modelled system dynamics namely friction, backlash, joint clearances, etc., and electrical and electronic errors caused due to resolution, delay and noise of the sensors, actuator responses, etc. Therefore, a simple joint space position tracking control scheme cannot ensure accurate tracking performance of the desired trajectory of the end-effector.

There are several established techniques have been proposed to overcome the above mentioned issues associated with the manipulator control [1–3]. Most of the recent proposed control schemes are either model based adaptive/robust control schemes [2] or task space control [3]. Model-based control, where the model used for control can be improved by incorporating issues such as friction, joint clearances, backlash, etc. in it [1, 2]. The adaptive control with model based may produce better performance in overall, but it could not reduce overcome the mechanical errors due to the joint space control scheme. In other hand, task space controller schemes are designed based directly on information in task space (no inverse kinematics solution is needed), their performance dependent on the manipulator Jacobian matrix and most of the schemes use inverse Jacobian matrix due to control inputs are in task space [3]. Moreover, the feedback gains in task space are not intuitive and joint space behavior is difficult to predict. There are several schemes come with robot calibration and parameter correction methods which are very common technique in mechanical systems to mitigate the mechanical accuracies [4]. However, this technique requires very sophisticated instrumentation and these instruments are expensive. Moreover, the parameter correction can correct the geometric errors; it may predict the joint clearance error range and not correct the effect of joint clearances in the task space. Therefore, the actual task space sensor information needs to be included in existing control loop which turns the dual-loop control system design. Several dual-loop control schemes proposed in the literature and few notable contributions are as follows:

Dual-loop control scheme has been successfully implemented in flexible robot manipulator trajectory tracking control along with adaptive sliding mode control [5] and in planar parallel manipulator for backlash error correction with a simple PID kinematic control (without considering the dynamics of the system) [6]. In addition, redundant sensor measurements of active and passive joints along with dual-loop control has successfully implemented in spatial parallel manipulator [7], a dual space adaptive control is implemented fin redundant parallel robots for payload adaption [8], and a dual-loop sliding mode control is implemented along with active vibration control in two link flexible robot to reduce the overall transient response of the system [9].

Each of the techniques described above have their respective advantages and limitations. Therefore, in the paper a dual-loop or cascade control scheme proposed with redundant sensor feedback inputs based on simple inner-loop proportional-integral-derivative (PID) control along with an integral outer-loop control which compensate the mechanical errors exist in the system. The outer-loop uses the task space sensor feedback and corrects the joint space through the help of inverse kinematics. The proposed control scheme is implemented on a planar 2-PRP+1-PPR parallel manipulator [10]. The manipulator is commanded to track a complex trajectory and a circular trajectory with and without outer-loop control. It is realized that the tracking performance is improved with the dual-loop control.

The remainder of the paper is structured as follows: in Sect. 2, the mechanical background is presented, in specific kinematic model of the manipulator and control scheme are discussed. The experimental arrangement is described in Sect. 3. Section 4, presents the complete results and their discussions. Finally, Sect. 5 presents the concluding remarks.

2 Mathematical Background

2.1 Kinematic Model

The kinematic arrangement of the manipulator is shown in Fig. 1. The fixed base, 0, and the moving platform (end-effector), 5 are connected through three legs. In these three legs: two of them having prismatic, revolute and prismatic joints which are arranged in a vertical plane and other leg has prismatic, prismatic and revolute joints which is fixed in a horizontal plane. In all three legs, the starting prismatic joint is actuated and other joints are passive. The geometry of the manipulator is identical to the reported in [10]. The actuator coordinates (joint displacements), $\mathbf{q} = [r_1 \ r_2 \ r_3]^T$, and the task coordinates of the end-effector, $\mathbf{\mu} = [x \ y \ \theta]^T$. The forward kinematic relation of the manipulator is as follows:

$$\mathbf{\mu} = [x \ y \ \theta]^T = \left[r_1 \ \ r_2 + \frac{r_1(r_3 - r_2)}{s} \ \ \tan^{-1}\left[\frac{r_3 - r_2}{s}\right] \right]^T \tag{1}$$

The inverse kinematic relations of the manipulator are given as follows:

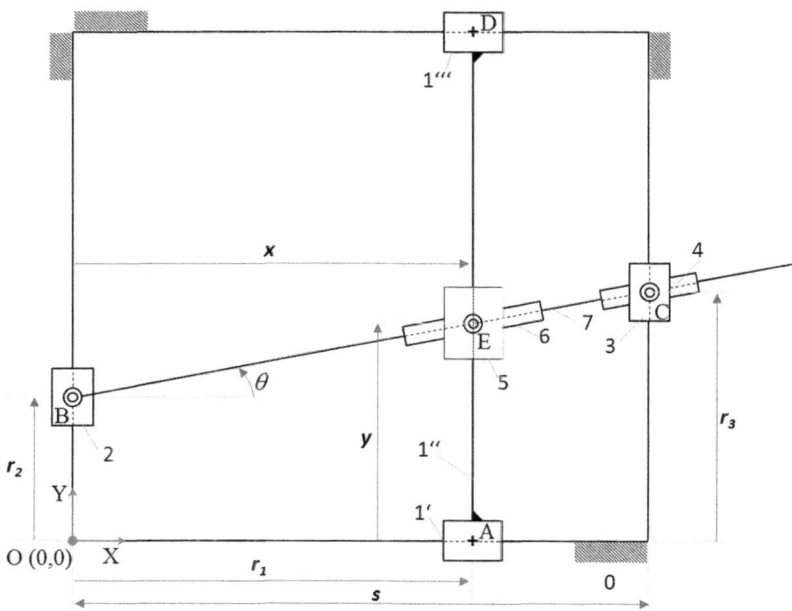

Fig. 1. Schematic representation of the 2-PRP+1-PPR planar manipulator.

$$\mathbf{q} = [r_1 \quad r_2 \quad r_3]^T = [x \quad y - x\tan\theta \quad y + [s-x]\tan\theta]^T \tag{2}$$

where, r_1, r_2 and r_3 are the joint displacements, and s is the horizontal span.

2.2 Dual-Loop Control Scheme

The block diagram of the proposed control scheme is presented in Fig. 2. The outer-loop uses the task space coordinates from the vision sensor outputs to find out the corrected joint space coordinates $\mathbf{q}_c$ through the help of inverse kinematics block. The corrected joint coordinates, $\mathbf{q}_c$ is compared with the desired joint coordinates, $\mathbf{q}_d$ and integrates this error, added with the desired joint coordinates along with outer-loop control gain and this value is considered as a virtual reference joint coordinates, $\mathbf{q}_r$. The virtual reference vector is considered as the desired values for the inner-loop and compared with the actual coordinate values, $\mathbf{q}$ comes from joint displacement sensors (linear potentiometers). The inner-loop has simple PID control and gives the control inputs to the DC motors as pulse width modulated (PWM) signals through the help of microcontroller. The proposed control law is given as follows:

$$\tau(t) = \mathbf{K}_P \tilde{\mathbf{q}}_I(t) + \mathbf{K}_I \int \tilde{\mathbf{q}}_I(t)dt + \mathbf{K}_D \frac{d}{dt}\tilde{\mathbf{q}}_I(t) \tag{3}$$

$$\tilde{\mathbf{q}}_I(t) = \mathbf{q}_r(t) - \mathbf{q}(t),$$
$$\mathbf{q}_r(t) = \mathbf{q}_d(t) + \mathbf{K}_O \int \tilde{\mathbf{q}}_O(t)dt, \tag{4}$$
$$\tilde{\mathbf{q}}_O(t) = \mathbf{q}_d(t) - \mathbf{q}_c(t)$$

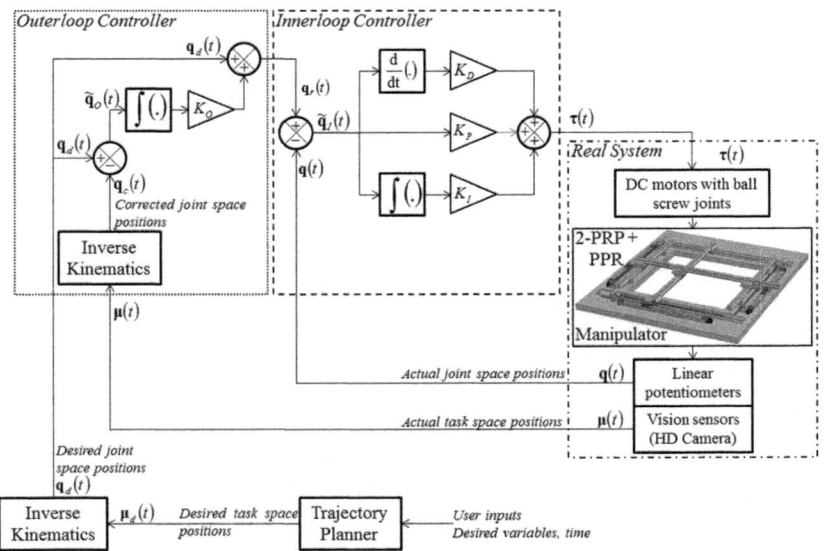

Fig. 2. Schematic representation of the proposed dual-loop control scheme.

3 Experimental Arrangement

To demonstrate the proposed control scheme, a complex trajectory tracking task is performed by running real-time experiments performed on an in-house fabricated prototype of the planar parallel manipulator and is presented in Fig. 3. MATLAB SIMULINK software package is used for programming the controller and Arduino Mega 2560 microcontroller board is used for feedback and hardware communication as shown in Fig. 3. The Arduino board is connected with a computer through a universal serial bus (USB) connection. It consists of two main interface modules: an analog input module and a digital input output module. There are three DC motors connected to the ball screw joints via servo-flex shaft couplers. The manipulator has a maximum workspace of 450 mm × 450 mm. To measure the task space displacements (angular and linear positions of the mobile platform centroid) of the manipulator, the experimental system which is equipped with digital camera with vision system (5 mega pixel with a maximum image size of 2560 × 2048 pixels, image capture capacity is 30 frames per second with a resolution of 1920 × 1080 pixels) are used. The linear actuator (ball screw with slider on a guide way) is coupled with a linear potentiometer (resolution of 100 microns) for determining the position of the actuated joint displacement. The linear ball joints can have a maximum velocity of 0.045 m/s. The complete hardware details both electrical and mechanical aspects can be found in reference [10].

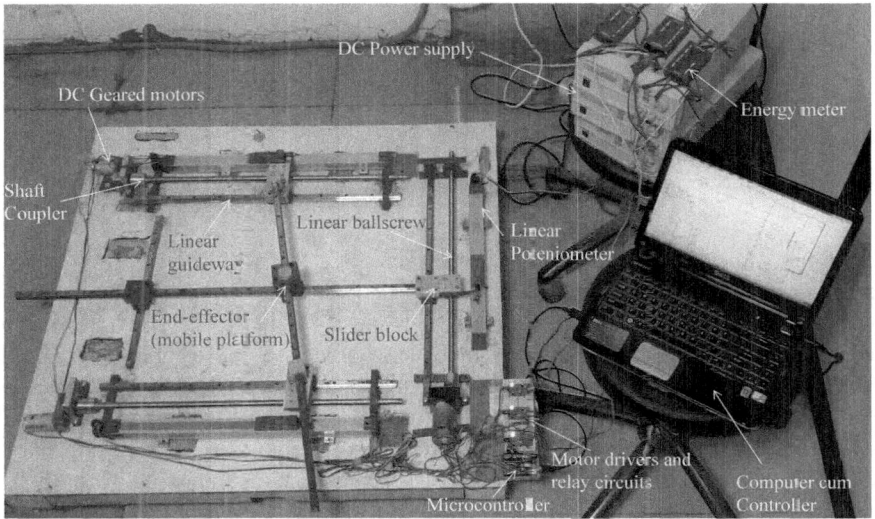

Fig. 3. Experimental arrangement of the 2-PRP+1-PPR planar manipulator.

4 Results and Discussion

The desired task space trajectory used in this paper is a loop comprising of a vertical rise, a circular path defined by two different arcs followed by a ramp and a vertical drop and ending with a horizontal span thus reaching the starting point in accordance with the time dependent functions. In prototype experiments, each segment took 40 s to complete the path which is well below the maximum speed of the each joint actuator. The desired task space trajectory tracking performance conducted separately with and without outer-loop for better comparison. Each experiment has conducted ten times and plotted their average values. The task space trajectories and time trajectories of norm of tracking errors are presented in Figs. 4 and 5. From these results, it is found that the tracking performance is improved when dual-loop applied.

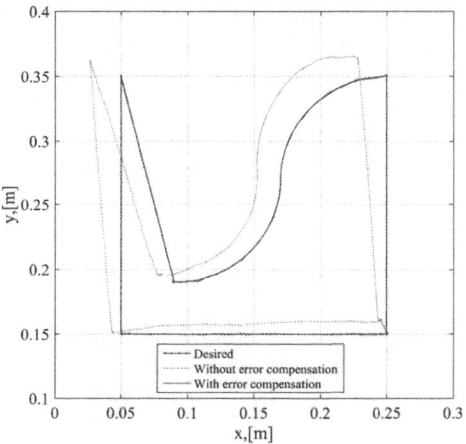

Fig. 4. End-effector (task space) trajectories: desired, with and without error compensation.

Further, time trajectories of joint displacements along with their desired and corrected (based on inverse kinematic solution of actual task space values) values is presented in Fig. 6. From these subplots, it is clearly noted that both controller performances are same in terms of their inner-loop objective and errors. The controller without outer-loop follows the desired joint space trajectory whereas with outer-loop follows the corrected desired joint space trajectory (virtual trajectory). It is also observed that $\mathbf{q}_c(t)$ follows the given desired joint trajectory $\mathbf{q}_d(t)$, inferring that the outer-loop integral control acting on the task space response attempts to nullify the errors not seen by the simple PID inner loop. Thus, from this plot of $\mathbf{q}(t)$, it is evident that the rectified joint position from integral control, $\mathbf{q}_r(t)$, is adjusted in order to correct the effect of mechanical errors.

In addition, the controller parameter sensitivity and robustness are also analysed. The root-square-mean error (RSME) is used as a performance measure for the analysis and it is given as:

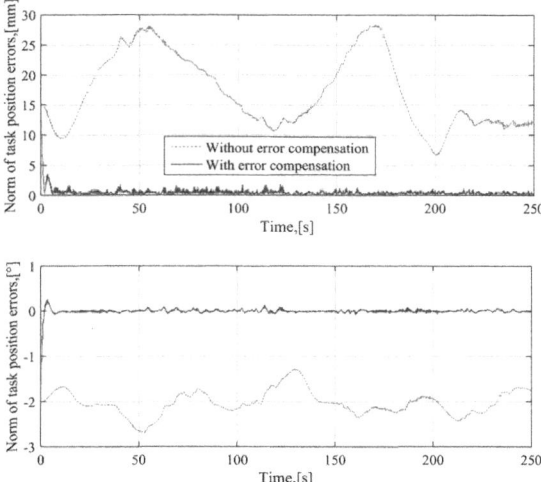

Fig. 5. Time trajectories of error norm of end-effector positions and orientation.

$$RSME_{xy} = \sqrt{\frac{\sum_{i=1}^{n}(x_{di}-x_i)^2+(y_{di}-y_i)^2}{n}}, RSME_{\theta} = \sqrt{\frac{\sum_{i=1}^{n}(\theta_{di}-\theta_i)^2}{n}}. \quad (5)$$

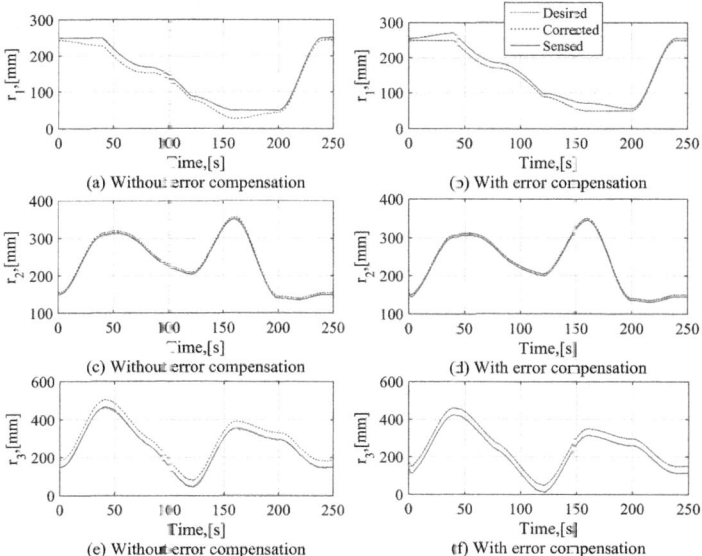

Fig. 6. Time trajectories of error norm of end-effector positions and orientation.

The controller parameter sensitivity analysis results are presented in Fig. 7. The robustness analysis is performed by tracking a circular trajectory of the end effector positions with different radii and average velocities at constant end-effector orientation ($\theta = 0°$). Figure 8 gives a comparison of the circular trajectory tracking results with and without outer-loop. The radius of the circular trajectory is varied from 0.05 m to 0.35 m. Similarly, there are two average velocity values 0.01 m/s and 0.03 m/s considered. Robustness results are presented in Table 1, are averaged over ten runs each.

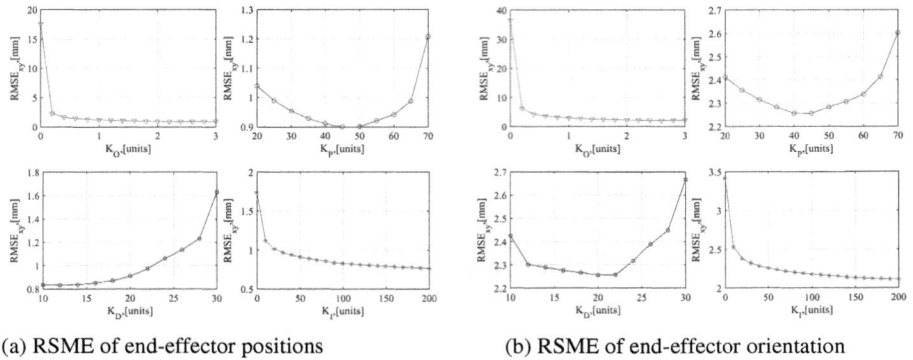

(a) RSME of end-effector positions   (b) RSME of end-effector orientation

Fig. 7. Controller parameter sensitivity results: parameter variations vs. RSME

From Table 1 and Fig. 8, it is clearly observed that at a lower velocity the tracking error reduces as the disturbance due to friction and inertial effects is not seen due to large mechanical advantage of the linear ball joints. It is also observed that on reducing the radius of the desired circular path from 0.35 m to 0.15 m, while keeping the velocity same, dual-loop control produces nearly the same values of RSME. For the complex trajectory, approximately 95% improvement in the average RSME is seen,

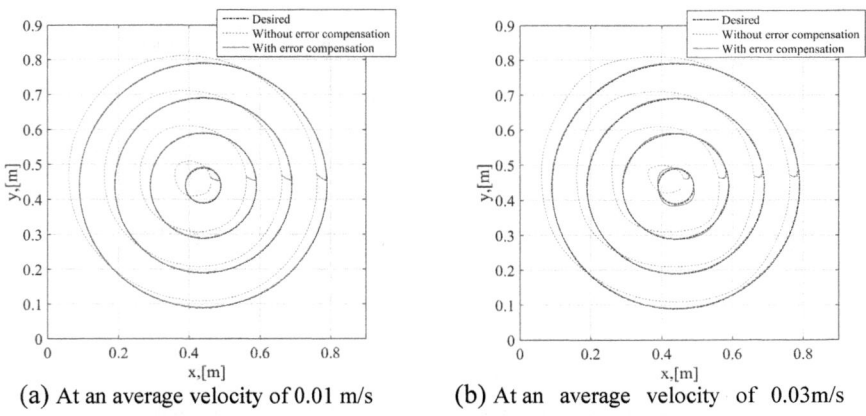

(a) At an average velocity of 0.01 m/s   (b) At an average velocity of 0.03m/s

Fig. 8. End-effector (task space) trajectories: desired, with and without error compensation.

Table 1. Comparative results of averaged root squared mean error (RSME) from experiments

Radius in m	Average Velocity in m/s	RSME (x, y) in mm		RSME (θ) in deg	
		Without compensation	With compensation	Without compensation	With compensation
Complex trajectory		18.21	0.90	2.04	0.09
0.05	0.01	36.06	4.17	2.01	0.23
0.15	0.01	35.60	2.52	2.02	0.14
0.25	0.01	37.44	2.09	2.03	0.10
0.35	0.01	40.15	1.91	2.03	0.09
0.05	0.03	41.64	9.55	2.01	0.41
0.15	0.03	36.76	5.05	2.05	0.24
0.25	0.03	37.50	3.85	2.05	0.20
0.35	0.03	39.94	3.54	2.05	0.17

while circle trajectory tracking it varies 77% to 95%. For the circular trajectory tracking with a smaller radius (0.05 m) has less improvement in error reduction due to large non-zero initial error. The error improvement is 77% and 88% at 0.03 m/s and 0.01 m/s, respectively. In overall, it presents a simple dual-loop control method with redundant sensing to evade calibration, and diminish the errors arising from mechanical inaccuracies.

5 Conclusions

The proposed dual-loop control scheme, envisioned at reducing tracking errors of the task space variables due to friction, geometric errors and joint clearances has been presented. Prototype experiments show that the inner-loop PID control with the outer-loop integral control improves the tracking performance significantly. Further, as expected, the enhancement of outer-loop is more significant at lower end-effector velocities rather than higher, as at higher speeds, inertial forces and friction cause greater disturbances which can overcome by introducing an additional nonlinear disturbance observer. The proposed dual-loop control scheme can potentially be applied to any robotic manipulators in the presence of geometric errors and joint clearances. The use of the proposed scheme in spatial manipulators conjunction with their dynamic models would be the next research objective.

Acknowledgments. This research was supported in part by the Humboldt Fellowship funded by the Alexander von Humboldt (AvH) Foundation, Germany and in part by the Council of Scientific and Industrial Research (CSIR), India.

References

1. Ghosal A (2009) Robotics-fundamental concepts and analysis. Oxford University Press, Oxford
2. Sabanovic A, Ohnishi K (2011) Motion control systems. Wiley, Singapore
3. Cheah CC, Li X (2015) Task-space sensory feedback control of robot manipulators. Springer, Singapore
4. He R, Zhao Y, Yang S, Yang S (2010) Kinematic-parameter identification for serial-robot calibration based on POE formula. IEEE Trans Rob 26:411–423
5. Lang J, Chen L (2011) Dual-loop integral sliding mode control for flexible space manipulator. Chin Mech Eng 22:1906–1912
6. Agarwal A, Nasa C, Bandyopadhyay S (2011) Dual-loop control for backlash correction in trajectory-tracking of a planar 3-RRR manipulator. In: Proceedings of 15th national conference on machines and mechanisms NaCoMM2011, Chennai, pp 189:1–189:8
7. Duarte F, Ullah F, Bohn C (2016) Modelling and dual loop sliding mode control of a two flexible-link robot to reduce the transient response. In: Proceedings of 24th Mediterranean conference on control and automation, Athens, pp 280–285
8. Agarwal A, Nasa C, Bandyopadhyay S (2016) Dynamic singularity avoidance for parallel manipulators using a task-priority based control scheme. Mech Mach Theory 96:107–126
9. Lamaury J, Gouttefarde M, Chemori A, Herve PE (2016) Dual-space adaptive control of redundantly actuated cable driven parallel robots. In: Proceedings of international conference on intelligent robots and systems (IROS 2013), Tokyo, pp 4879–4886
10. Singh Y, Santhakumar M (2015) Inverse dynamics and robust sliding mode control of a planar parallel (2-PRP and 1-PPR) robot augmented with a nonlinear disturbance observer. Mech Mach Theory 92:29–50

The Effects of Admittance Term on Back-Drivability

Ogulcan Işıtman, Orhan Ayit, and Mehmet İsmet Can Dede[(✉)]

İzmir Institute of Technology, İzmir, Turkey
{ogulcanisitman,orhanayit,candede}@iyte.edu.tr

Abstract. In the design of kinesthetic haptic devices, there are mainly impedance type and admittance type device. In a customary scenario, the human operator back-drives the haptic device by holding and providing motion to the handle of the haptic device. If the type of transmission system does not allow passive back-drivability, then the back-drivability is satisfied by the use of an admittance controller. This type of a haptic device is said to have admittance structure. The selection of the admittance term in this controller plays a critical part in the task execution performance. Determination of this term is not trivial and the optimal parameters depend on not only the key performance criteria but also on the human operator. An experimental study is carried out in this work to determine the effect of the admittance term parameters on the performance of human operators in terms of the energy efficiency and the best accuracy. In this paper, the experimental set-up and the results of the experiments are presented and discussed.

Keywords: Admittance control · Admittance term · Back-drivability · Haptics

1 Introduction

A typical kinesthetic haptic device acquires the motion of the targeted body part of the human operator. This information is used as the motion demand to drive the slave system in a teleoperation scenario. The physical interaction of the slave system with its environment is simultaneously measured as interaction forces/moments or if the slave system is a virtual reality scenario, then the virtual interaction forces/moments are modeled and calculated. The haptic device is used at this stage to display these forces/moments to the human operator.

In general, the targeted human body part is the hand and a handle is used to couple the human operator to the haptic device. Since the acquisition of the handle motion needs to be done at higher sampling frequencies [1] and with precision, external measurement techniques such as vision sensors or inertial measurement units cannot fulfill the task. The motion of the handle is typically calculated indirectly by using the joint sensor measurements and direct kinematics formulated for the haptic device's mechanism.

The haptic device mechanism is equipped with actuation systems in order to display forces/moments at the handle to the human. However, the human operator is required to back-drive this mechanism in order to issue motion demands for the slave system. If the actuation system is composed of a customary transmission system with larger reduction ratios (typically over 1:5), then the back-drivability of the system is affected adversely if not becomes almost impossible [2]. In this situation, a passive back-drivable system otherwise named as an impedance type system should not be used. A solution is to devise an admittance type structure for active back-driving by implementing an admittance controller.

An admittance controller used in an admittance type haptic device necessitates the use of a force/moment sensor to acquire the human operator interaction with the handle. An admittance term is used to convert this interaction information to the motion demand of the haptic device, which results in back-driving the haptic mechanism. This admittance term generates a virtual coupling between the human and the handle with the components of a virtual mass-spring-damper.

The selection of these mass-spring-damper components is not a trivial process. The selection process depends on the performance criteria set for the task such as task completion duration, task accuracy or human effort during the task, and the human operator's physical condition. The physical condition of the human operator can also be named as the impedance characteristics of the human arm for this specific case in which human hand motion is captured [3]. The human impedance varies from one individual to another and also it changes during the day and due to the psychological state of the same human operator [3].

In this paper, an experimental study is carried out to understand the effects of the virtual coupling components and their values on the task execution performance. A single degree-of-freedom (DoF) linear system with a handle is used and a task is designed to measure the performance of various human subjects. Different admittance terms are used and the performance results in terms of positioning accuracy and the human effort are obtained and discussed.

The next section provides a brief overview of the admittance type haptic devices and their controllers. The experimental set-up and the task are explained in the following sections. The experimental results obtained are presented and discussed to conclude the paper.

2 Overview of Admittance Type Haptic Systems

In the literature, admittance type devices are rarely seen as haptic systems, however, it is mandatorily used as the haptic system when there are mechanical limitations such as high inertia, high friction and as a consequence of these, non-back-drivability. The admittance type haptic systems are mostly preferred in surgical operations [4], rehabilitation systems [5], industrial operations [6] and haptic researches [7].

In this type of devices, forces or torques applied from user or environment, are transformed to states of the system such as position, velocity or acceleration by using admittance gain. Admittance gain is generally modeled by using different combinations of mass, spring and damper components. In [8, 9], admittance gain is designed as a mass/inertia to mimic free translational/rotational motion. By that way, the measured force/torque is converted to acceleration. In addition, other states are obtained by using integral operation to derive desired values for low-level controller in the system.

Another type of admittance gain model is used in an RML-glove which is a five-fingered admittance haptic interface. In this system, admittance gain is modeled as a virtual spring component for relating force to position [10]. To obtain a relation between velocity and applied force, only damper component is used to design admittance gain as implemented in [11, 12]. Also, a virtual mass-damper model can be used to obtain a desired physical coupling with the environment by calculating the velocity from the measured forces as used in [13]. Furthermore, in some works, an admittance gain is modeled by using all of three components [14, 15]. Besides the mentioned models, E. Faulring et al. use dynamic simulation to compute desired acceleration from an applied force for admittance type hand controller [16]. Also, in [17], a model free PID admittance controller for an upper limb exoskeleton is designed.

Even though mass-spring-damper compositions are mostly preferred as admittance term in literature, there are not enough studies on determination of the values of these components. In [18], an experiment is performed with the participation of 15 subjects to evaluate controllability of device. According to results of the experiment, the optimum values of mass-damper components are designated for each subject separately. Moreover, M. Nambi et al. prepared an experiment to investigate different damper values as an admittance term which minimizes the force tracking error [12]. In literature, the spring parameter is mostly chosen as zero in composing the admittance term. In this paper, the effects of choosing mass – damper components with spring and without spring investigated in compositions of the admittance term.

3 Experimental Set-up

A single degree of freedom admittance type device is used in the experiments as shown in Fig. 1. To obtain linear motion, a DC motor (HITACHI D06D401E) with an optical encoder (AEDA-3300AT) is assembled to a lead screw linear stage which has 220 mm length workspace. Since the resolution of the quadrature optical encoder is 16384 counts/rev and the lead screw stage has 5 mm pitch, the linear resolution of the system is obtained as 0.03 μm. In addition to these, a handle, which forces the user to hold it in precision grasp, is mounted on the top of the linear stage to interact with the subjects. Below the handle, a Kistler (type 9017B) 3 DoF force sensor is placed. An NI MyRIO 1900 data acquisition (DAQ) system is used for digital and analog data acquisition and controlling the system. DAQ system has a 12-bit ± 10 V analog input to get force data. Also, by using the device, analog outputs are issued as demands to the Maxon 4-Q-DC servo-amplifier that is used to drive the motor. To meet the power requirements of the system, a 20 V DC power supply is used. Control algorithm and graphical user interface are developed in NI LabView Software with Control Design and Simulation module and Real-Time module. Force readings are acquired and control algorithm is executed at a sampling rate of 1 kHz.

Admittance control is implemented on top of a low-level velocity controller, which is selected to be a proportional controller (P-controller). The implemented control scheme is shown in Fig. 2. Low-level controller gain K_p is experimentally found as 15 by observing minimal tracking error for given sinusoidal velocity input. In Fig. 2, G_c represents the low-level velocity controller. F_{ref} is reference force which is selected to

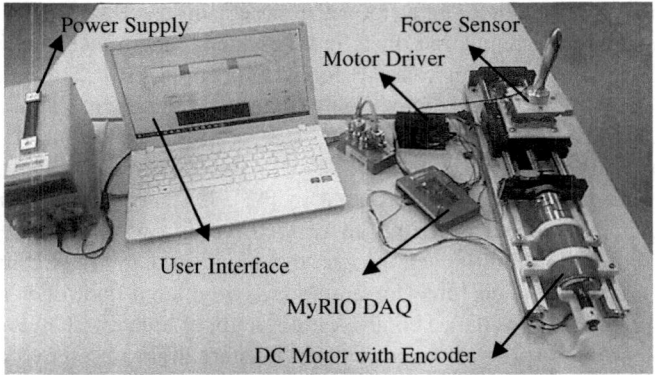

Fig. 1. Experimental set-up

be zero for free motion (full back-drivability), A is admittance term, X_m is measured position, X_e is the position vector of the contacted environment, V_s and F_s are the velocity and force sensor transfer functions, respectively.

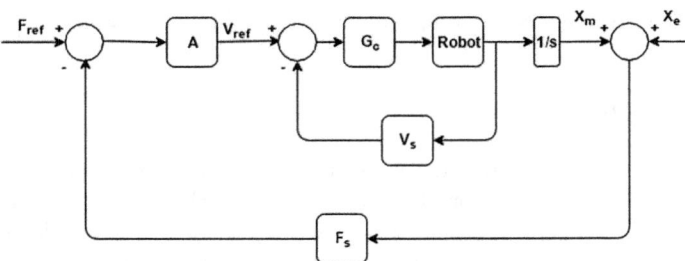

Fig. 2. Control scheme of single Dof admittance type device.

Admittance control requires measurements of the interaction forces between the robot and its environment; as well as accurate trajectory following capabilities. Hence, this implies that a robot with high power actuators and a stiff construction is preferred, which puts specific demands on the robot's design.

In an admittance controller, a force setpoint is specified and it is tracked by a force compensator given in Eqs. 1 and 2; where F, V, m, b, k and A parameters refer to force, velocity, mass, damper, spring and admittance term, respectively. Force compensator can be modeled as a mass, spring and damper system.

$$F(s) = m \cdot sV(s) + b \cdot V(s) + \frac{1}{s}k \cdot V(s) \tag{1}$$

$$\frac{V(s)}{F(s)} = \frac{s}{ms^2 + bs + k} = A \tag{2}$$

Determination of the parameters is constructed under consideration of tracking capability limit of the set-up and feasibility for the human force limits. The experimental setup, which is described above, has a 75 rad/s (≈12 Hz) velocity (V_{ref}) tracking capability as calculated by experimentally from its frequency response.

Characteristics of the admittance term depend on the chosen parameters. High admittance term causes a sudden reaction, while low admittance term gives a slow reaction to the applied force. Although the mass and the damper coefficients which are inversely proportional to the admittance term, they are chosen as low as possible to initiate the robot motion softly. However, selecting very low values for the coefficients cause the system to be very sensitive to the interaction forces.

4 Description of the Task

To determine the effect of the admittance term parameters on the performance of human operators, an experiment is designed. A task is organized to evaluate the accuracy and the energy usage of the subjects with different parameters.

In this task, the subject is asked to hold the handle, which is designed for the subject to have a precision type of grasp, with his/her dominant hand. As guided by the graphical user interface, the subject is instructed to move the handle to blue target and keep it there for four seconds as indicated by the user interface. After four seconds, the subject moves the handle by 116 mm to reach the red target and holds the handle at that location for four seconds. The subject repeats this process 2 times without releasing the handle. When the user reaches the target for the last time, it is requested to release the handle without any further action.

The visual information displayed to the human subject is shown in Fig. 3. The black bar demonstrates handle's position and targets are represented as a red (on the left) and blue (on the right) bars. The white region near the targets presents the scoreboard which is created to define a metric for comparison between the parameters. Figure 4 presents the score range used on both targets based on the position of the handle. When the black cursor is inside one of the targets, the indicator in green color by default, which is represented by the rectangular area above the targets, turns to red color for 4 s then turns back to green again. This gives the information to the user that he/she can move the handle to the other target. The accuracy score of users is calculated by taking the average value of gained points during 4 s. Since the human hand's movements may jiggle, the waiting duration provides us to obtain more reliable results on scoring the accuracy.

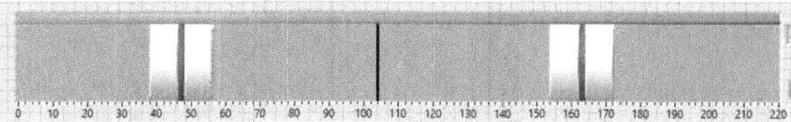

Fig. 3. Graphical user interface for the generated task (Color figure online)

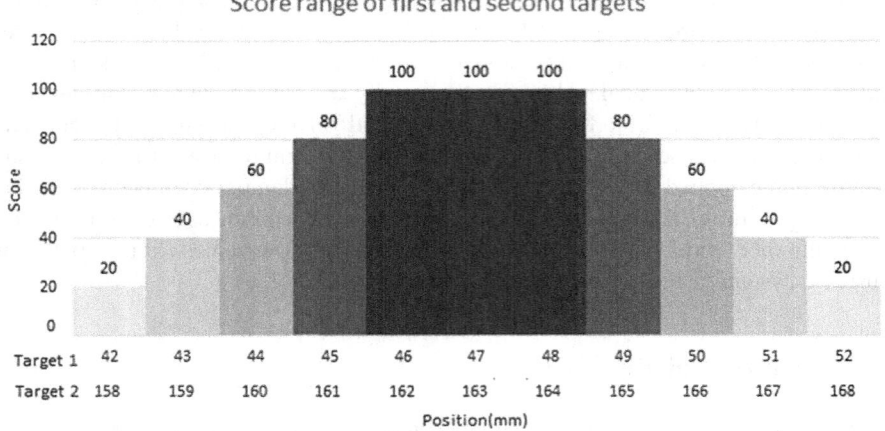

Fig. 4. Score range of the first target based on position

The evaluation of the effort of the subjects while moving the handle is carried out by using Eq. 3 where F, x, E_{kin}, E_{damp}, E_{spr}, and E_{lost} represent force, position, kinetic energy, dissipated energy due to virtual damper, potential energy stored in the virtual spring and energy lost due to viscous friction, respectively. Since the relative results are investigated, the energy loss of the system is neglected. At the end of the experiment, the handle comes back to its static condition and thus, sum of the kinetic energy terms is equal to zero. For each subject and each admittance parameter tested, locations of the targets are kept the same.

$$\begin{aligned}
F_1(x_1 - x_0) &= E_{kin1} + E_{damp1} + E_{spr1} + E_{lost1} \\
E_{spr1} + F_2(x_2 - x_1) &= E_{kin2} + E_{damp2} + E_{spr2} + E_{lost2} \\
&\vdots \\
F_n(x_n - x_{n-1}) &= E_{kin\,n} + E_{damp\,n} + E_{spr\,n} + E_{lost\,n} \\
E_{spr\,n} &= E_{kin\,n+1} + E_{damp\,n+1} + E_{lost\,n+1} \\
\hline
\sum_1^n F_n \Delta x_n &= \sum E_{kin} + \sum E_{damp} + \sum E_{lost}
\end{aligned} \quad (3)$$

As it can be observed from the energy equation, the potential energy of the spring term cancels out between each state. Due to these conditions, damping term dissipates all the energy in the experiments.

5 Experimental Results

The experiments are carried out with the participation of 4 women and 9 men subjects. The ages of these 13 subjects are varying between 24 and 32. Subjects are informed about the experiment and the task. Each subject is required to practice with the device before the experiment. For this practice, admittance parameters are chosen differently from the inspected ones not to affect the reliability of the comparison-based results.

Each subject participated in 6 experiments, which can be grouped into two main categories. In the first group of experiments, admittance term is modeled as spring - mass- damper and in the second group, it is modeled without the spring element. Table 1 presents chosen admittance parameters and designated corner frequency against the experiment number. Damping ratio ζ of the admittance term $(\frac{K\omega_n^2 S}{S^2 + 2\zeta\omega_n S + \omega_n^2})$ is kept constant to determine parameters with respective to desired natural frequency. For the both first order $\left(\omega_c = \frac{b}{m}\right)$ and second order $\left(\omega_n = \sqrt{\frac{k}{m}}\right)$ admittance gains, bandwidth frequency is taken as corner frequency in the rest of this paper.

Table 1. Mass-spring-damper parameters

Admittance parameters				
Experiment number	(m) Mass (kg)	(k) Spring (N/m)	(b) Damping (N.s/m)	(w_c, w_n) Corner Frequency (rad/s)
1	2	2	3,78	1
2	2	18	11,38	3
3	2	72	22,77	6
4	2	0	2	1
5	2	0	6	3
6	2	0	12	6

The spring parameter is adjusted to reach the desired corner frequency in first three experiments. For last three experiments, damper parameter is modified to obtain the desired corner frequencies. The value of the natural frequency for the first experiment and the value of the corner frequency for the fourth one are 1. In the same manner between the experiments 2-5 and 3-6, these values are 3 and 6, respectively. The effect of the spring term is investigated by comparing these experiment groups. The main reason to keep these the same is to have similar bandwidths for both types of admittance terms with and without the spring component.

The result of the accuracy experiments is given in Table 2 with the average values of the each set of experiments. The results can be investigated separately in each group of the experiment. With the increasing corner frequency, obtained accuracy scores also increase. Also, better accuracy is obtained by using the admittance term with spring term when corner frequency is kept constant.

The energy supplied by the user to the system in each experiment is listed in Table 3. The units for these results are in Joules. As observed in the accuracy experiments, energy consumption, which indicates the effort by the user, increases as the chosen corner frequency values are increased. The first three experiment group consumes more energy with respect to the latter one.

Table 2. Scores of subjects

Accuracy Subject	Exp # 1	Exp # 2	Exp # 3	Exp # 4	Exp # 5	Exp # 6
1	77,28	76,80	91,23	66,67	86,85	85,47
2	88,72	93,21	90,46	71,42	78,63	90,27
3	89,75	89,02	90,32	55,91	72,67	68,25
4	92,22	93,03	94,92	68,04	79,71	80,15
5	84,51	81,59	92,85	82,36	79,41	84,41
6	80,21	87,49	90,86	67,34	84,57	87,78
7	71,59	84,43	85,90	69,60	84,68	85,47
8	61,14	88,94	90,08	51,56	82,10	74,70
9	47,03	77,82	78,19	67,50	74,71	83,84
10	62,79	89,85	85,48	58,78	69,31	75,84
11	68,25	75,32	73,98	57,78	81,69	76,58
12	72,60	90,40	90,08	67,04	76,31	84,22
13	88,30	91,25	82,74	70,31	84,03	90,81
Average	75,72	86,09	87,47	65,71	79,59	82,14
Norm. Avg.	−0,328	0,582	0,816	−1,378	−0,013	0,320

Table 3. Energy consumption of subjects

Energy consumption (J) Subject	Exp # 1	Exp # 2	Exp # 3	Exp # 4	Exp # 5	Exp # 6
1	0,12	0,34	0,50	0,05	0,10	0,21
2	0,06	0,16	0,34	0,04	0,11	0,23
3	0,05	0,17	0,37	0,05	0,20	0,17
4	0,07	0,20	0,43	0,04	0,09	0,14
5	0,13	0,28	0,50	0,05	0,12	0,20
6	0,07	0,28	0,55	0,05	0,09	0,16
7	0,07	0,18	0,40	0,06	0,10	0,28
8	0,06	0,16	0,32	0,04	0,08	0,15
9	0,09	0,21	0,47	0,03	0,06	0,11
10	0,12	0,20	0,54	0,05	0,12	0,19
11	0,06	0,18	0,36	0,03	0,07	0,15
12	0,08	0,23	0,55	0,02	0,12	0,19
13	0,11	0,28	0,50	0,04	0,11	0,23
Average	0,08	0,22	0,45	0,04	0,10	0,18
Norm. Avg.	−0,665	0,249	1,779	−0,929	−0,495	0,06

Besides the given results, normalized averages of the accuracy and energy results are provided based on Eq. 4.

$$x_{normalized} = \frac{x - \bar{x}}{SD} \qquad (4)$$

This normalization is made for every subject with respect to their average values, $\bar{x}$, and standard deviations, SD. Then, for each experiment set, the average of the normalized data is calculated. This approach provides an objective comparison on the improvement of the performance.

6 Discussions and Conclusions

During the literature survey, it is observed that mass–spring–damper model is commonly used as admittance term due to the effective usage and simple design. For that reason, an experimental study is carried out to investigate the effects of spring, damper components and corner frequency for back-drivability success. An experimental setup is prepared and a task is defined to evaluate the effects on the human operator. The experiment results, for evaluating the accuracy and energy consumption according to the described task are presented in Sect. 5.

According to the results with spring and without spring, the corner frequency is one of the essential parameters for better accuracy. On the other hand, the comparison between the experiment groups 1-4, 2-5, 3-6, which have same corner frequencies, reflects that combination of mass-spring-damper terms causes better accuracy results than mass-damper terms. This implies the positive effect of the spring parameter on accuracy. As a consequence of higher corner frequency, spring and damper terms are increased. However, when the results of these groups are investigated, energy consumption is also increased as accuracy is enhanced since the mass term is kept constant and the energy calculations depend on damping term and velocity as shown in Eq. 3. The velocity demands coming out of the admittance term increases as a result of increasing corner frequency in each group and this effect also plays a role in energy increase. In addition to these, increasing the coefficients of admittance term requires higher force values and this increase is limited with the operating limits of the user. By interpreting the obtained results, mass – spring – damper parameters can be chosen optimally for specific tasks and these parameters should be experimentally tuned for different cases and set-up. For the tasks that require accuracy, the higher spring coefficient in the admittance term must be preferred such as in surgical robots. In contrast to that, the spring parameter might be unnecessary for rehabilitation tasks due to high-energy consumption. Further work will be focused on the variation of mass parameter effect on back-drivability in the sense of accuracy and operator effort.

Acknowledgements. This work is supported in part by The Scientific and Technological Research Council of Turkey via grant number 115E726.

References

1. Kern TA (ed) (2009) Engineering haptic devices: a beginner's guide for engineers. Springer Science & Business Media, Heidelberg, pp. 56–58
2. Ishida T, Takanishi A (2006) A robot actuator development with high backdrivability. In: 2006 IEEE conference on robotics, automation and mechatronics. IEEE
3. Podobnik J (2012) Haptics for virtual reality and teleoperation, vol 64. Springer Science & Business Media, Heidelberg, pp. 52–54
4. Osa T et al (2015) Hybrid rate—admittance control with force reflection for safe teleoperated surgery. IEEE/ASME Trans Mechatron 20(5):2379–2390
5. Ozkul F, Barkana DE (2011) Design of an admittance control with inner robust position control for a robot-assisted rehabilitation system RehabRoby. In: 2011 IEEE/ASME international conference on advanced intelligent mechatronics (AIM). IEEE
6. Tang T et al (2016) Teach industrial robots peg-hole-insertion by human demonstration. In: 2016 IEEE international conference on advanced intelligent mechatronics (AIM). IEEE
7. Wang H, Kosuge K (2012) Control of a robot dancer for enhancing haptic human-robot interaction in waltz. IEEE Trans Haptics 5(3):264–273
8. Peer A, Buss M (2008) A new admittance-type haptic interface for bimanual manipulations. IEEE/ASME Trans Mechatron 13(4):416–428
9. Morbi A, Mojtaba A (2016) Safely rendering small impedances in admittance-controlled haptic devices. IEEE/ASME Trans Mechatron 21(3):1272–1280
10. Ma Z, Ben-Tzvi P (2011) An admittance type haptic device: RML glove. In: ASME 2011 international mechanical engineering congress and exposition
11. Arbuckle TK et al (2016) Human velocity control of admittance-type robotic devices with scaled visual feedback of device motion. IEEE Trans Hum Mach Syst 46(6):859–868
12. Manikantan N, Provancher WR, Abbott JJ (2011) On the ability of humans to apply controlled forces to admittance-type devices. Adv Robot 25(5):629–650
13. Heidingsfeld M et al (2014) A force-controlled human-assistive robot for laparoscopic surgery. In: 2014 IEEE international conference on systems, man, and cybernetics (SMC). IEEE
14. Nabeel M et al (2015) Increasing the impedance range of admittance-type haptic interfaces by using Time Domain Passivity Approach. In: 2015 IEEE/RSJ international conference on intelligent robots and systems (IROS). IEEE
15. Baumeyer J (2015) Robotic co-manipulation with 6 DoF admittance control: application to patient positioning in proton-therapy. In: 2015 IEEE International Workshop on Advanced Robotics and its Social Impacts (ARSO). IEEE
16. Faulring EL, Colgate JE, Peshkin MA (2006) The cobotic hand controller: design, control and performance of a novel haptic display. Int J Robot Res 25(11):1099–1119
17. Yu W, Rosen J, Li X (2011) PID admittance control for an upper limb exoskeleton. In: Proceedings of the 2011 American control conference. IEEE
18. Ozkul F, Barkana Erol D (2016) Bir Robot Destekli Rehabilitasyon Sisteminin Admitans Filtre Parametrelerinin Adaptif Olarak Ayarlanması. Otomatik Kontrol Ulusal Toplantısı TOK, Eskişehir, 29 Eylül–1 Ekim 2016

Design and Proposed Model Reference Trajectory Control of a Snake like Robot

Sahin Yildirim[✉] and Kirakoya Abdoulaye Ben-Aziz

Erciyes University, Kayseri, Turkey
sahiny@erciyes.edu.tr, kirakoyaben@live.fr

Abstract. Despite remarkable innovation in robotic domain, there is still some problems to overcome sudden trajectory variation of 10 degree of freedom (DOF) snake like robots. In this investigation, two types of control system were employed to control prescribed trajectory of snake like robot. First control system is well known PID controller with constant gain parameters. Second controller is a proposed model reference adaptive control system. However, a proposed neural network controller is used for modeling reference adaptive controller. Finally, the proposed control system has superior performance to adapt the trajectory of snake-like robot.

Keywords: Snake robot · PID controller · Adaptive controller · Modeling · Control system · Results

1 Introduction

The inspiration for the construction of a snake like robot are biological snakes, which are populating a large territories on earth and are capable of moving using different locomotion gait depending on the environment they are surrounded by [1]. The proposed ten DOF snake-like robot, has the capacity of moving using two (02) different locomotion gait. The serpentine locomotion also known as lateral undulation, via the coupling of internal shape changes [2, 3] which utilizes passive wheels; and the crotaline motion known as sidewinding [4–6]. But in order to build an approximate motion similarity between the biological snake and the mechanical structured snake, the proposed control systems should adapt perfectly the body shape position of each revolute joint of the ten DOF snake-like robot.

Several tests were investigated on both PID and Model Reference Adaptive controller; and obtained results were analyzed in order to identify the most efficient control system that will be used on the snake robot. Currently the adaptive control has already been researched in the robotic domain [7–13]. The major challenges of adaptive control are the high coupling dynamic equation, many degrees of freedom and under actuated characteristic. The approach that we will use will be a modified Model Reference Adaptive Control (MRAC) [14], which is composed of reference model, adaptive mechanism, feedback controller, and plant. The structure of this paper is as follows. In Sect. 2, a brief overview of snake- like robot kinematics and dynamics. Section 3 introduces PID controller and the adaptive controller designed for the hyper-redundant multi-body robot on simmechanics. Section 4 gives simulations results and demonstrates the high performance of the proposed adaptive controller. Finally Sect. 5 presents some concluding remarks.

2 Snake-like Robot Kinematics and Dynamics

2.1 Snake-like Robot Kinematic

The proposed snake-like robot moves on a horizontal and flat surface, and has N−1 degrees of freedom. The following definitions are illustrated in Fig. 1.

Definition 2.1 (Link angle). The link angle of link $i \in \{1,\ldots,N\}$ of the snake robot is denoted by $\theta i \in R$ and is defined as the angle that the link forms with the global x axis with counterclockwise positive direction.

Definition 2.2 (Joint angle). The joint angle of joint $i \in \{1,\ldots,N-1\}$ of the snake robot is denoted $\varphi i \in R$ and is defined as

$$\varphi i = \theta i - \theta i + 1 \qquad (1)$$

A link angle is the orientation of a link with respect to the global x axis, while a joint angle is the difference between the link angles of two neighboring links. The link angles and the joint angles are assembled in the vectors $\theta = [\theta 1, \ldots, \theta N]^T \in R^N$ and $\varphi = [\varphi 1, \ldots, \varphi N-1]^T \in R^{N-1}$, respectively.

The snake robot has no explicitly defined orientation since there is an independent link angle associated with each link. A measure of the heading of the robot can be obtain as follows:

Definition 2.3 (Heading). The heading (or orientation) of the snake robot is denoted by $\bar{\theta} \in R$ and is defined as the average of the link angles,

$$\bar{\theta} = \frac{1}{N} \sum_{i=1}^{N} \theta_i \qquad (2)$$

Where θ_i is Angle between link i and the global x axis and N is the number of link.

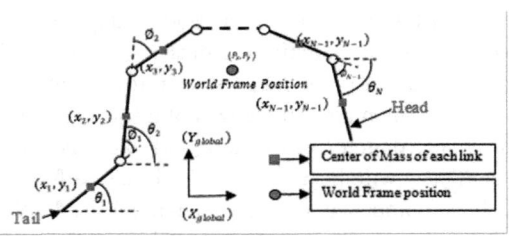

Fig. 1. The kinematic parameter of the snake-like robot

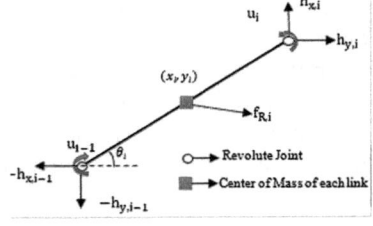

Fig. 2. Forces and torques acting on each link

2.2 Snake-like Robot Dynamic

The N -1 degrees of freedom of the snake-like robot are defined by the link angles $\theta \in R^N$ and the center of mass (CM) position $p \in R^2$. We now present the equations of motion of the robot in terms of the acceleration of the link angles $\ddot{\theta}$, and the acceleration of the CM position, $\ddot{p}$.

As illustrated in Fig. 2, link $i \in \{1,...,N\}$ is influenced by the ground friction force $f_{R,i} \in R^2$, which acts on the center of mass of the link, and also the joint constraint forces $-h_{x,i-1}, -h_{y,i-1}, h_{x,i}$, and $h_{y,i}$, which keep the link connected to link $i - 1$ and link $i + 1$. Using first principles, the force balance for link i in global frame coordinates is given by

$$m\ddot{\theta}_i = f_{R,xi} + h_{x,i} - h_{x\,i-1}, \tag{3}$$

$$m\ddot{y}_i = f_{R,yi} + h_{y,i} - h_{y,i-1}. \tag{4}$$

2.3 Coulomb Friction

The coefficients describing the Coulomb friction force in the tangential (along link x axis) and normal (along link y axis) directions of a link, respectively, are denoted by μt and μn. The proposed Coulomb friction force for the ten DOF snake-like robot was designed on simmechanics see Fig. 3. The coulomb friction is applied to all links of the snake robot. In order to mimic a real time simulation on ground, the value of friction force were choose as: μt = 0.8 and μn = 0.5.

$$f_{R,i}^{link,i} = -mg \begin{bmatrix} \mu t & 0 \\ 0 & \mu n \end{bmatrix} sgn\, v_i^{link,i} \tag{6}$$

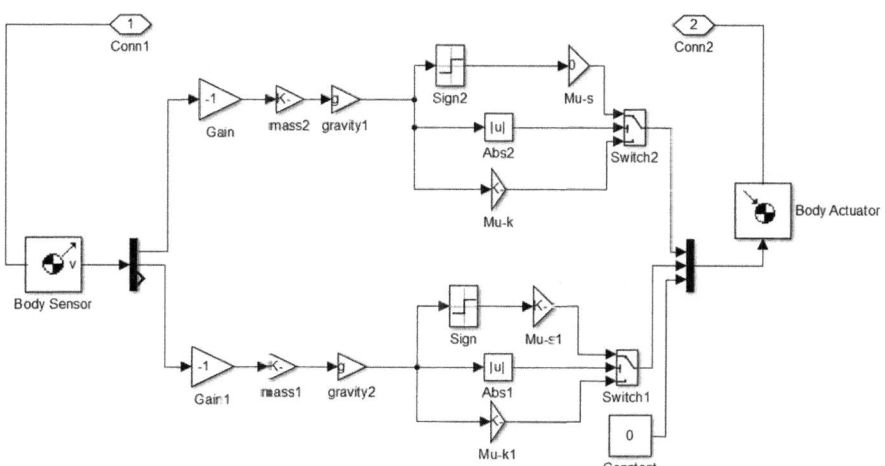

Fig. 3. Coulomb friction model of each link

3 Proposed Control System

3.1 PID Controller

The proposed PID controller continuously calculates the error value e(t) = yp − ym as the difference between the desired set point and the measured process variable and applies a correction based on proportional, integral and derivative parameters. The controller attempts to minimize the error over time by adjustment of a control variable See Fig. 4.

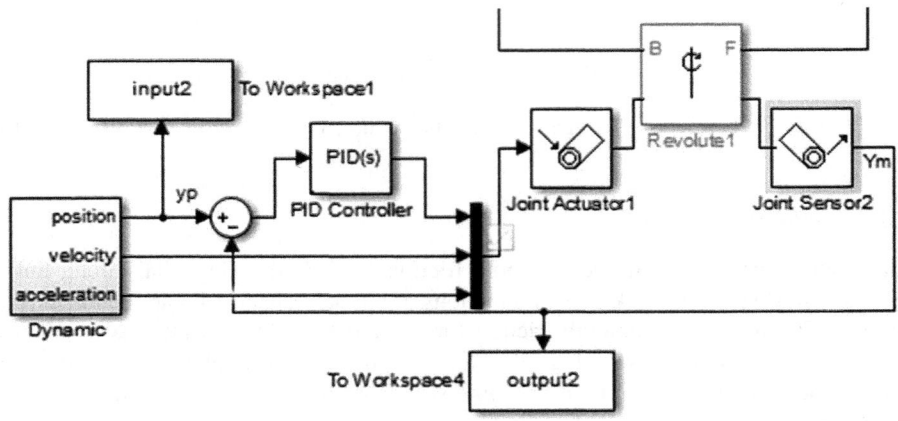

Fig. 4. PID controller representation on simmechanics

3.2 Adaptive Controller

3.2.1 Principle of Working

Model Reference Adaptive Control strategy is used to design the adaptive controller that works on the principle of adjusting the controller parameters so that the output of the actual plant tracks the output of a reference model having the same reference input [15].

3.2.2 Components

Reference Model: It is used to give an idyllic response of the adaptive control system to the reference input.

Controller: It is usually described by a set of adjustable parameters. In this paper only one parameter θ is used to describe the control law. The value of θ is primarily dependent on adaptation gain.

Adjustment Mechanism: This component is used to alter the parameters of the controller so that actual plant could track the reference model. The basic block diagram of

MRAC system is shown in the Fig. 5. As shown in the figure, ym(t) is the output of the reference model and y(t) is the output of the actual plant and difference between them is denoted by e(t).

$$e(t) = y(t) - ym(t) \qquad (7)$$

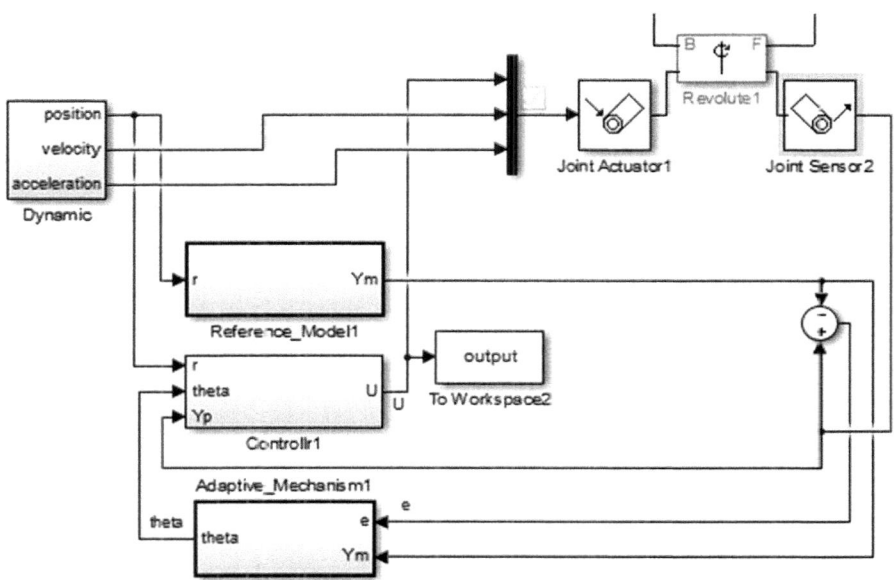

Fig. 5. Model reference adaptive controller simmechanics representation

4 Simulation Results

The simulations carried out are aimed at determining the capacity of performance of the proposed control systems. Both control systems were applied to all the revolute joints connecting each link of the robot. In order to mimic a real time simulation on ground and observed the performance of the proposed control systems, the tangential coefficient of friction was settled as mut = 0.8 and the normal coefficient of friction was settled as mun = 0.5. Information about the mass of each link and the gravitational acceleration were also provided respectively as m = 120 g and g = 9.8 m/s^2. Each observed figure contains two graphs representing the desired trajectory and the obtained trajectory, respectively of blue and red color. Four different tests were then investigated and are differentiated by the selected gains parameter. The gains parameter chosen in each case are respectively (kp = 1, ki = 10, kd = 0);(kp = 5, ki = 80, kd = 0); (Kp = 5, ki = 95, kd = 0.1) and (kp = 5, ki = 95, kd = 1.5). The experiments are carried out on both PID controller and adaptive controller and only the decisive results obtained are shown in the figures below. By observing Figs. 6 and 7, the results obtained are promising for the adaptive control system. In Fig. 6, a large difference is

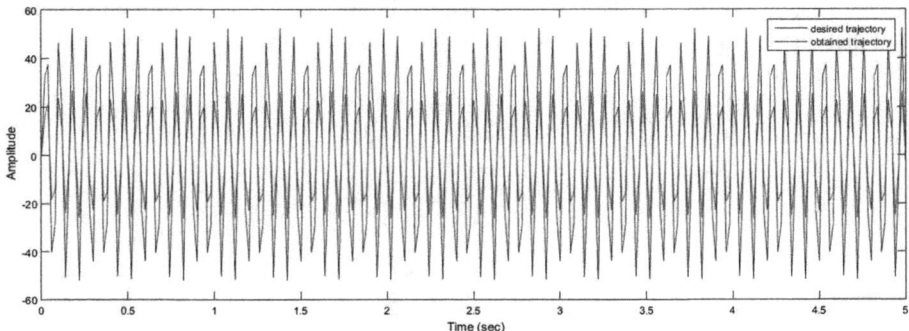

Fig. 6. PID controller observation at kp = 1, ki = 10, kd = 0

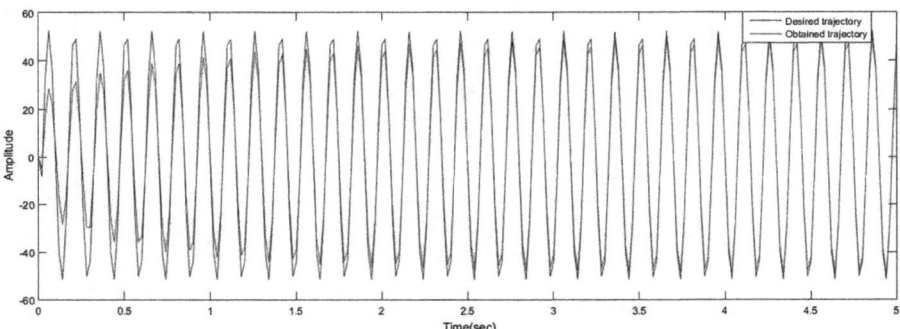

Fig. 7. Model reference adaptive controller observation at kp = 1, ki = 10, kd = 0

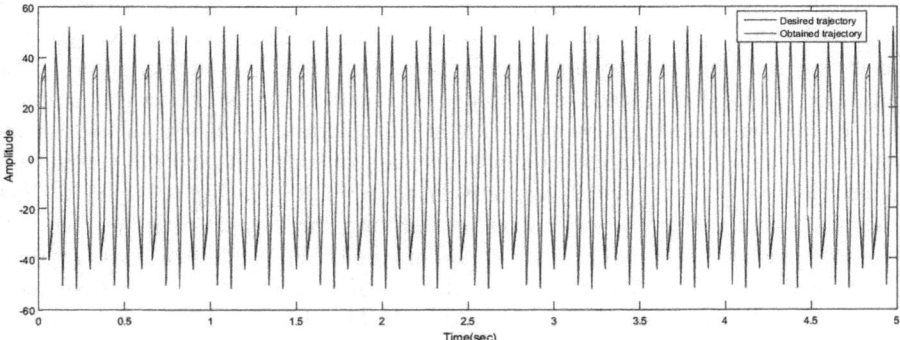

Fig. 8. PID controller observation at kp = 5, ki = 95, kd = 0.1

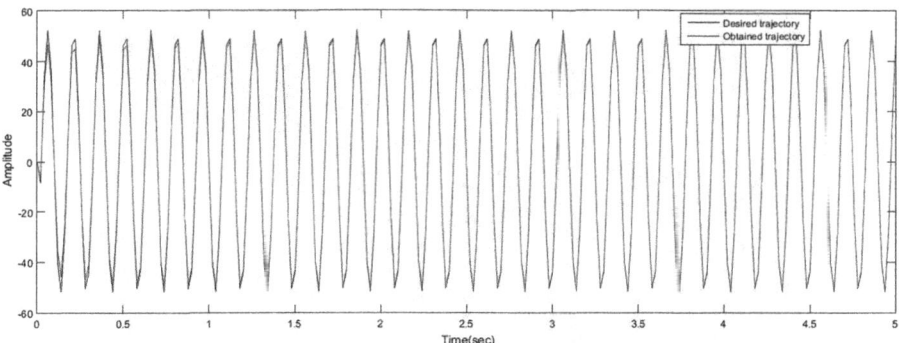

Fig. 9. Model reference adaptive controller observation at kp = 5, ki = 95, kd = 0.1

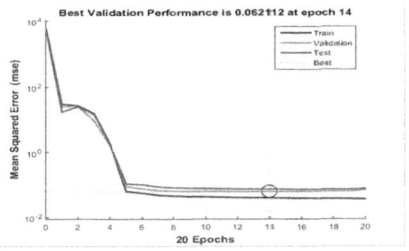

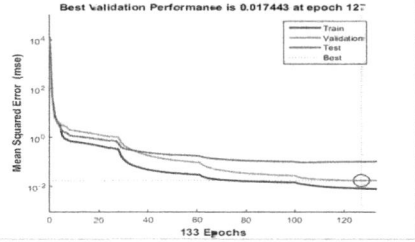

Fig. 10. Neural network training for PID-controller

Fig. 11. Neural network training for adaptive-controller

observed between the desired trajectory and the obtained trajectory. In the second case, the results obtained for the PID controller still show a great difference between the desired trajectory and the trajectory obtained while the adaptive controller performs better with a final trajectory greater than that of the PID controller. Finally by observing the results Obtained in the third cases, there is still a higher margin of error for the PID controller and lower for the adaptive controller see Figs. 8 and 9. Throughout the experiment, the results Have proved superior performance of the adaptive control system which has a higher accuracy and a strong capacity of achievement on trajectory control of the proposed ten DOF snake-like robot See Fig. 12.

In order to confirm the previous tests, a neural network training was investigated. The aim of this training is to compare the error margin between the desired trajectory and the trajectory obtained in each controller. The best performance of each controller was chosen (kp = 5, ki = 95, kd = 1.5) and the system was trained several times. The method of Levenberg-Marquardt method was then used and obtained results are represented on Figs. 10 and 11. The obtained mean squared errors ((target-output) ^2) were (0.062112 & 0.017443) respectively on Figs. 10 and 11. These results shows that the Adaptive controller performs with less error.

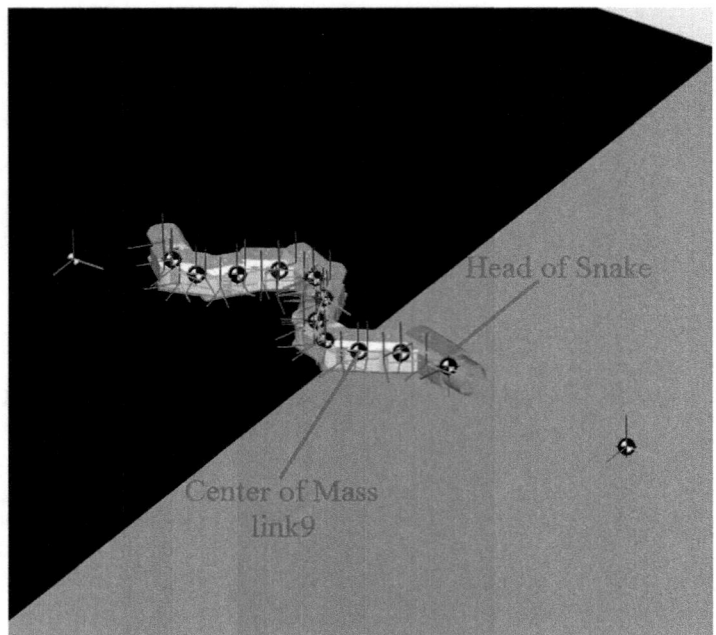

Fig. 12. Snake-like robot 3D representation on Simmechanics

5 Conclusions

Snake-like robot are very useful and important systems with multi-degrees of freedom in complex applications such as cluttered environment and rough terrain. Therefore it should be necessary to design the suitable control system that will run with high efficiency and allow a great motion control. The experimental work which was performed in simulation has given relevant results on the superior performances of the proposed adaptive controller for adapting large disturbances such as road roughness. Finally, this proposed adaptive control system will be employed in real time applications such as snake like robots with higher degrees of freedom.

Acknowledgements. The author would like to address his sincere thanks to Professor Dr. Şahin Yildirim for his availability and participation which have helped in the realization of this paper.

References

1. Sitar J, Racek V (2008) Snake robot: basic mechanical design, simulation and realization (movement variations of the snake robots). IEEE, pp 361–362
2. Levi P, Kernbach S (eds) (2010) Symbiotic multi-robot organisms reliability, adaptability, evolution. Springer, New York

3. Symbrion: Symbiotic Evolutionary Robot Organism (2008–2012) 7th Framework Program Project No FP7-ICT-2007.8.2. European communities
4. Replicator: Robotic Evolutionary Self-Programming and self-assembling Organisms (2003–2012) 7th Framework Program Project No FP7-ICT-2007.2.1. European communities
5. Wood GD, Kennedy DC (2003) Simulating mechanical system in simulink with simmechanics. Mathworks, The Mathworks, Inc. http://www.mathworks.com
6. Saito M, Fukaya M, Iwasaki T (2002) Serpentine locomotion with robotic snake. IEEE control Syst. Mag. 22:64–81
7. He W, Ge SS, How BE et al (2011) Robust adaptive boundary control of a flexible marine riser with vessel dynamics. Automatica 47(4):722–732
8. He W, Ge SS, Zhang S (2011) Adaptive boundary control of a flexible marine installation system. Automatica 47(12):2728–2734
9. He W, Ge SS (2012) Robust adaptive boundary control of a vibrating string under unknown time-varying disturbance. IEEE Trans Control Syst Technol 20(1):48–58
10. He W, Ge SS, Li Y et al (2014) Neural network control of a rehabilitation robot by state and output feedback. J Intell Rob Syst 80:1–17
11. Li Z, Ge SS, Ming A (2007) Adaptive robust motion/ force control of holonomic-constrained nonholonomic mobile manipulator. IEEE Trans Syst Man Cybern Part B Cybern 37(3): 607–616
12. Li Z, Li J, Kang Y (2010) Adaptive robust coordinated control of multiple mobile manipulators interacting with rigid environments. Automatica 46(12):2028–2034
13. He W, Chen Y, Yin Z (2015) Adaptive neural network control of an uncertain robot with full-state constraints. IEEE Trans Cybern PP(99):1–10
14. Hovakimyan N, Cao C, Kharisov E et al (2011) Adaptive control for safety-critical systems. Control Syst IEEE 31(5):54–104
15. Jain P, Nigam MJ (2013) Design of a model reference adaptive controller using modified MIT rule for a second order system. Adv Electron Electr Eng 3(4):477–484

Mechanical Transmissions

Free Vibration and Sensitivity Analysis of RV Reducer

Yuhu Yang[✉] and Chuan Chen

Tianjin University, Tianjin, China
yangyuhu@tju.edu.cn, chenchuan.1985728@126.com

Abstract. This work develops a dynamic model to investigate free vibration characteristics of RV reducer. The dynamic model considers key factors affecting vibration such as involute and cycloid gear mesh stiffness, crankshaft bending stiffness and bearing stiffness. For the model, equation of motion is derived. Numerical results of the associated eigenvalue problem reveal that the vibration modes are classified into three types: rotational, translational, and planetary component modes. Furthermore, sensitivity of natural frequencies associated with three vibration modes to stiffness and inertia parameters are studied. The investigation of the work is helpful for understanding the relationship between parameters and vibration modes.

Keywords: RV reducer · Free vibration · Vibration mode · Sensitivity

1 Introduction

RV reducer is a widely used in industrial robots as joint reducer due to many advantages such as compact structure, multi-teeth meshing and high carrying capacity, large transmission ratio and high transmission efficiency. However, vibration and noise generated by RV reducer remain key concerns among designers.

RV reducer is a two stage closed planetary gear train. The vibration of RV reducer has been studied by many researchers. Kahraman [1] observed three vibration modes with a four-planet gear train. Lin et al. [2] analyzed the free vibration and proved that they possess highly structured modal properties because of cyclic symmetry. There are exactly three types of modes: rotational, translational and planet mode. Furthermore, Kahraman [3] and Guo et al. [4] conducted the analytical study of the vibration of compound planetary gears with a purely rotational model. Kiracofe et al. [5] and Dhouib et al. [6] extended lateral-torsional model to compound, multi-stage planetary gears, and proved that they exhibited modal characteristics similar to simple, single-stage planetary gears. In addition to vibration modes, sensitivity of natural frequencies has also received research attention. Refs [7–10] examine the modal sensitivity based on typical free vibration modes of planetary gears. Although there are a lot of research about vibration of planetary gear, vibration of RV reducer has received little attention. Zhang et al. [11 – 12] calculate natural frequencies of RV reducer and made experiments with prototype.

Despite research above, systematic description of RV reducer's free vibration characteristics has not been considered. The objective of this paper is to investigate natural frequencies, vibration modes and sensitivity of natural frequencies of RV reducer. Knowledge of natural frequencies and vibration modes allows designers to adjust RV reducer to avoid resonances. Thorough calculation of sensitivity of natural frequencies is helpful for understanding the relationship between parameters and natural frequencies.

2 Dynamic Model and Equation of Motion

As shown in Fig. 1, RV reducer is a two-stage closed planetary gear train, which consists of involute gears and cycloid gears. The high-speed stage is K-H type differential gears, including the sun, planets and output wheel. The low-speed stage is K-H-V type planetary gears, including crankshafts, cycloid gears, pinwheel and carrier. The carrier and output wheel are fixed together as one component.

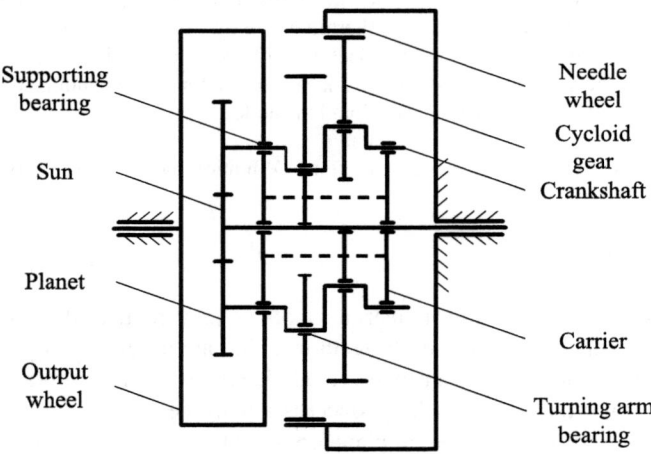

Fig. 1. Structure of RV reducer.

A lumped parameter model is shown in Fig. 2, where the pinwheel is considered as fixed, and the sun, M planets and crankshafts, N cycloid gears and output wheel are treated as rigid bodies. Component flexibility, bearings and gear meshes are represented by linear springs. The supports of the components are modeled as two perpendicular springs with equal stiffness. The transverse stiffnesses of the sun, planets, crankshafts, cycloid gears, and output wheel are designated as k_s, k_a, k_{Hb}, k_{cb} and k_o. The torsional stiffnesses of the sun, planet and output wheel are represented as k_{st}, k_H and k_{ot}. The gear meshes are modeled by springs acting along the line of action. The sun-planet and cycloid-pin mesh stiffnesses are k_{si} and k_{bj}. The stiffness of the pinwheel is not taken into account in the modeling since it is much larger. The external torques are T_s and T_o which are respectively applied to the sun and output wheel.

The coordinates are also illustrated in Fig. 2. Each component has two translational and one rotational motion. Thus the system has $6M + 3N + 6$ degrees of freedoms.

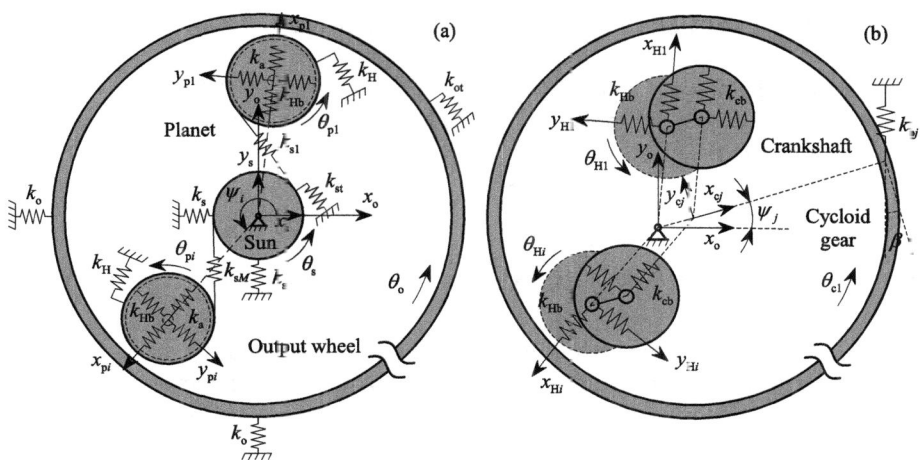

Fig. 2. Lumped parameter model of RV reducer and coordinates.

Using a fixed basis, translational coordinates (x_s, y_s) and (x_o, y_o) are assigned to the sun and output wheel, as shown in Fig. 2 (a). Coordinates (x_{pi}, y_{pi}), (x_{Hi}, y_{Hi}) and (x_{cj}, y_{cj}) are respectively assigned to planet i, crankshaft i, and cycloid gear j, where $i = 1, 2, \ldots M$ and $j = 1, 2, \ldots N$, as shown in Fig. 2 (b). Circumferential planet and cycloid gear locations are specified by the angles ψ_i and ψ_j. The x_{pi} and x_{Hi} coordinates are chosen to be positive from the geometric center of the sun to that of the arbitrarily chosen first planet. The x_{cj} coordinate is chosen to be positive from the geometric center of the pinwheel to that of the arbitrarily chosen first cycloid gear. All rotational coordinates are chosen to be θ which are positive in counterclockwise. This is illustrated in Fig. 2, where θ_s, θ_{pi}, θ_{Hi}, θ_{ci}, and θ_o are respectively assigned to the sun, planets, crankshafts, cycloid gears and output wheel.

According to the Newton's second law and theorem of angular momentum, the equations of motion for the sun, planets, crankshafts, cycloid gear and output wheel can be derived. Assembling the system equations and neglecting the gyroscopic effect, the governing equations of motion can be written in matrix form as

$$M\ddot{q} + (K_b + K_m)q = F. \quad (1)$$

where M, K_b and K_m are the inertia, support stiffness, and mesh stiffness matrices. F is the applied external torque. q is the generalized coordinate vector. The associated free vibration equation is

$$M\ddot{q} + (K_b + K_m)q = 0. \quad (2)$$

3 Natural Frequencies and Vibration Modes

To determine the nature frequencies and vibration modes the time-invariant system is considered. All externally applied moments are assumed to be zero. The associated eigenvalue problem is

$$\left(K_b + K_m - \omega_n^2 M\right)\phi_n = 0. \tag{3}$$

where ω_n are natural frequencies and the vibration modes have the form $\phi_i = [p_s, p_{p1}, \cdots, p_{pM}, p_{H1}, \cdots, p_{HM}, p_{c1}, \cdots, p_{cN}, p_o]^T$ with p_s for the sun, p_{Hi} for planet i, p_{Hi} for crankshaft i, p_{cj} for cycloid wheel j and p_o for the output wheel.

The eigensolution properties are illustrated by a numerical example with the parameters shown in Table 1. The parameters of mass and moment of inertia are obtained from simulations. The stiffnesses are calculated according to mechanics of materials. As in Ref [11], the meshing stiffness of cycloid gears is calculated and taken the mean value.

Table 1. Parameters of an example RV reducer.

Items	Sun	Planet	Crankshaft	Cycloid Gear	Carrier
Mass/kg	1.30	0.88	0.40	2.76	15.33
Moment of inertia/ (kgm^2)	4.44×10^{-4}	1.01×10^{-3}	7.56×10^{-5}	2.09×10^{-2}	1.06×10^{-1}
Base diameter/mm	10.57	48.63	2.20	85.80	63.50
Support stiffness/ (N/m)	$k_s = 4.19 \times 10^7$	$k_a = 2.33 \times 10^8$	$k_o = 1.51 \times 10^9$		
Bearing stiffness/ (N/m)	$k_{Hb} = 9.76 \times 10^8$	$k_{cb} = 9.84 \times 10^8$			
Mesh stiffness/(N/m)	$k_{sn} = 2.68 \times 10^8$	$k_{cr} = 8.35 \times 10^8$			
Torsional stiffness/ (Nm/rad)	$k_{st} = 1.16 \times 10^4$	$k_H = 6.99 \times 10^4$			

The nature frequencies and their multiplicities are shown in Table 2. All vibration modes for RV reducers can be classified into one of three types: rotational modes, translational modes and planetary component modes.

Rotational modes: There are exactly 11 rotational modes, each with an associated natural frequency of multiplicity one. In a rotational mode, the central components (the sun and output wheel) have pure rotation and no translation. Planet components (planets, crankshafts and the cycloid gears) have identical motion. A rotational mode has the form

Table 2. Natural frequencies (Hz) with multiplicity m

M	3	3	3	4	5
N	3	4	5	3	3
$m = 1$	456.56	499.63	532.20	454.13	449.47
	694.49	694.04	693.79	667.28	643.89
	2034.15	2025.39	2015.27	2118.41	2181.95
	2455.49	2407.62	2349.32	2501.34	2513.25
	2645.83	2589.33	2574.83	2932.20	3223.88
	4092.77	3979.59	3883.12	4330.62	4470.15
	5213.45	5213.52	5213.57	5802.57	5968.84
	5583.11	5598.21	5607.45	6017.18	6725.28
	6395.64	6578.29	6748.81	6858.17	7382.61
	16241.42	18055.34	19703.60	16241.42	16241.42
	16576.88	18375.22	20008.09	16688.60	16800.30
$m = 2$	776.81	791.89	805.35	750.85	729.21
	1999.34	2054.02	2088.99	2057.64	2098.51
	2262.19	2302.61	2337.30	2334.75	2372.00
	2573.12	2579.72	2585.18	2682.82	2818.86
	3713.51	3682.54	3657.23	4203.33	4565.57
	4209.01	4128.78	4071.80	4797.77	5304.61
	4270.38	4189.10	4132.08	4876.53	5404.39
	5255.98	5370.58	5465.94	5405.06	5599.51
	6080.19	6126.02	6183.41	6235.16	6402.26
	16554.35	18354.85	19989.19	16666.35	16782.25
	16585.97	18378.10	20007.22	16707.17	16831.56
$m = M - 3$				1971.55	1971.55
				2516.17	2516.17
				4808.66	4808.66
				5736.97	5736.97
				16241.37	16241.37
				16241.77	16831.10
$m = N - 3$		3877.22	3877.22		
		5203.98	5203.98		
		5280.81	5280.81		

$$\phi_n = \left[p_s, p_{p1}, \cdots p_{p1}, p_{H1}, \cdots p_{H1}, p_{c1}, \cdots p_{c1}, p_o \right]^T. \quad (4)$$

where $p_s = [0, 0, \theta_s]^T$, $p_o = [0, 0, \theta_o]^T$.

Translational modes: There are exactly 11 degenerate pairs of translational modes, where each pair has an associated natural frequency of multiplicity two. All central components have pure translational motion and no rotation. A pair of translational modes has the form

$$\phi_n = [p_s, p_{p1}, \cdots p_{pM}, p_{H1}, \cdots p_{HM}, p_{c1}, \cdots p_{cN}, p_o]^T.$$
$$\bar{\phi}_n = [\bar{p}_s, \bar{p}_{p1}, \cdots \bar{p}_{pM}, \bar{p}_{H1}, \cdots \bar{p}_{HM}, \bar{p}_{c1}, \cdots \bar{p}_{cN}, \bar{p}_o]^T. \quad (5)$$

where $p_s = [x_s, y_s, 0]^T$, $p_o = [x_o, y_o, 0]^T$.

Planetary component modes: There are exactly 6 or 3 degenerate pairs of modes, each with an associated natural frequency of multiplicity M-3 or N-3. The central components have no motion and the planet components deflect. A planetary component mode has the form

$$\phi_n = [0, p_{p1}, \cdots, \bar{p}_{p1}, \cdots, p_{pM}, p_{H1}, \cdots, \bar{p}_{H1}, \cdots, p_{HM}, 0, 0, \cdots, 0]^T.$$
$$\phi_n = [0, 0, \cdots 0, p_{c1}, \cdots, \bar{p}_{c1}, \cdots, p_{cN}, 0]^T. \quad (6)$$

4 Eigensensitivity to Parameters

One can obtain eigensensitivity by differentiation of Eq. (3) with respect to parameters. For a distinct eigenvalue, the eigensensitivitiy is

$$\lambda'_n = \phi_n^T \left(K' - \lambda_n M' \right) \phi_n. \quad (7)$$

where $\lambda_n = \omega_n^2$.

For a distinct eigenvalue, the eigensensitivities are

$$\left[\Gamma^T \left(K' - \lambda_n M' \right) \Gamma \right] a_n = \lambda'_n a_n. \quad (8)$$

where $\phi_i = \Gamma a_i$, $\Gamma^T M \Gamma = I_{m \times m}$.

4.1 Eigensensitivity to Stiffness

Eigensensitivity to stiffness is obtained by substitution of Eqs. (4)–(6) into Eqs. (7) and (8).

$$\begin{aligned}
\frac{\partial \lambda_n}{\partial k_s} &= x_s^2 + y_s^2, & \frac{\partial \lambda_n}{\partial k_{st}} &= \theta_s^2, & \frac{\partial \lambda_n}{\partial k_{sp}} &= \sum_{i=1}^M \delta_{si}^2, \\
\frac{\partial \lambda_n}{\partial k_H} &= \sum_{i=1}^M \left[(x_{pi} - x_{Hi})^2 + (y_{pi} - y_{Hi})^2 \right], & \frac{\partial \lambda_n}{\partial k_{Ht}} &= \sum_{i=1}^M \left(\theta_{pi} - \theta_{Hi} \right)^2, \\
\frac{\partial \lambda_n}{\partial k_{cb}} &= \sum_{i=1}^M \sum_{j=1}^N \left(\delta_{Hicjx}^2 + \delta_{Hicjy}^2 \right), & \frac{\partial \lambda_n}{\partial k_{Hb}} &= \sum_{i=1}^M \left(\delta_{iox}^2 + \delta_{ioy}^2 \right), & \frac{\partial \lambda_n}{\partial k_{cr}} &= \sum_{i=1}^M \delta_{cjr}^2, \\
\frac{\partial \lambda_n}{\partial k_o} &= x_o^2 + y_o^2, & \frac{\partial \lambda_n}{\partial k_{ot}} &= \theta_o^2.
\end{aligned} \quad (9)$$

On the basis of characteristics of three vibration modes, conclusions about sensitivity to stiffness can be drawn from Eq. (9):

(1) Substitution of Eq. (4) into Eq. (9), $\partial\lambda_n/\partial k_s$ and $\partial\lambda_n/\partial k_o$ are zero. Hence, natural frequencies of rotational modes are independent of the transverse support stiffness of the central components;
(2) Substitution of Eq. (5) into Eq. (9), $\partial\lambda_n/\partial k_{st}$ and $\partial\lambda_n/\partial k_{ot}$ are zero. Hence, natural frequencies of translational modes are independent of the torsional stiffness of the central components;
(3) Substitution of Eq. (6) into Eq. (9), $\partial\lambda_n/\partial k_s$, $\partial\lambda_n/\partial k_{st}$, $\partial\lambda_n/\partial k_o$ and $\partial\lambda_n/\partial k_{ot}$ are all zero. Hence, natural frequencies of planetary component modes are insensitive to stiffness parameters of the central components.

Figure 3 shows sensitivity of natural frequencies to stiffness parameters. In Fig. 3 (a), λ'_{t1} is larger when k_s is smaller. It means that sensitivity of the first natural frequency associated with translational mode to k_s is larger when the stiffness is smaller. In Fig. 3(b), λ'_{r1} is close to zero when k_{st} is 1.5×10^4 N·m/rad. It means that variation of k_{st} has little influence on λ_{r1} when it is 1.5×10^4 N·m/rad.

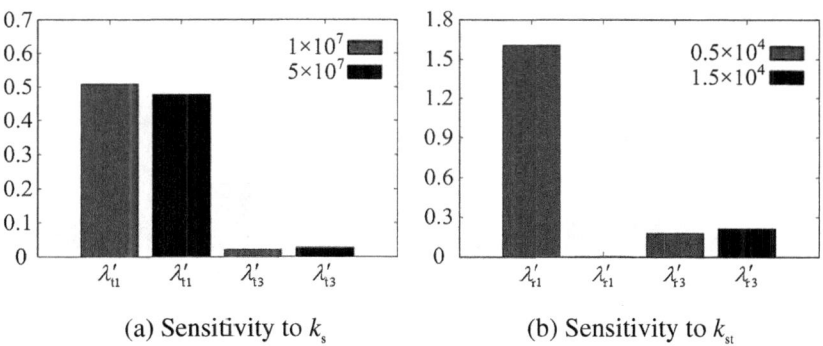

(a) Sensitivity to k_s (b) Sensitivity to k_{st}

Fig. 3. Sensitivity to stiffness parameters.

4.2 Eigensensitivity to Inertia Parameters

Similar to stiffness, eigensensitivity to mass and moment of inertia is

$$\begin{aligned}
&\frac{\partial\lambda_n}{\partial m_s} = -\lambda_n(x_s^2 + y_s^2),\ \frac{\partial\lambda_n}{\partial J_s} = -\lambda_n\theta_s^2,\ \frac{\partial\lambda_n}{\partial m_p} = -\lambda_n \sum_{i=1}^{M}\left(x_{pi}^2 + y_{pi}^2\right), \\
&\frac{\partial\lambda_n}{\partial J_p} = -\lambda_n \sum_{i=1}^{M}\theta_{pi}^2,\ \frac{\partial\lambda_n}{\partial m_H} = -\lambda_n \sum_{i=1}^{M}\left(x_{Hi}^2 + y_{Hi}^2\right),\ \frac{\partial\lambda_n}{\partial J_H} = -\lambda_n \sum_{i=1}^{M}\theta_{Hi}^2, \\
&\frac{\partial\lambda_n}{\partial m_c} = -\lambda_n \sum_{j=1}^{N}\left(x_{cj}^2 + y_{cj}^2\right),\ \frac{\partial\lambda_n}{\partial J_c} = -\lambda_n \sum_{j=1}^{M}\theta_{cj}^2, \\
&\frac{\partial\lambda_n}{\partial m_o} = -\lambda_n(x_o^2 + y_o^2),\ \frac{\partial\lambda_n}{\partial J_o} = -\lambda_n\theta_o^2\ .
\end{aligned} \quad (10)$$

On the basis of characteristics of three vibration modes, conclusions about sensitivity to inertia parameters can be drawn from Eq. (10):

(1) Substitution of Eq. (4) into Eq. (10), $\partial \lambda_n/\partial m_s$ and $\partial \lambda_n/\partial m_o$ are zero. Hence, natural frequencies of rotational modes are independent of the mass of the central components;
(2) Substitution of Eq. (5) into Eq. (10), $\partial \lambda_n/\partial J_s$ and $\partial \lambda_n/\partial J_o$ are zero. Hence, natural frequencies of translational modes are independent of the moment of inertia of the central components;
(3) Substitution of Eq. (6) into Eq. (10), $\partial \lambda_n/\partial m_s$, $\partial \lambda_n/\partial J_s$, $\partial \lambda_n/\partial m_o$ and $\partial \lambda_n/\partial J_o$ are all zero. Hence, natural frequencies of planetary component modes are insensitive to inertia parameters of the central components.

Figure 4 shows sensitivity of natural frequencies to inertia parameters. In Fig. 4 (a), λ'_{r1} is larger when m_s is 1 kg. It means that sensitivity of the first natural frequency associated with rotational mode to m_s is larger when the mass is smaller. From Fig. 4 (b), it suggests that the third natural frequency associated with rotational mode is more sensitive to J_s when J_s is 2×10^{-3} kg · m².

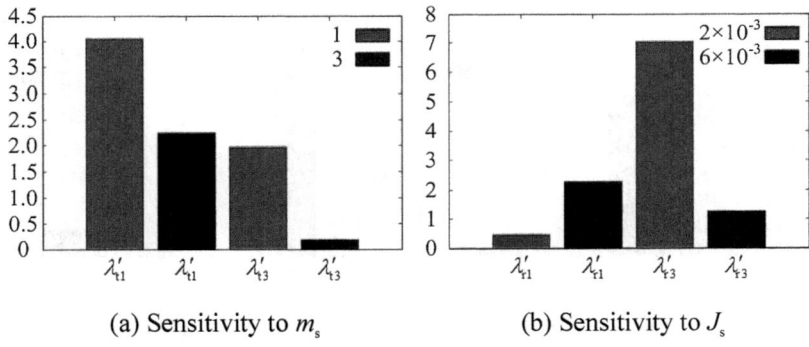

(a) Sensitivity to m_s (b) Sensitivity to J_s

Fig. 4. Sensitivity to inertia parameters.

5 Conclusions

The paper developed a dynamic model of general description to investigate the free vibration characteristics of RV reducer. The main results are as follows:

(1) All vibration modes can be classified into eleven rotational modes with distinct natural frequencies, eleven pairs of translational modes with natural frequencies of multiplicity two and three or six groups of planetary component modes with natural frequencies of multiplicity M-3 or N-3.
(2) In rotational modes, the central components have pure rotation and no translation. In translational modes, the central components have pure translation and no rotation. In planetary component modes, only planetary components have motion. The central components have no motion.

(3) Natural frequencies of rotational modes are independent of the transverse support stiffness and mass of the central components. Natural frequencies of translational modes are independent of the torsional stiffness and moment of inertia of the central components. Natural frequencies of planetary component modes are insensitive to the stiffness and inertia parameters of the central components.

Acknowledgments. The work is based upon work supported by the National High Technology Research and Development Program of China (863 Program) (No. 2011AA04A102).

References

1. Kahraman A (1994) Planetary gear train dynamics. J Mech Des 116(3):713–720
2. Lin J, Parker RG (1999) Analytical characterization of the unique properties of planetary gear free vibration. J Vibr Acoust 121:316–321
3. Kahraman A (2001) Free torsional vibration characteristics of compound planetary gear sets. Mech Mach Theor 36(8):953–971
4. Guo Y, Parker RG (2010) Purely rotational model and vibration modes of compound planetary gears. Mech Mach Theor 145(3):365–377
5. Kiracofe DR, Parker RG (2007) Structured vibration modes of general compound planetary gear systems. J Vibr Acoust 129(1):1–16
6. Dhouib S, Hbaied R, Chaari F et al (2008) Free vibration characteristics of compound planetary gear train sets. Proc Inst Mech Eng Part C J Mech Eng Sci 222(8):1389–1401
7. Lin J, Parker RG (1999) Sensitivity of planetary gear natural frequencies and vibration modes to model parameters. J Sound Vib 228(1):109–128
8. Guo Y, Parker RG (2010) Sensitivity of general compound planetary gear natural frequencies and vibration modes to model parameters. J Vib Acoust 132(1):011006
9. Qian B, Wu SJ (2014) The research on natural characteristic and eigensensitivity of ravingneaux compound planetary gear sets. Appl Mech Mater 446:590–596
10. Sun W, Ding X, Wei J et al (2015) A method for analyzing sensitivity of multi-stage planetary gear coupled modes to modal parameters. J Vibroengineering 17(6):3133–3146
11. Zhang YH, Xiao JJ, He WD (2009) Dynamical formulation and analysis of RV reducer. In: International conference on engineering and computation, pp 201–204
12. Zhang Y H, He W D, Wei J et al (2010) Dynamical model of RV reducer and key influence of stiffness to the nature character. In: Third international conference on information and computing, pp 192–195

Proof of Existence of a Gear Variator as Wheelwork with Constant Engagement of Toothed Wheels

Konstantin S. Ivanov[✉]

IMMASH MON RK, Almaty University of Power Engineering
and Telecommunications, Almaty, Kazakhstan
ivanovgreek@mail.ru

Abstract. Gear variator is a planetary train with constant gearing of toothed wheels and with the variable transfer ratio. Information on a gear variator occurs in the restricted number of patents and publications. The reliability of the gear variator calls a doubt as its kinematic chain has two degree of freedom at presence only one input. The additional constraint preserving a property of the two-mobile kinematic chain can provide reliability of motion. Such constraint is found. In the paper the existence of a gear variator with the brand new additional geometric constraint providing the variable transfer ratio is proved.

Keywords: Gear variator · Constant gearing · Additional constraint

1 Introduction

Subject of research is a wheelwork with constant cogging of toothed wheels and with the variable transfer ratio. Patents for inventions [1–3] and author's theory of a gear variator [4–7] are created.

The theory of a gear variator is based on science discovery "Effect of force adaptation". This discovery gives a possibility of creating a gear variator as continuously variable transmission [8–12].

The effect of force adaptation occurs in the kinematic chain containing an input link, the mobile four link closed contour and an output link. Such kinematic chain is not definable as it contains only one input link in the presence of two degree of freedom. The paradox of mechanics is that the closed contour imposes additional constraint on motion of links and secures definability. The closed contour creates brand new precedent – the circulation of energy which secures the equilibrium by a principle of virtual works. The interconnection of the kinematic and force parameters in the closed contour leads to redistribution of speeds and to receipt of the output energy flow with changeable parameters. The loading on the output link becomes the controlling factor defining speed of its motion without use of a control system.

The basic data on discovery are presented on the site http://www.adaptation.kz. On the site theoretical bases, the list of publications and patents, a video clip of animation model, a photo of experimental samples of adaptive transfers, wheelworks and reducers, test-beds are presented.

However reliability of the gear variator calls a doubt as its kinematic chain has two degree of freedom at presence only one input.

The earlier developed theory supposed that a mobile closed contour of wheelwork imposes additional constraint on motion of links and provides reliability of motion with the variable transfer ratio [8–12]. However the closed contour defines only possibility of additional constraint. Only the presence of the real additional geometric constraint should be a sufficient condition of reliability of a variator. A research statement of problem is following: to create the additional geometric constraint which conserves a property of variability of the transfer ratio and to prove possibility of existence of a gear variator as statically determinate mechanical system.

2 Brand New Constraint in the Kinematic Chain of a Gear Variator

The kinematic chain of a gear variator (Fig. 1) looks like two-row planetary wheelwork with two degree of freedom [3–5]. The mechanism contains a frame 0, input carrier H_1, satellite 2, block of the sun wheels 1–4, block of ring wheels 3–6, satellite 5 and output carrier H_2. Toothed wheels are forming the four-link mobile closed contour 1–2–3–6–5–4. Sizes of toothed wheels are defined through matching radiuses r_i $i = 1, 2, 3, 4, 5, 6$. Radiuses of carriers are $r_{H_1} = r_1 + r_2$, $r_{H_2} = r_4 + r_5$.

To the right of the mechanism the plot of linear speeds V_i $i = 1, 2, 3, 4, 5, 6$ of links of a mechanism is presented. Linear speeds V_i are expressed through angular

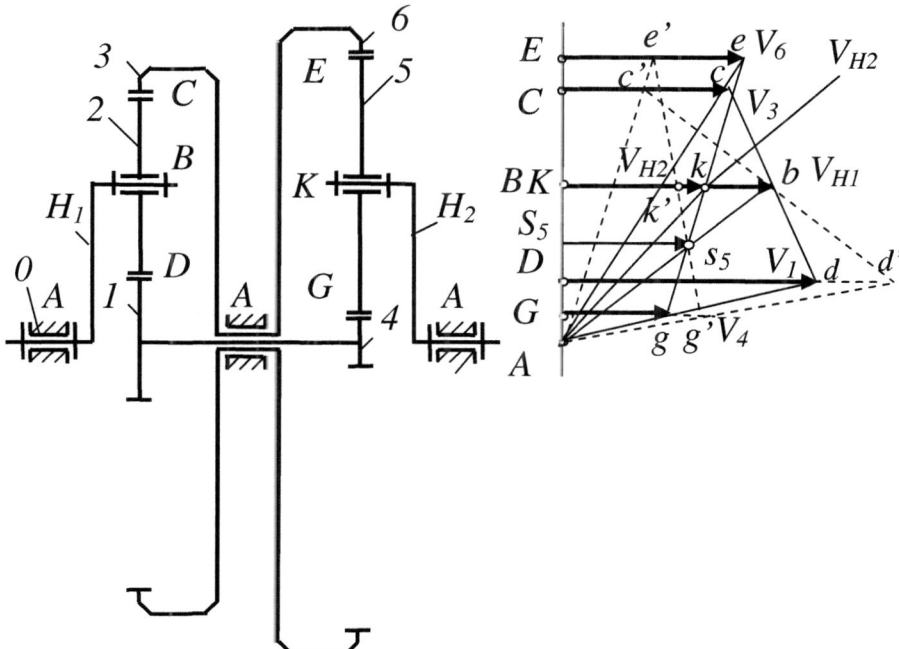

Fig. 1. The gear planetary train and its plot of linear speeds

velocities of links ω_i by Equation $V_i = \omega_i r_i$. Linear speeds of carriers are $V_{Hi} = \omega_{Hi} r_{Hi}$ $i = 1, 2$.

The plot of linear speeds in an initial position is shown by full lines. AE - a line of motionless points $B, C, \ldots$. The ends of vectors of speeds of mobile points are marked out through the matching small letters $b, c, \ldots$ Segment Bb defines input speed V_{H1}. Segment Kk defines output speed V_{H2}. The intermediate plot of speeds with the changed angular velocity of the satellite 2 (dashed line $c'bd'$) and output speed V_{H2} (segment Kk') is shown by dashed lines. Angular velocity of input carrier H_1 is defined by inclined line Ab. Angular velocity of the satellite 2 is presented by the inclined line cbd which is passing through point b. Angular velocity of the satellite 5 is presented by the inclined line ekg which is passing through point k.

Point s_5 of intersection of line ekg of angular velocity of the satellite 5 with line Ab of angular velocity of the carrier H_1 is the centre of coincidence of speeds of the satellite 5 and the carrier H_1. At the stopping of input carrier the centre of coincidence of speeds S_5 will be placed on a line of motionless points AB. In this case it represents the instant centre of relative turn of the satellite 5 relatively the input carrier. It is proved [7] that the instant centre of relative turn S_5 occupies an invariable position on lines AB and the centre of coincidence of speeds s_5 occupies a constant position on line Ab at constant input angular velocity. Line ekg of angular velocity of the satellite 5, passing through point s_5, can occupy any positions (it is shown by a dot line) at constant input angular velocity. Arbitrary position of line ekg and points k of the output carrier provides possibility of variable transfer ratio $u_{H1-H2} = V_{H1}/V_{H2} = \omega_{H1}/\omega_{H2}$ at constant input angular velocity.

The size $y = BS_5$ which determines position of a point S_5 on line AB can be univocal defined through critical buckling of the mechanism [7].

$$y = r_5 \frac{r_1 r_6 - r_3 r_4}{r_1 r_6 + r_3 r_4}. \tag{1}$$

Equation 1 expresses an additional geometric constraint analytically.

Thus the closed contour provides a necessary condition of existence of a gear variator - possibility of change of output angular velocity at constant input angular velocity of the mechanism.

The invariable position of point S_5 on carrier H_1 defines possibility of creation of brand new design with an additional geometric constraint which can occur in the kinematic chain with two degree of freedom. This additional constraint can be executed mechanically and can exist in the mechanism simultaneously (in parallel) with existing constraints not changing the plot of linear speeds. We will call this constraint as parallel constraint.

Parallel constraint introduces the real constructive changes to the mechanism creating conditions for transfer of forces. Introduction of forces in the analysis of the kinematic chain motion creates preconditions for provision of static determinacy.

The found regularity allows to build the equivalent one-mobile mechanism with parallel constraint in the form of high kinematic pair in cogging of additional toothed wheels 7 and 8 (Fig. 2). The wheel 7 is rigidly connected with the satellite 5 (the output satellite is executed in the form of the block of wheels 5–7). The wheel 8 is rigidly connected with the input carrier. The high pair is placed in point S_5. The plot of linear speeds of the equivalent mechanism with parallel constraint coincides with the plot of linear speeds of initial two-mobile mechanism. But thus a real possibility of change of a position of line *ekg* of angular velocity of the output satellite 5 appears.

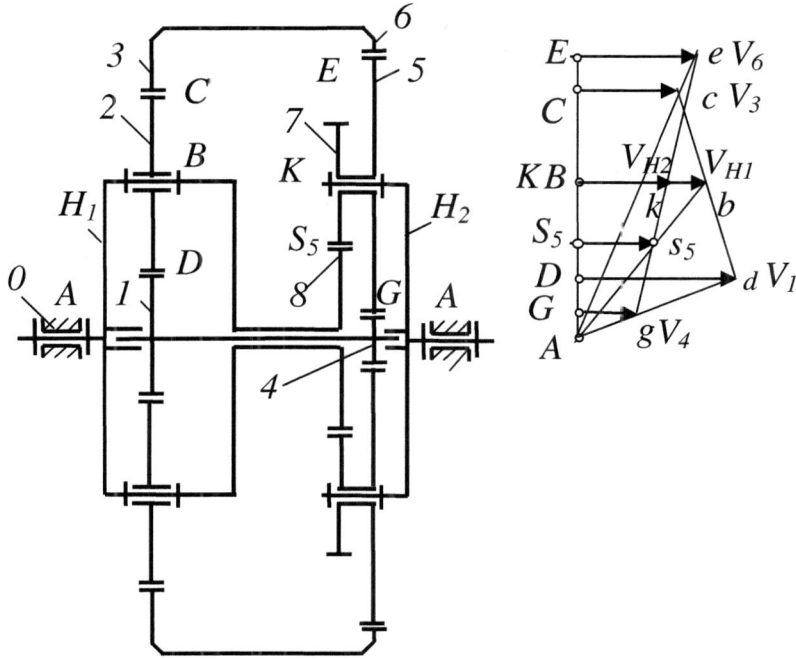

Fig. 2. The equivalent mechanism with one degree of freedom and its plot of linear speeds

Parameter y defined by Eq. (1), allows selecting the numbers of teeth of wheels 7, 8 on the preset sizes of wheels of a planetary kinematic chain

3 The Load Design of the Closed Contour

The load design of the mechanism without additional constraint is executed in [7].

The power analysis of the mechanism with parallel constraint is carried out under the same equations, that is without additional constraint, because the reactions counterbalanced in point S_5 (as well as the superposed forces that induce these reactions) can be eliminated from calculation according to applicable axiom of a statics. But the interacting of forces in a contact point of toothed wheels 7 and 8 will influence mechanism motion because a possibility of act of forces different in magnitude in points B and K of mechanism is admitted by different speeds of points application points of forces having one acting line (line Bkb on the plot of speeds).

The input motive force F_{H1} acts on the input carrier. The output force of resistance F_{H2} acts on output carrier. These forces are transferred to the closed contour as reactions R_{H1} and R_{H2} which can be expressed through external moments M_{H1} - the constant driving moment on the input carrier, M_{H2} - a variable moment of resistance on the output carrier $R_{H1} = M_{H1}/r_{H1}$, $R_{H2} = M_{H2}/r_{H2}$.

For the closed contour reaction R_{H1} is an motive force, reaction R_{H2} is force of resistance.

The output power is defined by equation $P_{H2} = R_{H2}V_{H2}$. Without a friction the output power is equal input power $P_{H2} = P_{H1}$. Hence

$$V_{H2} = P_{H1}/R_{H2}. \qquad (2)$$

Force of resistance R_{H2} acts from the output carrier on point K of the satellite 5 (Fig. 1) and within a constant output power can be a variable. According Eq. 2 variable force R_{H2} is capable to make a change of speed V_{H2} (the segment Kk on the plot of linear speeds). This situation can be modeled on the plot of linear speeds, as a change of a position of point k under acting of variable force R_{H2} applied to this point. Then line ekg of angular velocity of the satellite 5 is turning relatively motionless point s_5. Thus, the main property of parallel constraint - ability to change an output speed under the acting of variable force of resistance is carried out.

This result defines a sufficient condition of existence of a gear variator - possibility of change of an output speed by variable force of resistance.

From the load design [7] follows

$$\omega_{H2} = \frac{M_{H1} \cdot \omega_{H1}}{M_{H2}}. \qquad (3)$$

Equation 3 defines possibility of connection of force and speed parameters.

4 Features of Motion of the Kinematic Chain with Parallel Constraint

The closed contour of the mechanism placed between carriers has following two possible states of links in the motion:

(1) The joint state of links at the fixed position of links of a contour - motion with one degree of freedom.
(2) The disjoint state of links at relative motion of links of a contour - motion with two degree of freedom.

These two possible states of a contour are provided with the additional constraint executed in the form of contact point S_5 of an engagement of two additional wheels 7 and 8. On the plot of speeds (Fig. 1) the lines of angular velocities of the input carrier Ab and output satellite ekg always pass through point s_5. Line Ab of angle speed of the input carrier $\omega_{H1} = const$ has a constant angle of inclination. Line ekg угловой of speed of output satellite $\omega_5 \neq const$ can have a variable angle of inclination. Parallel constraint admits the joint and disjoint states of the kinematic chain. Functions of parallel constraint depend only on acting forces or the moments. We will displace the external contour forces R_{H1} and R_{H2} to the plot of speeds. They are acting along one line in the point k.

The joint condition occurs at $0 \leq M_{H2} \leq M_{H1}$. For a joint condition according to the Eq. (3) at $M_{H2} = M_{H1}$ we have $\omega_{H2} = \omega_{H1}$. The same result will occur at $M_{F2} < M_{F1}$ at the expense of a frictional force. On the plot of speeds motive force R_{H1} will be equal to resistance force R_{H2}. Application point k remains motionless.

The joint state will occur if $M_{H1} < M_{H2} \leq M_{H2\max}$. On the plot of speeds in point k the motive force R_{H1} will be less than the resistance forces R_{H2}. Point k will change its position (will move to the left) depending on magnitude of resistance force. When the maximum resistance force occurs the point k will occupy the position K, and the output carrier will be stopped.

Let's mark out $u_{H1-H2} = \frac{M_{H2}}{M_{H1}}$ - the variable force transfer ratio of the mechanism in the absence of a friction. Then from the Eq. (3) we will receive

$$\omega_{H2} = \frac{1}{u_{H1-H2}} \omega_{H1}. \tag{4}$$

In the disjoint state the output angular velocity depends on the variable force transfer ratio. Only the superposed resistance force in point K of the mechanism is capable to force the motion of application point k on the plot of speeds and turn the line ekg of angle speed ω_5 around point s_5. And it leads to change of the transfer ratio.

The Eq. 4 defines existence of a gear variator as mechanism with the variable transfer ratio.

5 Analysis of the Researches Executed Before

In earlier executed researches [7] the parallel constraint was absent and sufficient condition has not been carried out. But in the absence of parallel constraint the mechanism will appear disabled as it will have two degree of freedom.

It will lead to the following:

(1) At start-up (the motionless output carrier) one degree of freedom will occur. The mechanism will move in the disjoint state without ability to overcome a resistance starting torque.
(2) In motion operating conditions (at the moving output carrier) two degree of freedom will occur. The mechanism will move in the disjoint state without ability to overcome the operational moment of resistance exceeding the input driving moment. If the operational moment of resistance appears less than the driving moment, the mechanism will pass in a joint state of motion.

6 Modeling of Act of Variable Force of Resistance by Means of the Plot of Speeds

1. At $R_{H2} < R_{H1}$ parallel constraint does impossible the disjoint state of motion and provides joint motion of links of a contour. In this condition a line of angular velocity of output satellite ekg (Fig. 1) coincides with a line of angular velocity of the input carrier Ab. Equality of angular velocities $\omega_5 = \omega_{H1}$ occurs.

2. At $R_{H2} > R_{H1}$ force R_{H2} retards the motion and displaces point k to the left. Line ekg under acting of force R_{H2} applyed in point k, starts to turn round motionless point c_5. Parallel constraint provides the separated condition of motion of links of a contour at which the line of angular velocity ekg deviates line Ab. Turn of line ekg under the influence of force R_{H2} depends on magnitude of force. At greatest possible force $R_{H2} = R_{H2\max}$ line ekg will pass through point K, and the output carrier will be shut down.
3. Mechanism start occurs at motionless point K. Parallel constraint admits start possibility both in separated and in a joint condition depending on magnitude of force of resistance R_{H2}.

7 Check of Existence of the Gear Variator

The design of the operable gear variator has been developed, made and tested on the plant "Electric drive" (Almaty, Kazakhstan). On Fig. 3 the developed variator and the test-bed knocked-down are presented.

Results of tests have confirmed existence of a gear variator.

Fig. 3. Details of the gear variator and the test-bed

8 Check of Reliability of the Theoretical Proof of the Gear Variator Existence

The theoretical law proving existence of a gear adaptive variator can be confirmed by a numerical illustration of calculation of the kinematic and force parameters by statics method.

The adaptive gear variator has a compound kinematic chain which contains the basic chain with two degree of freedom and the parallel chain with one constraint. The main criterion of authenticity of the found regularity is a preservation of equilibrium of all links.

Conditions of equilibrium of the adaptive gear variator should take place for each set output moment of resistance in a regime of steady motion at uniform motion of the mechanism and constant input power.

Let's consider a numerical illustration of the gear variator in one of motion regimes.

The adaptive gear variator has the set constant parameters of power of an engine ω_{H1}, M_{H1} on input link H_1 and the set intermediate value of variable output moment of resistance M_{H2} on the output link H_2 (Fig. 2).

It is required to define the force and kinematic parameters of a variator.

It is given: $\omega_{H1} = 100\ s^{-1}$, $M_{H1} = 100\ Nm$, $M_{H2} = 200\ Nm$
Numbers of wheel teeth:

$$z_1 = 40,\ z_2 = 10,\ z_3 = 60,\ z_4 = 10,\ z_5 = 40,\ z_6 = 90.$$

Engagement module $m = 8$.

Basic equations of calculation of parameters of the gear adaptive variator are taken from works [6, 7].

Definition of the geometrics
Radiuses of toothed wheels

$$r_1 = mz_1/2 = 8 \cdot 40/2 = 160,\ r_2 = 40.$$
$$r_3 = 240,\ r_4 = 40,\ r_5 = 160,\ r_6 = 360.$$

Radiuses of the input and output carriers

$$r_{H1} = (r_1 + r_3)/2 = (160 + 240)/2 = 200,\ r_{H2} = (r_4 + r_6)/2 = 200.$$

Radius of the wheel 7 is equal to size $y = BS_5$ defined by Eq. 1

$$r_7 = r_5 \frac{r_1 r_6 - r_3 r_4}{r_1 r_5 + r_3 r_4} = 160 \cdot \frac{160 \cdot 360 - 240 \cdot 40}{160 \cdot 360 + 240 \cdot 40} = 114.3$$

The radius of the wheel 8 $r_8 = r_{H2} - r_7 = 200 - 114.3 = 85.7$.

The transfer ratio of wheels 1 and 3 at motionless carrier H_1

$$u_{13}^{(H1)} = -z_3/z_1 = -60/40 = -1.5.$$

The transfer ratio of wheels 4 and 6 at motionless carrier H_2

$$u_{46}^{(H2)} = -z_6/z_4 = -90/10 = -9.$$

Definition of force and kinematic parameters
To define: ω_{H2}, ω_1, ω_3, M_{12}, M_{32}, M_{45}, M_{65}

Power from the input carrier on the output carrier is transferred through the basic planetary kinematic chain with two degree of freedom and through the parallel chain in the form of toothed wheels 8, 7 with additional constraint. Therefore each of the specified chains transfers half of all power of the mechanism. Accordingly for each chain the input and output moment should be equal to half of magnitudes of the full moments.

Kinematic parameters of a planetary chain at $M_{H1} = 50$ Nm $M_{H2} = 100$ Nm
Output angular velocity

$$\omega_{H2} = M_{H1}\omega_{H1}/M_{H2} = 50 \cdot 100/100 = 50 \ s^{-1}.$$

Angular velocities of blocks of wheels 3–6 and 1–4:

$$\omega_3 = \frac{\omega_{H2}(1 - u_{46}^{(H2)}) - \omega_{H1}(1 - u_{13}^{(H1)})}{u_{13}^{(H1)} - u_{46}^{(H2)}}$$

$$= \frac{50(1+9) - 100(1+1.5)}{-1.5+9} = 33.33 \ s^{-1}.$$

$$\omega_1 = u_{13}^{(H1)}(\omega_3 - \omega_{H1}) + \omega_{H1} = (-1.5)(33.33 - 100) + 100 = 200 \ s^{-1}.$$

Moments on toothed wheels:

$$M_{12} = 0.5 M_{H1} r_1/r_{H1} = 0.5 \cdot 50 \cdot 160/200 = 20 \ Nm,$$
$$M_{32} = 0.5 M_{H1} r_3/r_{H1} = 0.5 \cdot 50 \cdot 240/200 = 30 \ Nm,$$
$$M_{45} = 0.5 M_{H2} r_4/r_{H2} = 0.5 \cdot 100 \cdot 40/200 = 10 \ Nm,$$
$$M_{65} = 0.5 M_{H2} r_6/r_{H2} = 0.5 \cdot 100 \cdot 360/200 = 90 \ Nm.$$

The moments on toothed wheels 8 and 7

$$M_8 = M_{H1} = 50 \ Nm,$$
$$M_7 = M_{H2} r_7/r_{H2} = 100 \cdot 114.3/200 = 57.1 \ Nm.$$

The main demonstration of reliability of the gained results is equilibrium of blocks of wheels 1–4 and 3–6. Each separate block of wheels appears unbalanced. However the moments on these blocks are internal. Hence the joint consideration of these blocks must confirm the equality of powers of internal forces

$$(M_{21} - M_{54})\omega_1 = -(M_{23} - M_{56})\omega_3.$$

After substitution of numerical values we will gain performance of balance of powers

$$(20 - 10) \cdot 200 = -(30 - 90) \cdot 33.33, 2000 = 2000.$$

The check shows a presence of balance of positive power on the block of wheels 1–4 and negative power on the block of wheels 3–6.

Thus equilibrium is carried out. The force and kinematic reliability of the variator take place. It confirms the reliability of the developed theoretical regularity.

Equation 3 allows defining output angular velocity in the set range of change of output moment of resistance (Table 1).

Table 1. Parameters of a tractive characteristic of adaptive variator

M_{H2} [Nm]	100	150	200	250	300	350	400
$\omega_{H2}[s^{-1}]$	100	66.7	50	40	33.3	28.6	25

On the data gained in Table 1 the theoretical tractive characteristic of a gear adaptive variator is built (Fig. 4). It is showing the effect of force adaptation - at a constant input power the output angular velocity is inversely proportional to the output moment of resistance.

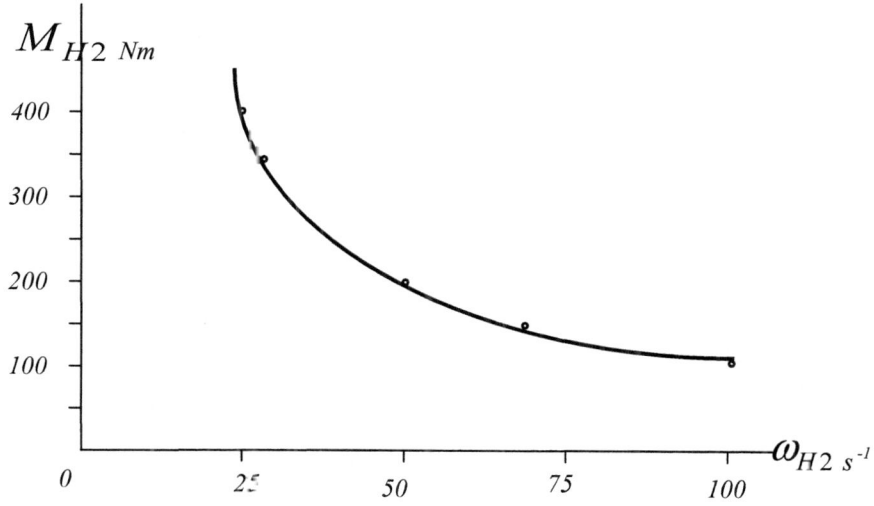

Fig. 4. Tractive characteristic of the gear adaptive variator

Experimental testing of the gear variator has been executed by M. Ceccarelli, G. Balbaev and K. Ivanov on the test-bed in LARM of university Cassino [13, 14]. The experimental tractive characteristic conforms to the theoretical one.

9 Conclusions

It is proved, that the two-mobile kinematic chain containing a mobile closed contour, has a number of brand new regularities. These regularities are following: (1) possibility of motion of an output link with a variable speed at constant input speed; (2) possibility of overcoming of variable force of resistance at the constant input motive force. These two new properties define necessary and sufficient conditions of existence of a gear variator. The necessary condition is provided with the kinematic chain with two degree of freedom. The sufficient condition is provided with brand new parallel constraint. This regularity defines possibility of creation of a gear variator with constant cogging of toothed wheels. Existence of a gear variator with constant cogging of toothed wheels is proved.

The use of gear variators in the industry opens the broadest areas of their application. Gear variators can be used for creation of automobile gear boxes, drives of manipulators and cars with variable technological resistance.

References

1. Crockett SJ (1990) Shiftless, continuously-aligning transmission. Patent of USA 4,932,928, Cl. F16H 47/08, U.S. Cl. 475/51; 475/47, 9 p
2. John H (1991) Power transmission system comprising two sets of epicyclic gears. Patent of Great Britain GB2238090 (A), 11 p
3. Ivanov KS, Almaty KAZ (2012) Owner of the registered sample. The name - Device of automatic and continuous change of rotation moment and change of rotation speed of output shaft depending on resistance moment. The deed on registration of the registered sample № 20 2012 101 273.1. Day of Registration 02.05.2012. The German patent and firm establishment. Federal Republic Germany. 12 p
4. Ivanov KS (1995) The question of the synthesis of mechanical automatic variable speed drives. In: Proceedings of the ninth world congress on the theory of machines and mechanisms, vol. 1. Politecnico di Milano, Italy, 29 August–2 September, pp 580–584
5. Ivanov KS (2004) Discovery of the force adaptation effect. In: Proceedings of the 11th world congress in mechanism and machine science, vol 2, 1–4 April, Tianjin, China, pp 581–585
6. Ivanov KS (2012) Synthesis of toothed continuously variable transmission (CVT). In: Proceedings of the First Conference MeTrApp 2011. Mechanism, Transmissions and Applications. Mechanism and Machine Science 3. Springer, pp 265–272
7. Ivanov KS, Dinasylov AD, Yaroslavseva EK (2013) Adaptive-mechanical continuously variable transmission. mechanism and machine science. In: New Advances in mechanics, transmissions and applications. Proceedings of the second conference MeTrApp 2013, vol 17, pp 83–90. Springer
8. Ivanov KS (2010) The simplest automatic transfer box. In: WCE 2010 world congress on engineering 2010 (ICME), London, UK, pp 1179–1184

9. Ivanov KS (2012) Paradox of mechanics – a basis of creation CVT. In: Transactions of 2-d IFToMM asian conference on mechanisms and machines science, Tokyo, Japan. 7–10 November, pp 245–264
10. Ivanov KS (2012) Theory of continuously variable transmission (CVT) with two degrees of freedom. In: Paradox of mechanics. Proceedings of the american society of engineers mechanics (ASME) international mechanical engineering congress and exposition (IMECE 2012). Houston, Texas, USA, pp 543–562
11. Ivanov KS (2013) Creation of adaptive-mechanical continuously variable transmission. In: 5th International conference on advanced design and manufacture (ADM 2013), Valencia, Spain, pp 63–70
12. Ivanov KS (2013) Continuously variable transmission: adaptive gear stepless mechanical CVT. In: International conference of gears with exhibition. VDI wissensforum GmbH. Technical University of Munich (TUM), Garching (near Munich), Germany, pp 984–987
13. Ceccarelli M, Balbaev G, Ivanov K (2014) An experimental test validation of a new planetary transmission. Int J Mech Control 15(02):3–7
14. Ivanov KS, Dinasilov AD, Yaroslavceva EK (2015) Gear variator – scientific reality. In: MeTrApp 2015. Springer International Publishing, Switzerland, pp 169–176

Parameterized Substructure Model of PGT for Finite Element Contact Analysis

Huimin Dong$^{(\boxtimes)}$, Chu Zhang, Jili Zhang, and Delun Wang

Dalian University of Technology, Dalian, China
donghm@dlut.edu.cn

Abstract. This paper develops a parameterized substructure modeling method for planetary gear train (PGT) quasi-static finite element contact analysis. The substructure models consist of two kinds of parameterized substructure, i.e. a pair of external gearing and internal gearing with a carrier. The models can be equivalent to the overall PGT system on account of the system equilibrium and the carrier elastic deformation. What's more, the kinematic synchronization between the substructures can be achieved due to the parameterized assembly of the models. Thus, the contact characteristic analysis of any kind of PGTs with different parameters can be conducted by using the equivalent substructure models. A spur gear planetary transmission system is taken as an example for finite element (FE) contact analysis to demonstrate the model feasibility. Comparing the results between using the substructure models and the complete model, it shows that this substructure method has enough calculating accuracy, and the calculating time is significantly reduced.

Keywords: Planetary gear train · Parameterized modeling · Substructure · Finite element method

1 Introduction

Planetary gear trains (PGTs) are widely used in various mechanical products due to their advantages of large transmission ratio, compact structure, strong carrying capacity, high transmission efficiency and stable transmission. Many scholars are attracted to study on their transmission characteristics for improving their properties by using theoretical method [1–5] and finite element (FE) method [6–16, 17]. Theoretical method [1–5] is based on rigid theory without considering elastic deformation of gears, which could not satisfy the high precision gear drive requirement. Finite element can deal with elastic deformation of the bodies and contact pairs. Several scholars adopted the overall system of a PGT for finite element model [6, 7] to investigate the characteristics. The solution to the overall system may be more precise and closer to the actual situation, however, sparse mesh of gear will cause analysis result not accurate, or the dense mesh will bring a huge amount of computation. Thereby, many scholars investigated the sub-system of a PGT, such as in Refs [8–10]. In available literature, almost all of them, the substructure finite element method ignoring the effect of the carrier was not parameterized model and without enough density of mesh, thus it is much difficult to meet the needs of accurate analysis of gears.

For the accuracy of gear drive FE analysis, the main influence factor is the modeling method which directly impacts accuracy due to the complicated shape of the tooth surfaces and the import error from other aided design software. The grid quality is an important part of the FE modeling, and some mesh generation methods are commonly used [11–14] to improve the grid quality. The grid density is required to balance the accuracy and the computation time.

At present, a PGT analysis using the substructures rarely operates at the same time; the mechanical characteristics of the substructures are not synchronous. And the modeling method is rarely parametric. In this paper, the substructure modelling method is established to simplify the contact analysis of PGTs. The phase difference between the substructures is considered to achieve time synchronization. The parameterized modeling method is adopted to program gear mesh, assemble PGT and separately analyze each substructure. A spur gear PGT is as a case study to display the substructure modeling method and process, and to compare the results with those of the overall system for verifying the accuracy of the substructure model.

2 Substructure of PGT

A PGT is much more complex than a conventional gear set, especially difficult for FE analysis, which is time-consuming and not easy to be convergent. In this section, for simplifying the structure of PGT, the substructures of PGT are introduced to be equivalent to the overall system according to composition principles.

2.1 Composition and Substructure of PGT

A pair of gear set, an external set or internal set, consists of three components, two gears and relative frame (carrier), as shown in Fig. 1 (a) and (b). A basic PGT (Fig. 1 (c)) can be regarded as the form of the two gear sets with the same planet gear on the carrier as shown in Fig. 1. Therefore, the two gear sets shown in Fig. 1 (a) and (b) are defined as the substructures of PGT, i.e. sub-1 and sub-2.

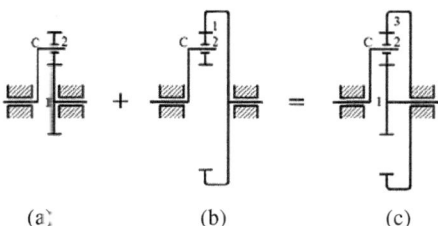

Fig. 1. Compositions of a PGT: (a) sub-1, external set; (b) sub-2, internal set; (c) PGT-1

The configurations of PGTs usually has four types shown in Fig. 2, however, they share the same substructure systems. PGT-2 consists of a sub-1 and sub-2, PGT-3 of

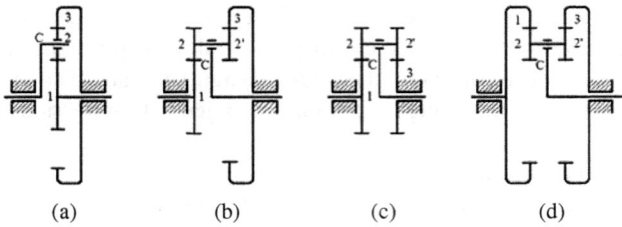

Fig. 2. Four kinds of PGTs: (a) PGT-1; (b) PGT-2; (c) PGT-3; (d) PGT-4

two sub-1 and PGT-4 of two sub-2, with two compound planet gears. Thus, the FE analysis of a PGT can be simplified through introducing the substructures approach.

For the substructure models, they should be equivalent to the overall system. Thereby, we propose a method of the force equilibrium equivalent, and the time synchronization between the substructures.

2.2 Force Equilibrium Equivalent

In order to make the substructure systems equivalent to the overall system, the transmission force and torque should be equivalent. For the force equivalent, the component elastic deformation can be ignored because it is much small. The equilibrium relationship of a planet of PGT overall system is shown in Fig. 3. F, T, I, α', δ_y and ω denote the force, torque, moment of inertia, engagement angle, deformation of carrier and angular velocity, while subscript s, p, r and c denote the sun, planet, ring gear and carrier respectively, subscript b and $i = 1,2$ denote the base circle and sub-1,2. According to the equilibrium, forces in Fig. 3 satisfy

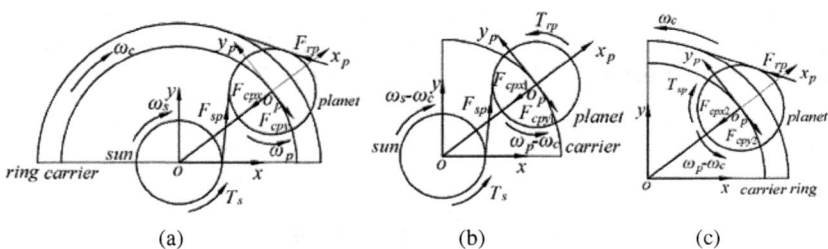

Fig. 3. The force diagram of the planet gear: (a) the overall system; (b) the sub-1; (c) the sub-2

$$\begin{cases} F_{sp} = (T_s/N)/r_{sb} \; F_{sp} \cos \alpha'_{sp} + F_{rp} \cos \alpha'_{rp} = F_{cpy} \; F_{rp}r_{pb} = F_{sp}r_{pb} \\ F_{sp} \cos \alpha'_{sp} = F_{cpy1} \; F_{rp} \cos \alpha'_{rp} = F_{cpy2} \; T_{rp} = F_{rp}r_{pb} \; T_{sp} = F_{sp}r_{pb} \end{cases} \quad (1)$$

$$\delta_y = T_c a/k_\theta = F_{cpy}a^3/(3EI_c) = F_{cpyi}a^3/(3EI_{ci}), \; i = 1,2 \quad (2)$$

where a, k_θ and E are the center distance, torsional stiffness and elastic modulus.

For the equilibrium of substructures, the torque T_{rp}/T_{sp} is applied to the sub-1/sub-2 to be equivalent to the torque from impact of the dissected component—ring/sun gear. The carrier is equivalent to a beam and the elastic deformation of carrier should also be consistent in the substructures and overall system. According to the change of torsional force, the moment of inertia for the carrier is distributed to the substructures. The rotational velocity of each component meets the transmission ratio conditions with the tooth number of gears Z. as

$$(\omega_s - \omega_c)/(\omega_r - \omega_c) = -Z_r/Z_s \qquad (\omega_p - \omega_c)/(\omega_r - \omega_c) = Z_r/Z_p \qquad (3)$$

3 Precise PGT FE Contact Model

For contact analysis, since the mesh generation is the basis of precise FE analysis, especially for the mated position, we propose a parametric modeling and assembly method to program the nodes and operate the assembly in ANSYS to achieve a precise FE contact model.

3.1 Parameterized Modeling

For convenience of gears and carrier modeling, the process is parameterized, which the discrete points on the tooth curve are calculated by tooth profile equations, and the mesh in the FE model is established with the half width of Hertz theory [15, 16]. The parameter of the carrier which the model cares is the moment of inertia by means of beam model without considering the complex structure. The steps of the external and internal gear FE modeling are:

(1) Get the coordinates of tooth curve directly by the equations, where the coordinate systems are established at gear center, and y-axis is on the symmetry line of the tooth thickness, shown in Fig. 4 (a).
(2) Generate nodes of gears in ANSYS and adopt the node connection element method to generate the plane elements of gears. The mesh of tooth profiles at

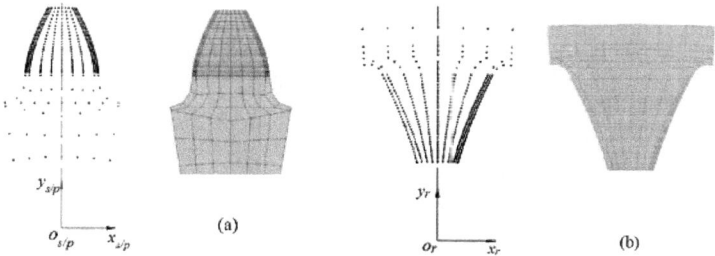

Fig. 4. End surface nodes and elements: (a) external gear tooth; (b) internal gear tooth

mating area is refined, with which planet gears on both sides and of sun gear opposed to that of ring gear. Based on the Hertz contact theory, the thickness of the gear contact area is double of half width shown in Fig. 4.

(3) Stretch the 2D model into 3D model, and then rotationally copy the 3D model to generate a complete gear model.

3.2 Parameterized Assembly

After parametrically modeling for each gear, the parameterized assembly is followed. A fixed (global) coordinate system is defined at PGTs center, in which Y-axes is coincident with the tooth space symmetric line of the ring gear. Letting r, $T = (x, y)^T$, and $R(\theta)$ be the point vector of gears, the translational and rotational coordinate transformation matrix from the gears' to fixed system, subscript f be the fixed system, N be number of the planets, based on the coordinates definition and the assembly condition of PGT, the assembly is operated parametrically by means of the point vector of ring/planet/sun in the global coordinate system, which are

$$\begin{cases} r_{f(r/s)} = R(\theta_{r/s})r_{r/s}, \; r_{fp,j} = R(\theta_{p,j})r_p + T_{p,j}, \; j = 1,\ldots,N \\ \theta_r = \pi/z_r, \theta_s = 0|_{z_p \text{isodd}}, \text{ or } \theta_s = \pi/z_s|_{z_p \text{iseven}} \end{cases} \quad (4)$$

For N planets, the interval angle of the carrier is $\theta_c = 2\pi/N$. For the jth planet,

$$T_{p,j} = (-a\sin(\theta_c(j-1)), a\cos(\theta_c(j-1)))^T, \; \theta_{p,j} = \theta_{c,j} - \frac{z_r}{z_p}\theta_{c,j}, j = 1,\ldots,N \quad (5)$$

This parameterized assembly different from that in the 3D modeling software is more precise for the FE analysis. The carrier is equivalent to the beam element. One side is fixed on the center axis of the sun gear while the other side is connected to the planet gear with MPC184 revolute.

3.3 Quasi-Static FE Modeling

Quasi-static FE analysis is to simulate the process of gear pair engagement with lower speed. The processes of the quasi-static FE modeling method need to hold:

(1) Establish the contact pair for each gear tooth. All the teeth are engaged in the meshing process, thus the nodes on the contact and target region are respectively defined as one group. One group chooses CONTA 174 element type while the other chooses TARGE 170 element type to establish contact pairs.

(2) Use the mass21 element to establish the revolute joints at the center of external gear and the nodes of gear inner surface are connected with revolute joints by the rigid beam of MPC184 as the Fig. 5 shows. In this way, the gear inner surface moves as a rigid body with the six degrees of freedom.

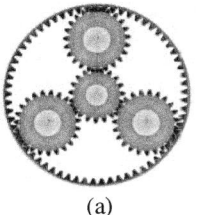  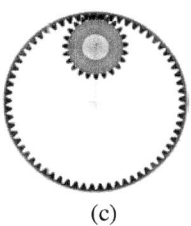

Fig. 5. FE model: (a) overall system; (b) the sub-1; (c) the sub-2;

(3) Apply the torque and angular velocity shown in Fig. 2 to the subs FE model (Fig. 5) with the value that can be calculated by Eqs. 1 and 3.
(4) Set the time interval Δt to run the quasi-static FE model.

After each operating at Δt, we can extract the contact stress of the gear pairs.

The phase difference between the sub-1 and sub-2 is directly embedded as long as the PGT is assembled. The initial phase and time interval for substructures are consistent with the overall system which ensures the time synchronization.

4 Case Study

We take a 3-pinion spur gear PGT with the configuration of PGT-1 as a case study with $Z_s = 23$, $Z_p = 22$, $Z_r = 67$, $a = 56.25$ mm, modulus 2.5 mm and tooth width 10 mm. T_s, ω_c, E and k_θ are 10000 N·mm, 0.2 rad/s, 206 Gpa and 5400 N·mm/rad. For the consistence to the carrier deformation, the moment of inertia of the equivalent carrier I_c is 490.8 mm^4.

The equilibrium relationship between overall system and substructures are calculated by the Eqs. 1 and 2. The T_{rp} and I_{c1} for sub-1 are 3188 N·mm and 245.4 mm^4 while the T_{sp} and I_{c2} for sub-2 are 3188 N·mm and 245.4 mm^4. The angular velocity of the components for the overall system and substructures is consistent with ω_s, ω_p, ω_r 0.7826 rad/s, -0.4091 rad/s, 0 rad/s respectively obtained by Eq. 3.

The force, torque and velocity are applied to the quasi-static model, and the time interval Δt is set at 0.01 s. The instantaneous contact stress of external and internal engagement for overall system are 655.4 MPa and 398.9 MPa respectively, while 653.8 MPa and 422.7 MPa for the substructures as shown in Fig. 6. The elastic deformation for the carrier of the overall system is 0 008266 mm while the sub-1 and sub-2 are 0.008485 mm and 0.008511 mm respectively, which verifies the rationality of the equivalent method.

The gears rotate and the contact stress is extracted at the same interval as shown in Fig. 7. The result of substructure FE model is similar to that of overall PGT FE model. The biggest error of the contact stress appears on the position of the single and double tooth alternate engagement. In addition to the alternating regions the greatest deviation is 12.6% which is acceptable.

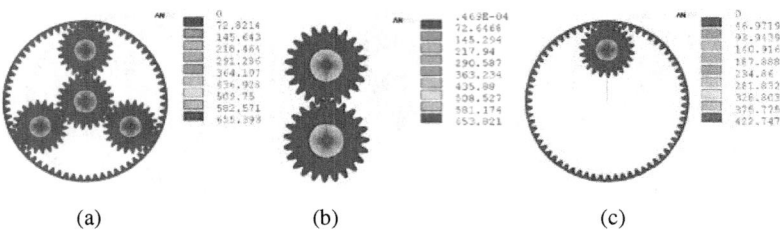

Fig. 6. Contact stress: (a) overall PGT model; (b) the sub-1; (c) the sub-2

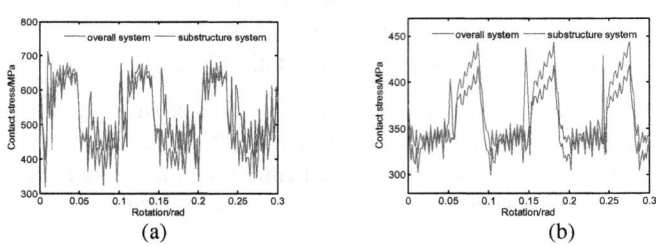

Fig. 7. Contact stress for the two model: (a) the external gear pair; (b) the internal gear pair

5 Conclusions

A rigorous substructure modelling method is proposed for PGTs FE contact analysis, in this paper, through explicitly dissecting the composition of PGTs. The combination of the substructure models can be effective equivalent to overall system for any type of PGTs, by means of the force equilibrium relationship and parameterized modeling and assembling. With this method, the kinematic synchronization between the substructures can be achieved. A case study of PGT-1 demonstrates the models feasibility, whose results have enough calculating accuracy, and the calculating time is significantly reduced. The work presented in this paper will provide a simplified method for further research of PGTs system transmission and contact characteristics.

Acknowledgement. This work is financially supported by the National Natural Science Foundation of China (No. 51375065).

References

1. Velex P, Maatar M (1996) A mathematical model for analyzing the influence of shape deviations and mounting errors on gears dynamic behavior. J Sound Vib 191(5):629–660
2. Zhao J, Li R (1984) Gears Strength Design Data. Machinery Industry Press, Beijing
3. Hertz H (1882) Uber die beruhrung fester elastischer korpe. J Fur Die Reine Undangewandte Math 92:156–171

4. Kahraman A (1994) Load sharing characteristics of planetary transmissions. Mech Mach Theor 29(8):1151–1165
5. Velex P, Flamand L (1996) Dynamic response of planetary trains to mesh parametric excitations. J Mech Des 118(1):7–14
6. Parker RG, Agashe V, Vijayakar SM (2000) Dynamic response of a planetary gear system using a finite element/contact mechanics model. J Mech Des
7. Ko KE, Lim DH, Kim PY, Park J (2010) A study on the bending stress of the hollow sun gear in a planetary gear train. J Mech Sci Technol 24(1):29–32
8. Ambarisha VK, Parker RG (2007) Nonlinear dynamics of planetary gears using analytical and finite element models. J Sound Vib 302(3):577–595
9. Wang XZ (2013) Optimization design and finite element analysis of the 2 K-H planetary gear reducer based on UG parameterization model. Taiyuan University of Technology
10. Wang J, Wang Y, Huo Z (2013) Finite element residual stress analysis of planetary gear tooth. Adv Mech Eng 27(2):127–132
11. Jun LV, Wang Z, Wang Z (2001) Generation of finite element hexahedral mesh and its trend of development. J Harbin Inst Technol 33(4):485–490
12. Wei H, Zhou B (1997) The choice of adaptive finite element mesh automatic generation method. J Southwest Jiaotong Univ 32(5):477–482
13. Yu H, Zhao K (2004) Finite element mesh automatic subdivision. High Voltage Eng 30(5):4–5
14. Blacker T (2001) Automated conformal hexahedral meshing constraints, challenges and opportunities. Eng Comput 17(3):201–210
15. Hao D (2012) Research on numerical analysis modeling method for gears meshing and its applications. Dalian University of Technology
16. Dong H, Zhang C, Wang X, Wang D (2017) A precise FE model of a spur gear set considering eccentric error for quasi-static analysis. Springer, Singapore

On the Impact of Transmission Error on the Dynamic Behavior of Geared-Linkages

Domenico Mundo[1(✉)], Shadi Shweiki[1,2], and Piervincenzo Catera[1]

[1] Department of Mechanical Energy and Management Engineering,
University of Calabria, Rende, CS, Italy
domenico.mundo@unical.it

[2] Siemens Industry Software NV, Interleuvenlaan 68, 3001 Louvain, Belgium

Abstract. Geared-linkages have been extensively studied for their capability to generate prescribed output motions. The combination of linkages with higher-pairs, in fact, enables the achievement of complex kinematic requirements, linked to function, path or motion generation tasks. At high operational velocities, the dynamic behavior of such mechanisms can be significantly different from the kinematic one. One of the sources of unwanted dynamics is represented by the Dynamic Transmission Error (DTE) between meshing gear pairs in the mechanism. In this paper, a Multi-Body simulation approach is proposed to model and analyze the dynamic behavior of a geared five-bar linkage, subject to the internal excitation due to the DTE. It is shown that unwanted vibrations, which are negligible at low speeds, become more and more significant while the gear velocity increases, which may cause an unacceptable behavior of the motion generating mechanism.

Keywords: Geared five-bar linkage · Multi-Body simulation · Transmission Error · Path generating mechanisms

1 Introduction

In several industrial applications, especially in the area of automation industry, the design of mechanical systems involves the type and dimensional synthesis of mechanisms that are able to generate wished output motions, while meeting a set of specified kinematic requirements. The objective of the mechanism can be to guide a rigid body through a series of prescribed configurations (*rigid body guidance*), to generate a wished input/output relationship (*function generation*), or to move a point of the coupler along a specified trajectory (*path generation*). A wide variety of methods for the dimensional synthesis of such mechanisms have been proposed by several authors, including graphical (Tao [16]), analytical (Hartenberg and Denavit [4]), numerical (Roth and Freudenstein [11]) and computer-aided (Erdman [2]) approaches. The proposed methods can be classified into two classes: precision point methods, which allow to exactly satisfy a set of specified configurations (Sandor and Erdman [12]), and optimal synthesis methods, which aim at minimizing the structural error, defined as a measure of the difference between the wished and the generated motion. In the latter case, the synthesis process is typically based on a global optimum search (Ullah and Kota [17]).

Several studies focused on the synthesis of a specific category of motion generating systems, represented by geared five-bar mechanisms (GFBMs). In 1963, Freudenstein and Primrose [7] analysed the kinematic properties of the curve generated by a point belonging to the coupler of a GFBM. Roth and Freudenstein [3] proposed a numerical method for the synthesis of a GFBM for path generation tasks defined by nine precision points. A GFBM was proposed by Oleksa and Tesar [9] as function generating mechanism, while Dimaragonas et al. [1] developed a generalized geared n-bar function generator, by using a synthesis method based on complex algebra. Zhang et al. [19] proposed an algorithm for the optimal synthesis of symmetric GFBMs as path generating mechanisms. Starns and Flugrad [14] used continuation methods to synthesize a GFBM for a path generation task defined by seven precision positions. Nokleby and Podhorodesky [8] proposed a method for the optimal synthesis of GFBM, based on a quasi-Newton optimization routine. A GFBM with non-circular gears has been proposed by Mundo et al. [7] to achieve a path generation task precisely, while Modler et al. [6] proposed a general method for the synthesis of geared linkages with non-circular gears. Recently, Visa et al. [18] proposed a Multi-Body approach for the structural synthesis of planar geared linkage mechanisms.

In mechanical transmissions, the variable mesh stiffness is one of the major sources of the so called Transmission Error (TE), which represents a deviation of the real transmission from the ideal one [13]. By using the Multi-Body simulation approach proposed in [10] and an FE-based methodology for the static characterization of the TE and of the mesh stiffness in a pair of cylindrical gears [5], in this paper, a Multi-Body simulation model of a GFBM is developed, in which the dynamic behaviour of the meshing gears and its impact on the output motion of the mechanism are analysed under different working conditions and compared with the kinematic behaviour.

The outline of the paper is the following. Section 2 provides a brief description of the proposed Multi-Body modelling approach for the dynamic simulation of gear meshing and related phenomena, while considering the time-varying mesh stiffness. In Sect. 3, a case study, consisting of a GFBM for approximate straight-line generation, is described, along with the results of the dynamic simulations for different values of gear velocity. Concluding remarks are provided in Sect. 4, along with an outlook of future developments.

2 Gear Dynamic Simulation in Multi-Body Environment

The modelling approach used for the dynamic simulation of gear meshing in Multi-Body environment is based on the use of a User Defined Force (UDF) element, which is defined for each gear pair in the model. Such a dedicated element derives the meshing forces mutually exchanged by engaging gears from the instantaneous conditions and determines the proper location of load application on the teeth flanks at the subsequent step of the dynamic simulation.

The process of meshing force determination starts from the computation of the DTE between the gears and its time derivative. It is derived by comparing the position of two reference systems connected to the gear bodies with the position of a third reference system for each time step of the simulation. The DTE is computed in a transverse plane, which is tangential to the pitch circles of the meshing gears, as the relative displacement between the two reference frames in the tangential direction. It accounts for two different contributions: the relative rotation of the two meshing gears (DTE_r) and the relative translation in the tangential direction (DTE_t), according to the following Eq. 1:

$$DTE = DTE_r + DTE_t \tag{1}$$

The gear contact force defined by the UDF element is derived from the DTE according to the following formulation:

$$\begin{cases} F_{tt}T = F \\ F_{tt} = F_{tt_k} + F_{tt_c} \\ F_{tt_k} = kDTE \\ F_{tt_c} = c\dfrac{dDTE}{dt} \end{cases} \tag{2}$$

where F is the normal contact force, F_{tt} is the magnitude of the tangential contact force and T is the transformation vector from the tangential direction to the direction normal to the tooth surface in the force application point. F_{ttk} and F_{ttc} represent the elastic and the damping components of the tangential contact force. The constants k and c are the instantaneous meshing stiffness and the viscous damping coefficient respectively. The transformation vector, which allows for computing the radial and tangential components of the gear contact force for a pair of spur gears, is defined as:

$$T = \begin{Bmatrix} \tan \varphi \\ 1 \\ 0 \end{Bmatrix} \tag{3}$$

where φ is the pressure angle.

The meshing force is applied at the location of the operating contact point, which is computed starting from the geometry of the gears and from their actual position. For the case study analysed in this work, the operating contact point is considered to be always located in the middle plane of each gear, since gears are spur and no misalignment is taken into account.

In case of constant transmitted load, the equivalent mesh stiffness can be derived from the Static Transmission Error (STE) curves, which in turn can be obtained through non-linear static Finite Element simulations, as shown in [5, 15], according to Eq. 4:

$$k = \dfrac{F_{tt}}{STF} \tag{4}$$

where F_{tt} is the tangential component of the static load transmitted by the driving to the driven gear in the static scenario.

The mesh stiffness is variable with respect to the relative angular position of the gears, given the time-varying nature of the STE. For axisymmetric gear pairs, the angular period of the STE is coincident with the length of the meshing cycle. If one of the gears is non-axisymmetric, due, for instance, to the presence of holes in the blank, the STE angular period does not coincide with the meshing cycle and must be calculated based on the number of teeth and of holes. The position along each STE cycle (PC) is derived as:

$$PC = \frac{1}{\theta_{P1}} \left[\theta_1 - floor\left(\frac{\theta_1}{\theta_{P1}} \theta_{P1}\right) \right] \quad (5)$$

where θ_1 is the actual rotational angle and θ_{P1} is the angular period of the STE curve. The obtained value of PC is between 0 and 1.

3 Case Study Description

With the aim of assessing the impact of the DTE of the gears on the dynamic behaviour of a geared-linkage, a case study is analysed, consisting a GFBM, which was proposed by Zhang et al. as approximate straight line generator in [19]. Figure 1 shows the kinematic schematic of a generic five-bar linkage, while the design specifications of the mechanism analysed in this work are reported in Table 1.

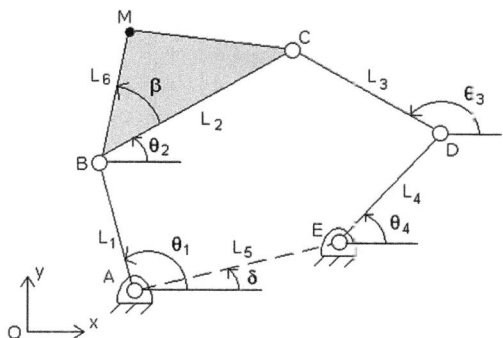

Fig. 1. Kinematic scheme of the five-bar linkage

Figure 2 shows the Multi-Body model of the GFBM, which was built by using the Motion module of the LMS Virtual.Lab software, along with the trajectory followed by the coupler point M, which approximates a straight line for a significant portion of the motion period. It is worthy to notice that the two cranks of the five-bar linkage are operated at the same angular velocity by a gear train consisting of two identical gears, meshing with an idler gear that has the same tooth geometry, but with a lightweight

Table 1. Design parameters for the mechanism of Fig. 1 as specified in [19]

Parameter	x_A	y_A	L_1	L_2	L_3	L_4	L_5	L_6	$\theta_{1,in}$	$\theta_{4,in}$	β	δ
Units	mm								Degrees			
Value	0.0	8.38	25.4	101.6	101.6	25.4	127.0	101.6	238	163	0	0

design. The latter is purposely used to increase the overall flexibility of the gear train and, hence, the TE in the geared system. The other two gears have a solid design, since a transmission with three extremely lightweight gears could result in a system with unrealistic flexibility. In the Multi-body model, ideal revolute joints are used to restrain the Degrees of Freedom on the linkage, while the meshing forces exchanged by the engaging gear pairs are calculated by using the UDF element described in Sect. 2. The mesh stiffness of each gear pair has been derived from the STE curve shown in Fig. 3, which has been calculated through static non-linear FE simulations, as described in [15].

Fig. 2. Multi-body model of the GFBM with the design specifications of Table 1. The path generated by the coupler point in kinematic conditions, approximating a straight line, is shown as well.

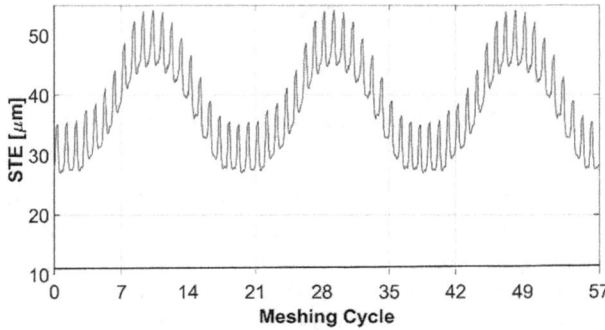

Fig. 3. STE curve estimated for each gear pair through non-linear FE simulations [15].

A simulation campaign has been executed by using different values of the angular velocity of the input gear in the range between 25 rpm and 100 rpm, in order to assess how the transmission error affects the dynamic response of the GFBM under different working conditions. In all the simulations, a constant torque of 350 Nm, which is the load used in the static simulation to derive the STE curve of the gear pair, was applied to the driven gear, while the driving gear was controlled in velocity.

The simulation results are represented in Fig. 4 which shows the trajectory followed by the coupler point in the straight-line region under the different dynamic conditions. Figure 5 shows the dynamic acceleration of the coupler point in x and y directions for the different angular velocities of the input gear along with the kinematic one, thus providing visual information on the unwanted vibrations generated in the mechanism by the gears' DTE. It is worthy to notice that the dynamic curves generated at 25 turn/min and 50 turn/min are very close to each other, showing small differences that are slightly appreciable in Fig. 5.

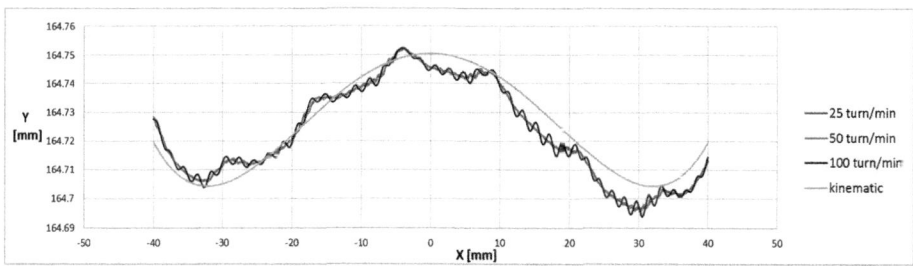

Fig. 4. Trajectory of the coupler point in the straight-line region in the kinematic and in the different dynamic conditions.

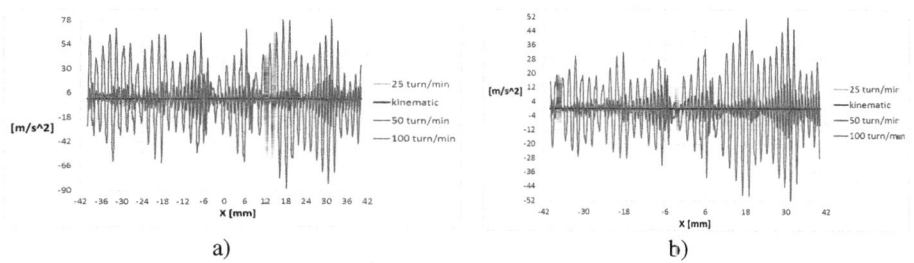

Fig. 5. Dynamic and kinematic acceleration of the coupler point in x and y direction for different values of the input angular velocity, estimated in the region of the straight-line trajectory.

4 Conclusions

In this paper, the impact of the gears' DTE on the dynamic behavior of a geared-linkage has been investigated by means of a Multi-Body simulation approach. By deriving the contact forces exchanged by the gears from the time-varying mesh stiffness, dynamic simulations are enabled, from which an analysis of the unwanted dynamics can be obtained.

The simulation results demonstrate that significant deviations of the system from the kinematic behavior can occur, especially at high operational velocities, which makes a dynamic study of such mechanisms a necessary step to prove their capability to generate the prescribed output motion with sufficient accuracy.

Future research activities are planned to include in the Multi-Body model additional vibration sources, such as the flexibility of the link and the compliance of bearings, with the aim of developing a simulation approach that enables accurate predictions of the dynamic behavior of geared-linkages under realistic working conditions.

Acknowledgments. The research leading to these results has received funding from the People Programme (Marie Curie Actions) of the European Union's Seventh Framework Programme FP7/2007-2013/under REA grant agreement no. 324336 DEMETRA: Design of Mechanical Transmissions: Efficiency, Noise and Durability Optimization.

References

1. Dimaragonas AD, Erdman GN, Sandor AG (1971) Synthesis of a geared N-bar linkage. Trans ASME J Eng Ind 93:157–164
2. Erdman AG (1995) Computer-aided mechanism design: now and the future. Trans ASME J Mech Des (Special 50th Anniversary Design) 117:93–100
3. Freudenstein F, Primrose EJF (1963) Geared five-bar motion. Trans ASME J App Mech 30:161–175
4. Hartenberg RS, Denavit J (1964) Kinematic synthesis of linkages. McGraw-Hill Inc., New York
5. Korta JA, Mundo D (2017) Multi-objective micro-geometry optimization of gear tooth supported by response surface methodology. Mech Mach Theory 109:278–295. doi:10.1016/j.mechmachtheory.2016.11.015
6. Modler K-H, Lovasz E-C, Bär GF, Neumann R, Perju D, Perner M, Mărgineanu D (2009) General method for the synthesis of geared linkages with non-circular gears. Mech Mach Theory 44:726–738
7. Mundo D, Gatti G, Dooner DB (2009) Optimized five-bar linkage with non-circular gears for exact path generation. Mech Mach Theory 44:751–760
8. Nokleby SB, Podhorodesky RP (2001) Optimization-based synthesis of Grashof geared five-bar mechanisms. Trans ASME J Mech Des 123:529–534
9. Oleksa SA, Tesar D (1963) Multiply separated position design of the geared five-bar function generator. Trans ASME J Eng Ind 298–305
10. Palermo A, Mundo D, Hadjit R, Desmet W (2013) Multibody element for spur and helical gear meshing based on detailed three-dimensional contact calculations. Mech Mach Theory 62:13–30

11. Roth B, Freudenstein F (1963) Synthesis of path-generating mechanisms by numerical methods. Trans ASME J Eng Ind 30:298–305
12. Sandor GN, Erdman AG (1991) Mechanism design: analysis and synthesis, vol 1, 2nd edn. Prentice Hall, Englewood Cliffs
13. Smith JD (2003) Gear noise and vibration, Cambridge
14. Starns G, Flugrad DR (1993) Five-bar path generation synthesis by continuation methods. Trans ASME J Mech Des 115:988–994
15. Shweiki S, Korta J, Palermo A, Mundo D, Heirman HKG (2016) On the effects of blank lightweighting on gear dynamics. In: 2016 international conference on power transmissions ICPT2016, ChongQing University, China, 27–30 October 2016
16. Tao DC (1964) Applied linkage synthesis. Addison-Wesley Publishing Company Inc., Reading
17. Ullah I, Kota S (1997) Optimal synthesis of mechanisms for path generation using fourier descriptors and global search methods. Trans ASME J Mech Des 119:504–510
18. Visa I, Neagoe M, Moldovan MD (2017) Structural synthesis of planar geared linkage mechanisms as multibody systems. Mech Mach Sci 46:99–106
19. Zhang C, Norton RL, Hammonds T (1984) Optimization of parameters for specified path generation using an atlas of coupler curves of geared five-bar linkages. Mech Mach Theory 19:459–466

Effect of the Coil Shape on Magnetic Field of an Electromagnet for Contactless Power Transmission to Microrobots

Abdulkareem Alasli[1], Nail Akçura[2], and Levent Çetin[3(✉)]

[1] Mechanical Engineering Department, The Graduate School of Natural and Applied Sciences, Dokuz Eylul University, Izmir, Turkey
eng.asli@gmail.com
[2] Mechatronics Engineering Department, The Graduate School of Natural and Applied Sciences, Dokuz Eylul University, Izmir, Turkey
nail.akcura@deu.edu.tr
[3] Mechatronics Engineering Department, Faculty of Engineering and Architecture, Izmir Katip Çelebi University, Izmir, Turkey
levent.cetin@ikc.edu.tr

Abstract. This paper proposes a comparative study in order to investigate the effect of the coil geometry on the intensity and homogeneity of the magnetic field. For this purpose, a solenoid with rectangular prism core and two different coil shapes, cylindrical and cuboid, was modeled. Several finite element numerical studies at different structural design parameters combinations for two different distances between the coils were carried out. The obtained results were evaluated according to the flux intensity and the homogeneity at the center region between the coils. Consequently, it was found that cylindrical coils generate higher flux intensity with more homogeneous magnetic field.

Keywords: Electromagnetic based actuation (EMA) · Microrobot · Solenoid · Numerical analysis

1 Introduction

Huge developments in today's technology give the possibility of designing and implying micro and nano scale functional systems in different aspects of life. In medicine, for example, it is foreseen that these nano and micro electromechanical systems, named as micro or nanorobots, will play a critical role in the forthcoming years. Specifically, nano and microrobots floating within bodies, liquids and tissues, will have many uses such as drug delivery, sampling and surface disintegration at a certain point [1].

Nano and microrobots are generally picked as ferromagnetic or permanent magnet materials. Their shapes and dimensions are picked with different structures in studies considering the limitations of micro level. The application environment is generally located inside body fluid which is affecting the robot during motion in term of lifting and drag force. Two fundamental problems are reported about these robots; actuation and monitoring. The small structure of the robots makes the integration of on board actuator hard to be applied. There is a great effort to produce actuators for micro and nano scale applications based on electroactive materials [2]. Nevertheless, contactless power transmission is still the most emphasized solution for navigation of nano and microrobots [3]. The most appreciated means of untethered power transmission are the fields arising from heat, electrical and magnetic sources. It is also shown in literature that different types of motions can be observed by the aid of spatially distributed quantities like electrical potential and magnetic field density. Between these alternatives, the magnetic actuators are more popular due to their ability of generating force with higher strength by just using intentionally created magnetic field forms.

When a magnetic particle is positioned at any point within the magnetic field, its magnetization vector is aligned in the direction of the flux lines. Also, the gradient of the external field causes a force applied on the particle, which is translated into motion. So, motion specific characteristics of magnetic field can be recapped as a field vector and a field gradient for control of orientation and applied force respectively.

The needs for the specific magnetic field characteristic for robot motion can be satisfied by using Helmholtz and Maxwell coil pair configurations. Homogenous magnetic field, which is responsible for controlling robot orientation, can be generated by using Helmholtz coil pairs. On the other hand, constant gradient magnetic field can be generated by Maxwell coil pairs, whose responsibility is to generate force applied to the robot. These electromagnets are placed around the environment to cover the micro robot working space. Based on this idea, Nelson controlled the motion of a ferromagnetic particle in the plane using Helmholtz and Maxwell coil pairs, which are rotating around an axis parallel to work-plane by a motor [4]. Choi and his team improved the design of electromagnetic based actuator (EMA) by adding another pair of fixed Helmholtz and Maxwell coils on perpendicular axis to control motion of a microrobot on a plane without using motor [5]. However, in these mentioned studies, a major problem has arisen that using the actuators, EMAs, the generated magnetic field is not powerful enough due to their coreless designs.

The solenoid structure with core inside can increase the magnitude of the magnetic field several hundred times which is a major advantage, but on the other hand cores of the coils affects the symmetry of the flux lines because of concentrating most of the flux lines into the core and spreading between the cores through an environment with much lower magnetic permeability (vacuum or fluid) to close the magnetic flux contour during actuation. Along all the study, also with all referenced studies, the magnetic flux lines are assumed not to affect the flux lines distribution. Therefore, cooperation of multiple coils requires more rigorous design methodology [6].

The main idea of this study is to take advantage of solenoids and combine it with coaxial coil pair design. For this purpose, we analyzed the structural design parameters and present a comparison study for probable coil structures.

1.1 Problem Definition

It is obvious that a pole magnet pulls any ferromagnetic object around it. The fundamental observation of this experience is not only the existence of the magnetic force but also the difficulty to keep position of ferromagnetic object within the effective area of magnet. Therefore, the starting point for EMA design is picked as the design of the coil structure which will form a homogenous magnetic field to orient the object with very low gradient to cause no force excitation to it.

As an initial choice; one should define the volume of the core. The more core material inside the coil, the more field magnitude it will generate. So, the core volume should be as large as possible compatible with weight requirements. Another consideration is the core material which can be ferrite or magnetic steel with relative magnetic permeabilities of ca. 600 and 4000 respectively.

As mentioned above, the volume of the core is constrained with weight and the coil material which is defined according to the need of amplification of magnitude of magnetic field. With relatively higher amplification factor and relatively simple manufacturing possibilities, steel is often selected as core material. Basically, the ferromagnetic cores are rectangular prisms, made of steel, with definite width height and length as shown in Fig. 1.

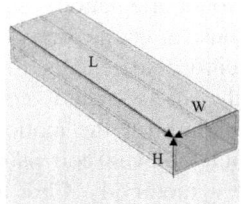
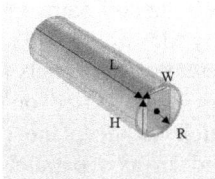

Fig. 1. Dimensions of the core for cylindrical and cuboid coils in term of W (width), H (height), and L (length).

Our previous studies showed that the longer cores create smaller magnetic field gradients [7]. Thus, selection length of core depends on the limit of footprint area of the EMA. The length of core can be determined as long as this spatial constraint allow. Eventually, the width and height of the core can be found using cross sectional area calculation for given weight and core length.

The ferromagnetic cores can be manufactured with square and rectangular. For the both cases, there are two possible geometries for coil structure: cylindrical or cuboid. This paper represents a comparative study to present the effects of the proposed geometries to the homogenous magnetic field generation.

2 Methodology

2.1 Basic Solution Methodology

In an EMA system, the purpose of the coil structure which will function like Helmholtz coils is to align the considered microrobot in the desired direction within the region of interest (ROI, work space of the micro-object). The alignment occurs when the coil pair generates a uniform magnetic field that causes the permeant magnetic object to rotate until its magnetic flux lines are parallel to the one generated by the coils. The generated torque τ [N.m] for alignment can be defined as in (1):

$$\tau = v(M \times B) \tag{1}$$

where, B [T] refers to the magnetic flux. v [m^3] and M [A/m] denote the volume and the magnetization of the microrobot. From the previous equation, v and M are constant and depend on the material of the considered object. Therefore, to adjust the rotation torque τ, the magnetic flux B should be adjusted. B can be controlled by changing either the distance d between the coils or the current I, which flows through the coils. However, observing the changes in B in the ROI by changing d or I can be challenging with the available experimental equipment. Therefore, another way must be found.

COMSOL is a simulation program used for understanding, prediction, and design problems in static and transient applications. For this steady state study, AC/DC module, integrated in COMSOL, is used for simulating electromagnetic fields by solving Maxwell's equations for different scenarios. In the stationary domain (the steady state), equations set (including Ampere law) are solved by COMSOL as in (2), (3) and (4):

$$\nabla \times H = J \tag{2}$$

$$B = \nabla \times A \tag{3}$$

$$J = \sigma.E \tag{4}$$

where, H [A/m] represents the magnetic field vector and B [T] the magnetic flux density, which is defined as $B = \mu_0.\mu_r.H$ where μ_0 [Wb/(A.m)] and μ_r are the magnetic permeability of free space and relative permeability, respectively. A [V·s/m] represents the magnetic vector potential while J [A/m^2] and E [V/m] are the vectors of the external current density and the electric field, respectively.

The work flow of each study can be described in the following steps: design the geometry, choose the materials, define boundary and initial conditions, define the finite elements mesh, select suitable solver, and visualize the results. Furthermore, in any numerical study, it is important to have sufficient resolution of the generated mesh. In this study, the mesh was divided into regions with different elements size to shorten the consumed time for calculation. The region of interest (which is located between coils and its size depends on the distance between them), the cores, and the coils were chosen to have the most intensive number of elements (with size less than 3 mm) as seen in Fig. 2. In these shown simulations, the default numerical stationary solver was chosen with default setting. Also, parametric sweep was added to initiate a sequence of studies to find the solutions to a sequence of varied parameters.

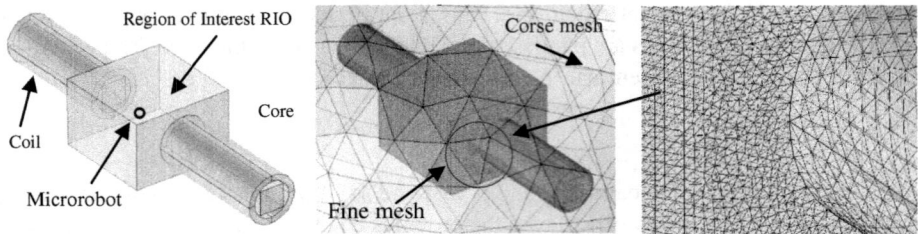

Fig. 2. The designed model for simulations (on the left) and the used mesh with multi-size regions (on the right).

2.2 Modeling and Evaluating

This part is devoted to propose a method for evaluating the effect of the coil's shape on the intensity and the homogeneity of the magnetic field. As discussed previously, two shapes (cuboid and cylindrical) were considered. According to that, a model with two coils facing each other on the same axis, as in Fig. 2, has been developed.

Six parametric simulations, three for each shape were carried out for two different distances: A close distance 50 mm (in between), and a wide one 100 mm. The close and the wide distance were selected in order to give us a better understanding of the effect of the geometry shape at two extreme points for two coils. The dimensions of the coil is related directly with the dimension of the core and since the electrical stainless steel cores are shaped generally as cuboid, the dimensions of the cuboid coil will be similar to the ones of the cores because it is winded directly around the core. However, the radius of the cylindrical coil can be defined as in (5),

$$R_c = \sqrt{\left(\frac{W}{2}\right)^2 + \left(\frac{H}{2}\right)^2} \tag{5}$$

where, W and H are the width and the height of the core, respectively as in Fig. 1.

For simplicity, the presented method will be built on the assumptions of a maximum weight of 1 kg per coil and a limitation of 500 mm for the considered footprint area. The W (width), the H (height), and the L (length) of the cuboid geometry, as shown in Fig. 1, can be calculated with the correlation in (6),

$$V_{Coil} = W \times H \times L \tag{6}$$

The maximum weight (mass) of the coil is related with its maximum volume by the density of the chosen materials. The largest part of the coil's weight is devoted for the electrical steel core and almost 2–5% is devoted for the wire and the plastic cylinder (holder). The density of electrical steel is $\rho = 7850\,\text{Kg/m}^3$ and based on that the maximum volume is calculated by using (7) to be 127.4 cm^3,

$$V_{\max} = \frac{m_{\max}}{\rho} \tag{7}$$

However, calculation based on maximum volume alone will give a huge number of W, H, and L combinations. Therefore, another design choice should be made. As it is known, the homogeneity of the generated magnetic field is proportional with the length of the used core. For the considered footprint area limitation, a maximum 150 mm as the length of the core was chosen. Based on the two previous assumptions the combinations of $W \times H$ showed in Table 1 will be simulated.

Table 1. Calculated combination of W and H according to the maximum volume and the $L = 150$ mm

W (mm)	20	25	30	35	40
H (mm)	41.7	33.3	27.8	23.8	20.8

Two parameters were used for evaluation. The first one is the intensity of the magnetic flux at the center point between the running coils B_0. The second one is the dimensions of the area in which the intensity of the magnetic field is 100%, 90%, and 80% of its maximum value as seen in Fig. 3. This parameter is important for evaluating the homogeneity of the magnetic field. Also, it is important to mention here that all of the simulations were carried out with current of 1 A flow through a wire of 1 mm diameter and 600 turns for each coil.

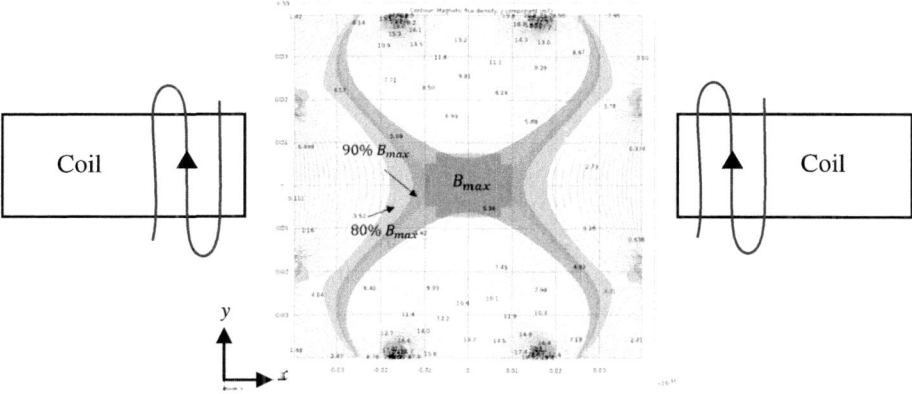

Fig. 3. The ROI divided according to the flux intensity regions of the magnetic field (Color figure online)

3 Results and Discussion

The flux intensity and the homogeneity of the magnetic field within the ROI are crucial criteria for designing any electromagnetic based actuation. The flux intensity indicates how much the generated magnetic field is powerful to move the considered microrobot and the homogeneity of the magnetic field is essential for directing the microrobot in the desired direction. Tables 1 and 2 present these two criteria as a function of the distance between the coils (d) at the suggested combination of W and H.

Table 2. Flux intensity of the magnetic field and B_{max} region at different distances for several combinations of W and H for cylindrical coils

W (mm)	H (mm)	d (mm)	B_0 (mT)	B_{max} region (cm)	d (mm)	B_0 (mT)	B_{max} region (cm)
20	41.7	50	14.427	1.5×1	100	3.659	2×1.5
30	27.8	50	14.763	2×1	100	3.607	2×1.5
40	20.8	50	14.479	3×1	100	3.646	2.5×1.5

For the both cases (cylindrical and cuboid coils), it can be obviously seen that the flux intensity at the center between the coils B_0 is increasing with the decrement of the distance d between the coils. This behavior is a neutral result of Biot–Savart law which indicates that the flux intensity is inversely propositional with distance. Furthermore, when the combinations with the aspect ratio close to 1 (i.e. the cross section is close to the shape of square), the generated flux intensity also shows higher values. This can be clearly seen at $W \times H = 3 \times 2.78$.

By comparing Tables 2 and 3, it can be observed that for all selected distances between the coils, it is obvious that the flux intensity of the magnetic field B_0 for the cylindrical coils is at least two times greater than the one of the cuboid coils. Moreover, both tables indicate that the B_{max} areas, in which the intensity is constant and equals to B_{max}, at the center of the working space between the coils are twice larger for the cylindrical coils. To get a better understanding about the effect of the coil's shape on the homogeneity of the magnetic field and the B_{max} region, the flux intensity distribution within the ROI for the both considered cases at distance $d = 50$ mm for all combinations were evaluated and presented in Fig. 4.

Table 3. Flux intensity of the magnetic field and B_{max} region at different distances for several combinations of W and H for cuboid coils

W (mm)	H (mm)	d (mm)	B_0 (mT)	B_{max} region (cm)	d (mm)	B_0 (mT)	B_{max} region (cm)
20	4.17	50	5.119	0.5×1	100	1.714	1×1
30	2.78	50	5.210	1×1	100	1.695	1×1
40	2.08	50	5.132	1×1	100	1.699	1.5×1

As mentioned previously and as shown in Fig. 3, the ROI was divided into three main areas, the red area in the middle represent the region where B_0 is constant and equal to B_{max}. The lager the area of B_{max}, the more homogeneous the magnetic field generated by the coils, which can be also translated as the more stable the microrobot at the defined work conditions. Furthermore, the length of the other two area, adjacent to B_{max} region (80% and 90%), give an indication about the gradient of the magnetic field inside the ROI. The longer the distance between those areas the less the gradient occurs within this region, this can be transformed as a long plateau at the middle of axis between two coils. So, by comparing all the figures in Fig. 4 given above, surely it can

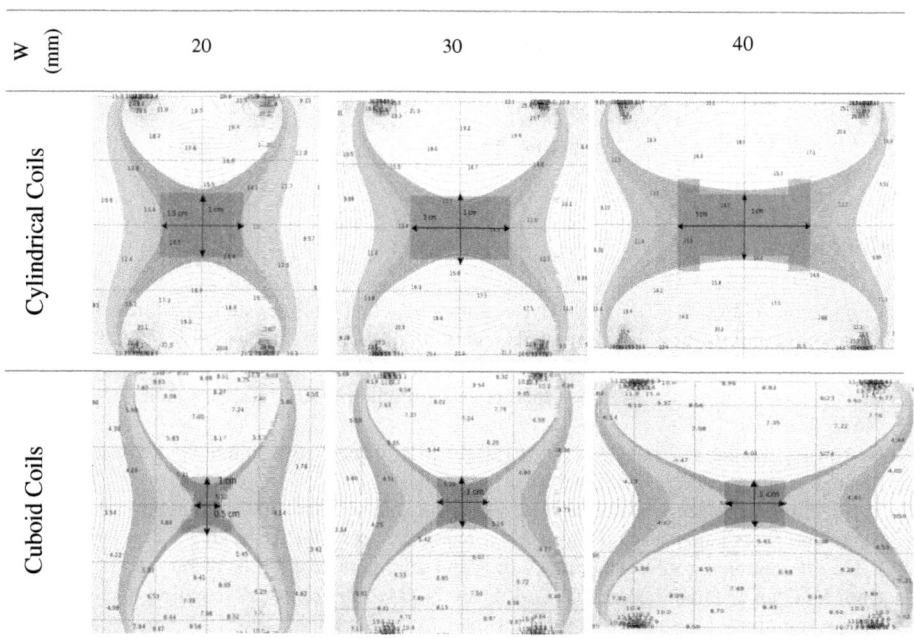

Fig. 4. ROI between the coils (cylindrical and cuboid) divided by the flux intensity regions for different values of W (All information within the graph is listed in Tables 2 and 3).

be said that the cylindrical coils at all values of W and H provide better B_{max} regions. It is important to mention here that the same phenomenon was also realized with the other distance ($d = 10$). Therefore, and for the reasons discussed above, choosing the circular geometry for synthesizing the electromagnetic coils seems to be more beneficial.

4 Conclusion

The presented study discussed the idea of taking the advantage of solenoids and combines it with coaxial coil pair design in order to use it in electromagnetic based actuator (EMA). For this purpose, two coils geometries (cylindrical and cuboid) were considered and an analysis on the structural design parameters was carried out. Consequently, a comparative study was presented to evaluate the effects of the proposed coil geometries on the homogenous magnetic field generation. It was concluded that choosing the cylindrical geometry for synthesizing the electromagnetic coils provides higher flux intensity B_0 at the center between the coils in addition to a more homogeneous magnetic field with larger B_{max} region.

Acknowledgement. We would like to acknowledge the financial support of the Scientific and Technology Research Council of Turkey (TÜBITAK, Grant no. 215M879). We would also like to thank Smart Structures Laboratory of Ege University Department of Mechanical Engineering for providing the support of using COMSOL program and for the very useful reviews of the anonymous reviewers.

References

1. Chungseon YU et al (2010) Novel electromagnetic actuation system for three-dimensional locomotion and drilling of intravascular microrobot. Sens Actuators, A 161(1):297–304
2. Shahinpoor M, Bar-Cohen Y, Simpson JO, Smith J (1998) Ionic polymer-metal composites (IPMCs) as biomimetic sensors, actuators and artificial muscles-a review. Smart Mater Struct 7(6):R15
3. Nelson BJ, Kaliakatsos IK, Abbott JJ (2010) Microrobots for minimally invasive medicine. Annu Rev Biomed Eng 12:55–85
4. Bell DJ, Leutenegger S, Hammar KM, Dong LX, Nelson, BJ (2007) Flagella-like propulsion for microrobots using a nanocoil and a rotating electromagnetic field. In: Proceedings 2007 IEEE international conference on robotics and automation, p 1128–1133
5. Choi H, Choi J, Jang G, Park JO, Park S (2009) Two-dimensional actuation of a microrobot with a stationary two-pair coil system. Smart Mater Struct 18(5):055007
6. Kummer MP, Abbott JJ, Kratochvil BE, Borer R, Sengul A, Nelson BJ (2010) OctoMag: an electromagnetic system for 5-DOF wireless micromanipulation. IEEE Trans Rob 26(6): 1006–1017
7. Çevik H (2015) Bir mikrorobotun düzlemsel hareketinin elektromanyetik aktüatör yardimiyla kontrolü. Dokuz Eylül University Mechatronics Engineering Department, Master Thesis, İzmir

Dynamics of Machinery

The Edge of Chaos in Kinematics and Dynamics of Mechanism

Zhaohui Liu, Jin Xie$^{(\boxtimes)}$, and Yong Chen

Southwest Jiaotong University, Chengdu, China
zhhliu@home.swjtu.edu.cn, xj_6302@263.net,
cykine163@163.com

Abstract. Edge of chaos is a new concept deriving from complexity. Its significance for mechanism science is described in this paper. It is demonstrated that one can obtain all solutions of nonlinear equations by Newton's method utilizing the edge of chaos. And in dynamic analysis and controlling of mechanism, where the models usually are nonlinear dynamic system, one is able to understand something about the arising of chaos, and find efficient measures to control and anti-control chaos through studying of edge of chaos. Following these, some methods to detect the edge of chaos are introduced briefly.

Keywords: Edge of chaos · Computational kinematics · Dynamics of mechanism · Controlling

1 Introduction

Complexity is a new field of science research [1]. In its development, many words and concepts from science and engineering are reused with new contents, such as entropy, self-adaptation, self-organization, among many others. And in the meantime, new words or phrases emerge from time to time. One of them is the "edge of chaos", coined by C. G. Langton in 1990 [2].

One definition of edge of chaos is that it is the region between the rigid order and chaos. At the edge of chaos, the system transits from rigid order to chaos in either smooth or abrupt fashion. Smooth transition suggests that there is an intermediate regime between rigid order and chaos, analogically fluid is the intermediate phase between solid and gas [2, 3]. Whereas an abrupt transition suggests that the system passes straight from the rigid order to chaos with no intermediate regime, only exists there a critical point, so-called chaos threshold [4–7].

Another definition of edge of chaos is based on the concept of basin of attractor. In a dynamical system, the basin of attractor refers to a set of initial points if starting with them the system will converge to the same attractor. While the edge of chaos is definite as the basin boundary, which separates the basins of the system [8, 9].

The edge of chaos is the source of complexity, and most interesting features can be found there. The study of it allows us to gain more knowledge about nonlinearity, chaos, complexity occurring both in nature and in human society, not only in theory but also in engineering practice. Therefore many scientists and engineers from different fields have been engaging in the studies of edge of chaos [10–18].

The history of mechanism science can be summed up as a history of struggling with nonlinearity, which arises from kinematics and dynamics of machines and mechanisms. To a certain degree, all kinds of methods explored for design or analysis of mechanism are aimed to improve the tools and methods to deal with nonlinearity.

In this paper, we review some researching outcomes achieved in recent years aiming at addressing the significance of edge of chaos for mechanism science and clarifying the direction of further research.

The organization of this paper is as follows. The significance of the edge of chaos for computational kinematics and dynamics and controlling of mechanism is illuminated by the examples in Sects. 2 and 3 respectively, and the methods to determine the edge of chaos are briefly introduced in Sect. 4, and the conclusions is provided in Sect. 5.

2 The Edge of Chaos Utilized in Computational Kinematics

Many problems in kinematics of mechanism lead naturally to a set of nonlinear equations as following

$$\mathbf{F}(\mathbf{X}) = 0 \tag{1}$$

where $\mathbf{X} = [x_1, x_2, \cdots, x_n]^T$.

Usually the nonlinear equations have more than one solution. And every real solution has its concrete meaning. It may correspond to a configuration of the mechanism, or a candidate for the designed mechanism. Therefore, the methods to find out all the solutions of Eq. (1) are so critical that lots of methods have been proposed [19].

Newton's method is widely utilized for its simplicity. Let $\mathbf{X}^*$ denote one of the solution of Eq. (1) and $\mathbf{X}_0$ the initial point, the following sequence will approximate to $\mathbf{X}^*$.

$$\mathbf{X}_{n+1} = \mathbf{X}_n - \mathbf{J}(\mathbf{X}_n)^{-1}\mathbf{F}(\mathbf{X}_n) \tag{2}$$

where $\mathbf{J}(\mathbf{X})$ is the Jacobian matrix of $\mathbf{F}(\mathbf{X})$. Unfortunately, Newton's method is of quadratic local convergence [20]. According to this property, it seems that Newton's method cannot be employed to find out all solutions of Eq. (1).

However, things are not as simple as it appears if taking Newton's method as nonlinear discrete dynamical system.

Assuming Eq. (1) has more than one real solution. Then each solution of Eq. (1) possesses a basin of attraction, and the basin boundaries, defined as edge of chaos in Sect. 1, will come into being.

It is definite that no one is able to predict which solution the sequence of Eq. (2) will converge to if the initial point $\mathbf{X}_0$ is chosen at edge of chaos. The reason for it is that at the edge of chaos two sequences may converge to quite different solutions even if the initial points are very close to each other. In other words, at the edge of chaos a little change in initial point may cause great change in the final result. So, Newton's method behaves chaotically at the edge of chaos. It is this property that allows us to find out all solutions of Eq. (1) by Newton's method.

To illustrate this idea, we here discuss the case with one-variable. As an example, we take

$$f(x) = x^3 - x = 0 \tag{3}$$

Obviously Eq. (3) has three solutions. They are -1, 0, and 1. The Newton's method can be written as

$$x_{n+1} = x_n - \frac{x_n^3 - x_n}{3x_n^2 - 1} \tag{4}$$

As done in Ref. [21], Newton's method is extended to the complex plane, and the basin of attraction for each solution is depicted in Fig. 1. The three attractors of solutions, -1, 0, and 1, are shaded with blue, red and black color respectively.

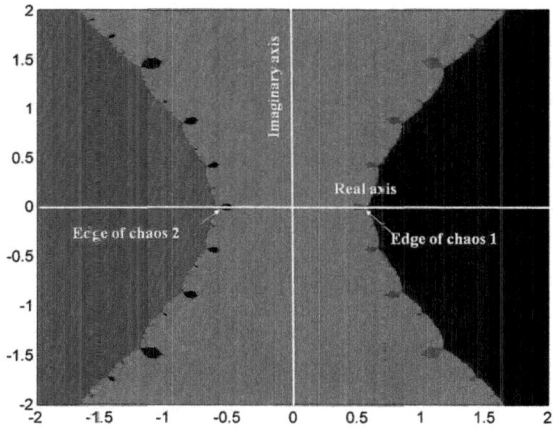

Fig. 1. Edge of chaos (Julia set) by the definition of boundary of the basins of the three solutions of $f(x) = x^3 - x = 0$ (Color figure online)

In Fig. 1, the set bounding the basins of attractions is named as the edge of chaos or Julia set. From this figure, it can be seen that the domains of attraction for each solution are intertwined. In the neighborhood of any point at the edge of chaos, there are domains of attraction for each and all solutions. Based on this fact, one can find out all solutions using Newton's method if picking many enough initial points in the vicinity of any point at the edge of chaos.

This paves the way for finding out all solutions using the edge of chaos. In real domain, Newton's method is carried out alone the real axis, the white horizontal line shown in Fig. 1. There are two edge of chaos on this real axis. Of course, one can find out all real solutions if picking many enough initial points in the vicinity of any point at edge of chaos.

It should be noted that the most crucial step in obtaining all solutions with Newton's method is to find out the edge of chaos. Based on the period-doubling route to chaos, Jovanovic et al. proposed a sensitive inverse iterating [22] and Feng et al. employed optimization [23] to find periodic point. Xie et al. find that the intersection of boundary of attraction with real axis is located in the neighborhood of the singular points of Newton's method, say $\dot{f}(x) = 0$, i.e. $x = \pm\sqrt{\frac{1}{3}} = 0.577$, as shown in Fig. 1 [24].

Comparing with Newton's method with one-variable, Newton's method with multi-variables are more complicated, and the studies on it are rare. Motyka et al. presented some examples to illustrate the basins of attraction in multivariable case [25].

Inspired by the edge of chaos in the Newton's method with one-variable, Xie et al. extended their finding to multi-variables [24]. They assumed that the edge of chaos is in the neighbourhood of a singular point, which satisfied the condition $|\mathbf{J}(\mathbf{X})| = 0$. Many numerical experiments prove this assumption, and show that they arrived at all solutions of Eq. (1) if picking many enough initial points at the edge of chaos, or nearby point of singular point.

3 The Edge of Chaos Utilized in Dynamics and Controlling of Mechanism

The dynamics analysis of mechanism is to determent motion of a mechanism starting from an initial point under action of some kinds of forces. They are of initial-value problem, and can be mathematically expressed as following

$$\dot{\mathbf{X}} = \mathbf{F}(\lambda, \mathbf{X}), \quad \mathbf{X}(0) = \mathbf{X_0} \tag{5}$$

where λ is parameter of the system.

In recent years, high-fidelity modeling of mechanism becomes popular. In such modeling, more factors are taken into account, such as nonlinear elastic or spring elements, nonlinear dapping, backlash, play, or bilinear springs. All these factors can be regarded as nonlinear elements, and lead Eq. (5) to be nonlinear. In a nonlinear dynamic system, chaotic motion occurs under certain circumstances [26]. And in Refs. [27, 28], another kind of mechanism, namely underactuated mechanism, exhibiting chaotic motion is presented.

Chaotic motion is a double-edged sword. It can be used to increase the efficiency of some kinds of mechanism, and at meantime cause undesirable vibration, wear and damages. So the controlling and anti-controlling of chaos are significant for the mechanism design.

Here, we will present one example to demonstrate the edge of chaos and its utilizations in controlling and anti-controlling of chaos.

A PR manipulator supported on a flexible foundation is shown in Fig. 2.

Chaos arises from this system under some parameters. A PD controller is employed to control chaos. The mathematical model of the controller is

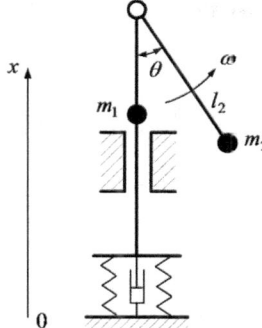

Fig. 2. A PR manipulator supported on a flexible foundation

$$\begin{cases} \dot{x}_1 = k_1 x_2 + k_2 \dot{x}_2 \\ \dot{x}_2 = -\alpha x_2 - \beta x_1 - \gamma x_1^3 + \lambda \cos x_3 + \mu \\ \dot{x}_3 = \Omega \end{cases} \quad (6)$$

where k_1 and k_2 are the proportional and derivative action coefficients respectively

Usually k_1 and k_2 are determine by trial and error, which is blindness and require more computational effort. However, we can regard both of them as normal parameters of the dynamic system, and depict the edge of chaos, as shown in Fig. 3 [29], where the point assigned with blue color if the parameters lead to chaotic motion and with white color if the parameters lead to periodic motion. With helping of the information drown from Fig. 6, it is easy to determine the controlling parameters k_1 and k_2. Firstly to fix parameter k_1, say $k_1 = 1$, and then we can choose k_2 for different purpose. k_2 should be

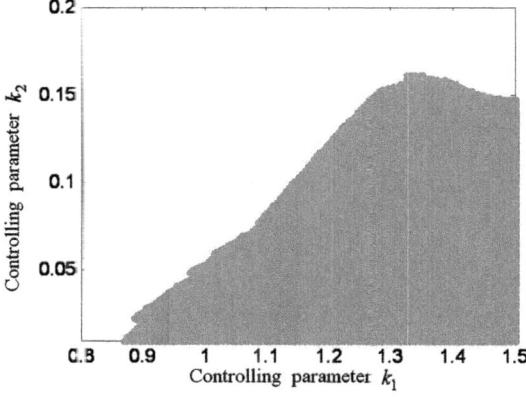

Fig. 3. The edge of chaos for the parameters k_1 and k_2 (Color figure online)

chosen from the interval [0.046, 2] for suppressing chaotic motion and from the interval (0, 0.046) for anti-controlling chaotic motion.

4 The Approaches to Finding Out the Edge of Chaos

Since the edge of chaos can be represented by some kinds of figures, all the numerical methods used to detect the chaos occurrence can be an alternative of the approaches to find out the edge of chaos.

Here we speculate on the bifurcation diagram, Lyapunov exponent, entropy, and Melnikov method. Due to the limitations of paper length, we omit the dynamical models and relevant information of the systems.

It is well known that period-doubling and torus bifurcation are the routes to chaos. Bifurcation diagram depicts such behaviors of dynamic system with the respect to one of parameters of the system.

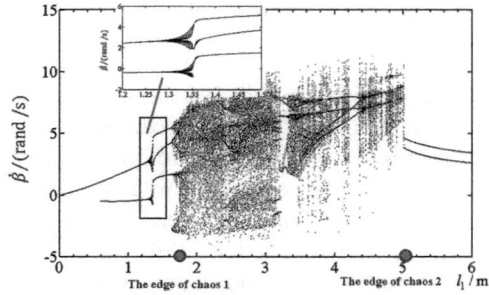

Fig. 4. A bifurcation diagram

Figure 4 is a bifurcation diagram of a dynamic system. There are two edges of chaos. At the edge of chaos 1 the system changes from periodic motion into chaotic motion, and at the edge of chaos 2, from chaotic motion into periodic motion.

Since chaos is of sensitive dependence on the initial conditions, a small change in initial conditions leads to a great change in the location at a later time. To replace this criterion, we can use a more mathematical definition based on Lyapunov exponent. That says that the dynamic system is chaos if at least one of the Lyapunov exponents is positive. According to this rule, one edge of chaos can be found in Fig. 5.

Lyapunov exponent is one of rare computable parameters to reflect the characteristics of chaos.

From the information theory point of view, chaos generates unpredictable messages which cannot be coded in a concise way and can be quantified by positive entropy. There are different algorithms to calculate entropy, namely different entropies, for examples, Shannon entropy, Kolmogorov-Sinai entropy, topological entropy, and so on.

Figure 6 presents the edge of chaos for Logistic mapping determined through permutation entropy.

In the normal case of mechanical engineering, there is more than one parameter which affects the dynamics behaviors of mechanism. So the edge of chaos versus two

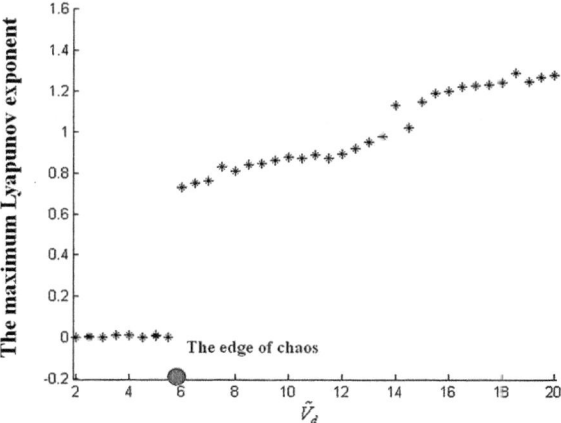

Fig. 5. The maximum Lyaponov exponent versus a parameter of dynamic system

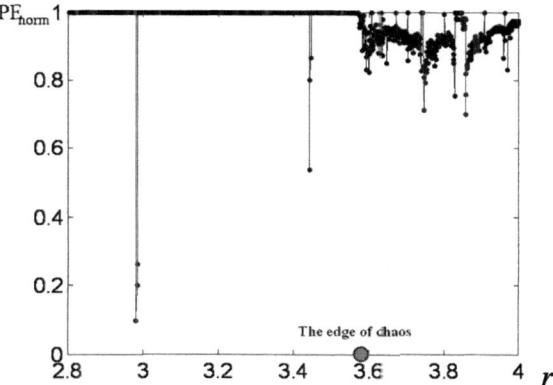

Fig. 6. The permutation entropy for Logistic mapping

parameters, like that shown in Fig. 3, will be more helpful for analyzing chaotic motion and controlling chaos.

Figure 3 is plotted by a time-consuming method. Firstly, the plane of parameters (k_1, k_2) is uniformly sampled in rectangle parallelepiped by 100×100 nodal points. Then the values of parameters at each nodal point are put into the dynamic equations and simulation is carried out. With the result of simulation and criterion of chaotic motion, whether the state of motion is chaotic or not is determined and each of nodal point is assigned with blue or white color depending on the state of motion being chaotic or periodic.

Melnikov method is applied to predict the chaos occurrence in a Hamiltonian system. It can generate a boundary between chaos and regular motion. For example, Duffing system is written as

$$\begin{cases} \dot{x} = y \\ \dot{y} = x - x^3 + \varepsilon f \cos(\omega t) - \varepsilon k_1 y \end{cases} \quad (7)$$

The boundary generated by Melnikov method for Duffing system is shown in Fig. 7 [30], where the nodal points assigned with pale blue are the ones with these values the motion states are chaos.

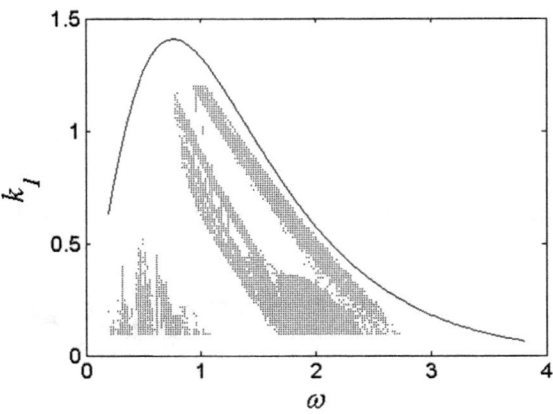

Fig. 7. The boundary generated by Melnikov method (Color figure online)

From Fig. 7, it can be seen that in some area the onset value of chaos by Melnikov method is almost same as that by simulation technique, but in other area, the value by Melnikov method is too conservative. As a whole, applying Melnikov method to detecting the edge of chaos is far from perfect method and some modifications are needed.

5 Conclusions

In the field of mechanism science there exist many nonlinear dynamics phenomena in which chaos is unavoidable.

This paper reviews some progresses on utilizing the edge of chaos. And some approaches to detecting the edge of chaos are introduced briefly.

Although it has been confirmed that using Newton's method can obtained all real solutions of nonlinear equations arising from computational kinematics with the help of the edge of chaos, as shown in Sect. 2, an efficient method to locate the edge of chaos is not yet established. It requires more understanding of nonlinear discrete dynamics.

Referring to dynamics of mechanism, the researching on the edge of chaos shines new light on the controlling and anti-controlling chaos, as described in Sect. 3. With the researching going on, we believe a new mechanism of chaos occurrence or tout to chaos might be revealed in the near future.

The edge of chaos is a new concept. Its study and utilization are in their infant stage. We argue that the edge of chaos is a new fundamental in science, including mechanism science, and hope that this paper evokes much more interest in it.

Acknowledgment. The authors would like to acknowledge the financial support of NSFC (National Natural Science Foundation of China) under the grant No. 51175437 and No. 51575457.

References

1. Horgan J (1995) From complexity to perplexity. Sci Am 272:74–79
2. Langton CG (1990) Computation at the edge of chaos: phase transition and emergent computation. Phys D 42:12–37
3. Hiett PJ (1999) Characterizing critical rules at the 'edge of chaos'. BioSystems 49:127–142
4. Borges EP, Tirnaki U (2004) Mixing and relaxation dynamics of hénon map at the edge of chaos. Phys D 193:148–152
5. Mayoral E, Robledo A (2004) Multifractality and nonextensivity at the edge of chaos of unimodal maps. Phys A 340:219–226
6. Borges EP, Tirnakli U (2004) Two-dimensional dissipative maps at chaos threshold: sensitivity to initial conditions and relaxation dynamics. Phys A 340:227–233
7. Celikoglu A, Tirnakli U (2006) Sensitivity function and entropy increase rates for z-logistic map family at the edge of chaos. Phys A 372:238–242
8. Skufca JD, Yorke JA, Eckhardt B (2006) Edge of Chaos in a parallel shear flow. Phys Rev Lett 96:174101
9. Muñoz PR, Barroso JJ, Chian AC-L, Rempel EL (2012) Edge state and crisis in the pierce diode. Chaos 22:033120
10. Mahmoudabadi A, Seyedhosseini SM (2014) Developing a chaotic pattern of dynamic hazmat routing problem. IATSS Res 37(2):110–118
11. Moglia M, Pascal P, Stewart B (2010) Modelling an urban water system on the edge of chaos. Environ Model Softw 25(12):1528–1538
12. Osborn RN, Hunt JG (2007) Leadership and the choice of order: complexity and hierarchical perspectives near the edge of chaos. Leadersh. Q. 18(4):319–430
13. Oxley L, George DAR (2007) Economics on the edge of chaos: some pitfalls of linearizing complex systems. Environ Model Softw 22(5):580–539
14. Rai V, Upadhyay RK (2006) Evolving to the edge of chaos: chance or necessity. Chaos, Solitons Fractals 30(5):1074–1087
15. Upadhyay RK (2009) Dynamic of an ecological model living on the edge of chaos. Appl Math Comput 210(2):455–464
16. Robledo A (2004) Aging at the edge of chaos: glassy dynamic and nonextensive statistics. Phys A 342(1–2):104–111
17. Xu D-Y, Yu C-W, Cheng Q-M, Bao Z-Y (2011) Application of the chaos domain of the Zhabotinski CNN to explore insights to hydrothermal deposit-forming processes. Comput Geosci 37(12):1928–1934
18. Hu Y-D, Zhang Z-Q (2011) Bifurcation and chaos of thin circular functionally graded plate in thermal environment. Chaos, Solitons Fractals 44(9):739–750

19. Raghavan M, Roth B (1995) Solving polynomial systems for the kinematic analysis and synthesis of mechanisms and robot manipulators. ASME Spec 50th Anniv Des Issue, 117:71–79
20. Dennis JE Jr, Schnabel RB (1996) Numerical methods for unconstrained optimization and nonlinear equations. SIAM: Society for Industrial and Applied Mathematics, Philadelphia
21. Hoppensteadt FC (2000) Analysis and Simulation of Chaotic Systems, 2nd edn. Springer, New York
22. Jovanovic VT, Kazerounian K (1998) Using chaos to obtain global solutions in computational kinematics. ASME J Mech Des 120:299–304
23. Feng C, Xie J, Chen Y (2004) Using chaos and fractals to synthesis planar mechanism. China Mech Eng 15(9):753–756 (in Chinese)
24. Xie J, Yan K-Y, Chen Y (2006) On global aspects of real newton's method and applied to synthesis of linkage. In: Proceedings of 2006 ASME international design engineering technical conferences and computers and information in engineering conference, DETC2006-99087
25. Motyka MA, Reiter CA (1990) Chaos and Newton's method on systems. Comput Graph 14(1):131–134
26. Moon FC (1987) Chaotic vibrations an introduction for applied scientists and engineers. Wiley, Hoboken
27. Nakamura Y, Suzuki T, Koinuma M (1997) Nonlinear behavior and control of a nonholonomic free-joint manipulator. IEEE Trans Robot Autom 13(6):853–862
28. Ravishankar AS, Ghosal A (1999) Nonlinear dynamics and chaotic motions in feedback-controlled two- and three-degree-of-freedom robots. Int J Robot Res 18(1):93–108
29. Cui Q-Y, Xie J, Wei W, Yi Z-Q (2016) Chaotic motion and its control of rotary machinery. J Mech Transm 40(3):12–16 (in Chinese)
30. He S-H (2015) On the application of melnikov method to determining edge of chaos of mechanical dynamic system with multi-parameters. Mater Dissertation, Southwest Jiaotong University

Dynamics of Orthogonal Mechanism of Vibrating Table in View of Friction

Zharilkassin Iskakov[1(✉)], Kuatbay Bissembayev[2], and Nutpulla Jamalov[2]

[1] Institute of Mechanics and Machine Science,
Almaty University of Power Engineering and Telecommunications,
Almaty, Kazakhstan
iskakov53@mail.ru
[2] Institute of Mechanics and Machine Science, Almaty, Kazakhstan
kuat_06@mail.ru, nutpulla@mail.ru

Abstract. The article deals with dynamics of orthogonal mechanism of vibrating table in view of friction. The emphasis is placed on the influence of friction on the oscillatory motion of the mechanism at its interaction with non-ideal energy source. As a result of analysis of numerical solutions of nonlinear motion equation has been determined that the coefficient of sliding friction affects the average angular speed of the driving link, and at that its growth in the first half cycle, when the vibrating table moves upward will cause reduction of the average angular speed, but in the second half-cycle, conversely, will cause increase of this value of angular speed. A coefficient of rolling friction influences on the vibration amplitude of the angular speed of the driving link: with its increasing the amplitude of vibrations increases. Therefore, sliding friction may affect the coefficient of non-uniformity of rotation in two ways, i.e. one half cycle of the mechanism rotation can reduce it and increase it in the other half cycle, and only rolling friction destabilizing effect has been found.

Keywords: Oscillating motion · Orthogonal mechanism · Vibrating table · Friction coefficient · Non-ideal source

1 Introduction

In recent years the vibration equipment in the mechanical engineering practice is developed based on the lever mechanisms. These mechanisms have a unique opportunity to create a vibratory motion of the working member. Development of vibration mechanisms on the basis of mathematical modeling gives the good results acceptable for the practice. The structural scheme of vibrating machines usually are not complicated, but for the successful operation it is necessary to determine accurately their parameters that can only be done on the basis of studies of the dynamics of vibrating machines and processes performed by them.

Any engine has original limitations, determined by the characteristics of its power. Such limited power energy source is named "non-ideal source" and the system "non-ideal system". In the event of the restricted power of energy source strong interaction between the dynamic system and the motor leads to fluctuations in the engine speed with the amplitude having a fairly large value.

A complete overview of the various theories of non-ideal oscillatory systems is presented and discussed in the papers [1, 2, 7].

One type of the vibrating equipment is vibrating table with flat lever mechanism has a broad prospect of using in the construction industry, in the chemical, pharmaceutical and food industries and in the mining industry.

In the papers [3–6] the dynamic and mathematical models of the mechanism of the orthogonal vibrating table with non-ideal energy source were built. The rotational and librational motions of the mechanism were considered. The influence of mechanism parameters and energy source parameters on the mean value, the frequency and amplitude of vibrations of the angular speed of the driving link were determined. The conditions imposed on the amplitude of the vibratory motion of the driving link were found. Criteria were defined for stability of rotational and librational motion of the orthogonal mechanism. The article [5] focuses on the oscillatory motion of the mechanism in the low-speed rotation of the motor. In the paper [3] a damless hydro-turbine with inclined blades was used as a non-ideal energy source. It also has been found the dependence of the coefficient of rotation non-uniformity on the system parameters.

The objective of this work is to study of the dynamics of the orthogonal mechanism of vibrating table with low-speed motor in view of friction.

2 Kinematic Correlations

Computed model of orthogonal mechanism is shown in the Fig. 1. We place the origin of the coordinate system in the axis of rotation of the crank. Here, by X and Y we denote the coordinates of the hinge joint C (Fig. 1). From the closedness equation of vector contours in projections to the coordinate axes the following kinematic relations [5, 6, 9] can be written

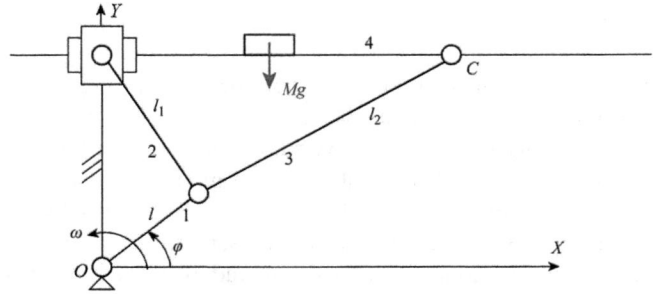

Fig. 1. Diagram of orthogonal mechanism of vibrating table.

$$X = l\cos\varphi + l_2\sqrt{1 - (l_1/l_2)^2 + (l/l_2)^2\cos^2\varphi};$$
$$Y = l\sin\varphi + l_1\sqrt{1 - (l/l_1)^2\cos^2\varphi},$$
(1)

where l, l_1, l_2 are length of links 1, 2, 3, respectively, φ - the crank rotation axis angle (Fig. 1).

Expressions of maximum, minimum and average values of the coordinates X and Y are given in articles [5, 6].

As the calculations [5, 6] show, the amplitude of the horizontal and vertical vibrations of hinge joint C of orthogonal mechanism with respect to the mean values of the coordinates are equal to each other and equal to the length of the crank l.

Structural model of the orthogonal mechanism of vibrating table in the different ratios of the lengths of links performed by specialized software package [8] is shown in Fig. 2, and motion paths of knee joints, including hinge joint C are also shown here.

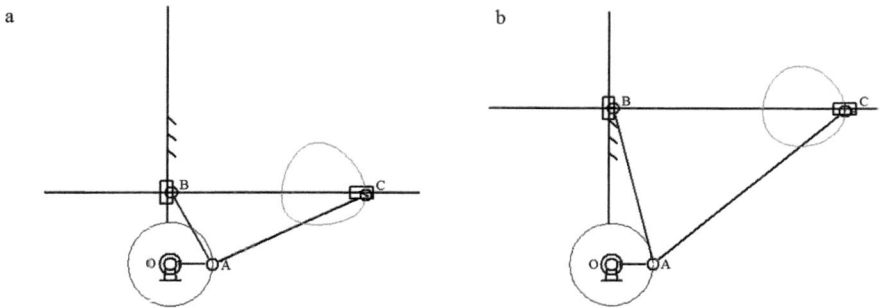

Fig. 2. Structural model of orthogonal mechanism performed by a computer: a - for the ratio of the lengths of links $l:l_1:l_2 = 1:2:4$; b - for the ratio of the lengths of links $l:l_1:l_2 = 1:5:6$.

Figure 3 shows graphs of variance projection components of analogs of speed and acceleration of hinge joint C. The motion begins from the right horizontal position of the crank. From here, it is easy to observe that the change in the ratio of the lengths of links not only affects the values of speed and acceleration, but also their graphical behavior. The developers of orthogonal mechanism of vibratory table will select the appropriate option of size and ratio of the lengths of links depending on the process requirements.

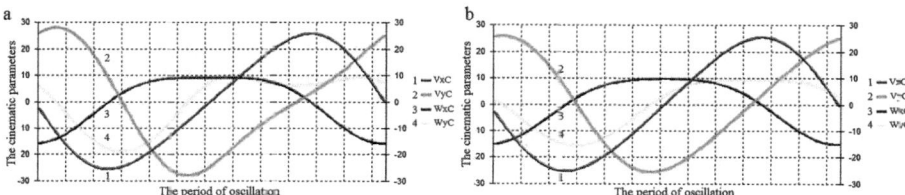

Fig. 3. Graphs of variance of projection components of the speed and acceleration of the working point C: - for the ratio of the lengths of links $l:l_1:l_2 = 1:2:4$; b - for the ratio of the lengths of links $l:l_1:l_2 = 1:5:6$.

3 Dynamics of Mechanism Motion

The motion equation of orthogonal mechanism in view of the friction in the presence of fixed load in the horizontal links (link 4) is obtained in the paper [4]. It is given by

$$A_0 \ddot{\varphi} + \left[A_0 \left(\frac{1}{2} \cos^3 \varphi - \sin^2 \varphi \cos \varphi - \sin \varphi \cos \varphi \right) + fMl^2 \sin \varphi \cdot sign\dot{X} \right] \dot{\varphi}^2$$
$$+ P \cos \varphi + kM_D \cos \varphi \sin \varphi = M_D - fMgl \cdot sign\dot{X} - kM_D. \quad (2)$$

At this point

$$A_0 = M(1 + m_3/M)l^2 + J; A_3 = 2Ml^2 l/l_1;$$
$$P = M[1 + m_1/M + m_2/M + m/(2M)]gl; J = ml^2/3, \quad (3)$$

where M - load mass m, m_1, m_2 and m_3 - mass of the links 1, 2, 3 and 4; J - inertia moment of link 1, f - the coefficient of sliding friction in the prismatic pair, k - rotational friction coefficient in the revolute pair; M_D - moment of driving forces.

Influence of non-ideal energy source on the vibratory system have to be expressed in the form of the $M_D(\varphi, \dot{\varphi})$, where φ - coordinate the motion of energy source. Torque on the shaft of some motor, for example, of the DC motor is determined by the formula

$$M_D = a - b\dot{\varphi}, \quad (4)$$

where a and b - constant coefficients, depending on the motor parameters.

The nonlinear differential Eq. (2) in view of (4) was solved by using the Simulink (Fig. 4). Calculation of the coefficients (3) was performed using Coefficients block diagrams (Fig. 5). The input signals for the model are the mass of the load, the mass of the links, the friction coefficients, as well as the initial position and the initial angular velocity of the driving link.

Calculations were performed for the following values of parameters:

$a = 600N \cdot m, b = 300N \cdot m \cdot s, = 70kg, m = 2kg, m_1 = 2,5kg, m_2 = 3kg, m_3 = 3,5kg, l = 0, 1m, l_1 = 1m, l_2 = 2m.$

The results of the solution of Eq. (2) in the absence of friction are completely consistent with the results of paper [5]. The graphs of dependence of the angular speed of the driving link (motor shaft) $\dot{\varphi}$ and dependence of motor torque M_D on the rotation angle φ at different values of the coefficient of sliding friction f and for various values of coefficient of rotational friction k are shown in Figs. 6 and 7. The graphs in the Fig. 6a shows that the average angular speed depends on the coefficient of sliding friction, and at that in the first half cycle of rotation of the mechanism under the influence of sliding friction this value increases, but in the second half-cycle - decreases due to reversal of sign the sliding friction. The amplitude of the angular speed of the driving link depends on rotational friction, it increases with increasing of the rotational friction coefficient (Fig. 7a). Figures 6b and 7b show the motor torque waveforms with significant amplitude and influence of sliding and rotational friction on them.

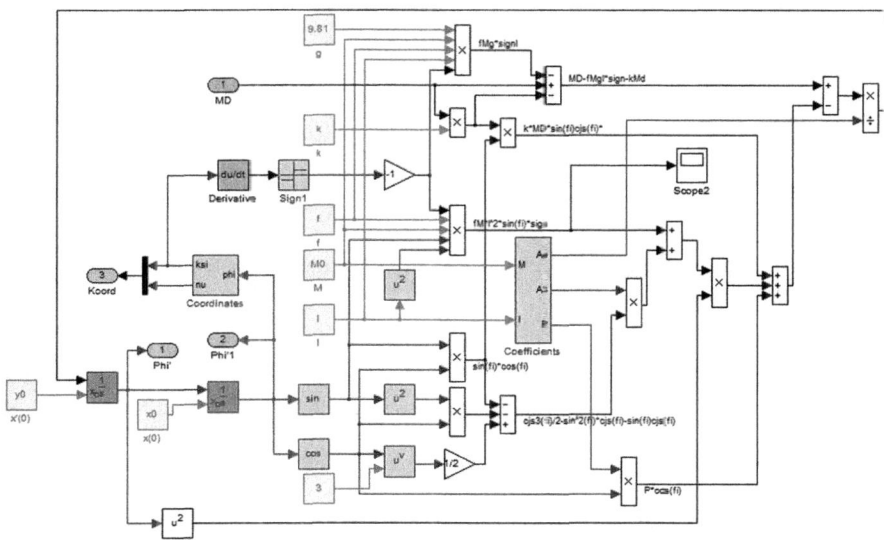

Fig. 4. Simulink model of the vibrating table (solution of Eq. (2)).

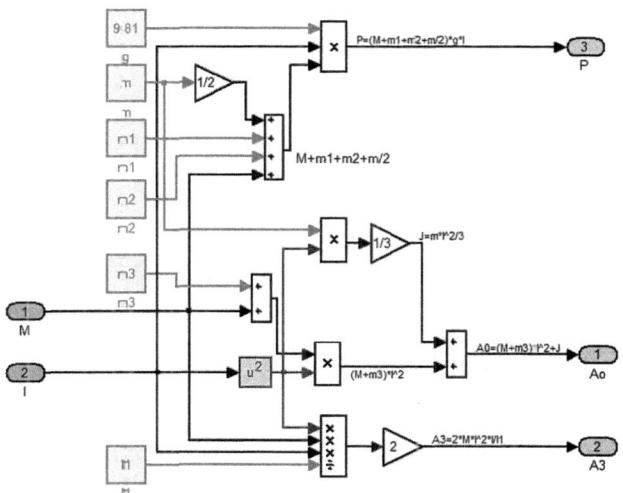

Fig. 5. Coefficients block diagrams for calculation of the coefficients (3).

Dependence of the rotation angle of the driving link φ on time t is represented in the Fig. 8, which shows the similar upward shift of the dependence diagram $\varphi(t)$ along the whole line under the influence of the coefficient of sliding friction. Rotational friction coefficient due to the smallness of its value does not affect practically the behavior of the dependence $\varphi(t)$.

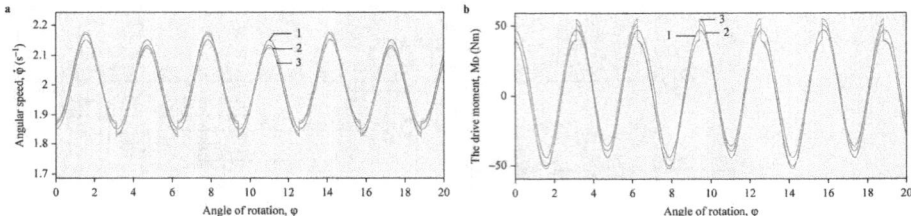

Fig. 6. Dependencies of angular speed (a) and motor torque (b) on the rotation angle of the driving member (the motor shaft) at $1 - f = 0$; $2 - f = 0.15$; $3 - f = 0.2$ and $k = 0$.

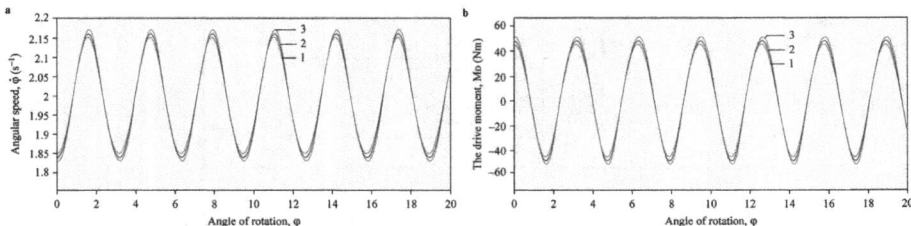

Fig. 7. Dependence of the angular speed (a) and motor torque (b) on the angle of rotation of the driving link (motor shaft) at $1 - k = 0$; $2 - k = 0.05$; $3 - k = 0.1$ и $f = 0$.

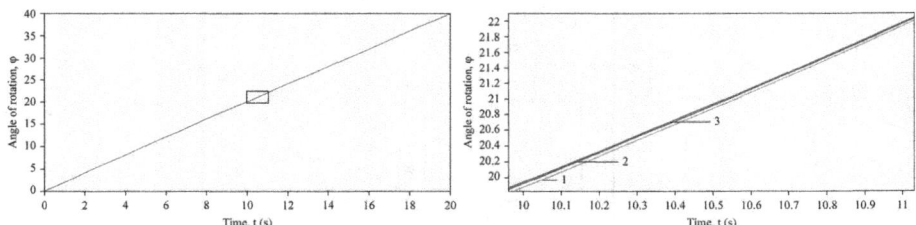

Fig. 8. Dependence of rotation angle of the driving member (motor shaft) on time in $1 - f = 0$; $2 - f = 0.15$; $3 - f = 0.2$ and $k = 0$.

At steady state $\dot{\varphi}$ the speed of the driving member even if remains constant at the average, but it changes within the cycle, passing through the maximum $\dot{\varphi}_{max}$ and minimum $\dot{\varphi}_{min}$ values. The rotation non-uniformity estimated by the coefficient of non-uniformity.

$$\delta = (\dot{\varphi}_{max} - \dot{\varphi}_{min})/\dot{\varphi}_a, \qquad (5)$$

where $\dot{\varphi}_a$ - average speed per cycle. Ratio δ describes the amplitude of speed fluctuations with respect to its mean value. The smaller δ, the relatively smaller the speed fluctuations, the calmer the driving member rotates. From this point of view, if the sliding friction in one half cycle of the mechanism rotation may increase the coefficient of non-uniformity of rotation, in the other half-period it may reduce it, and the rotational friction only has a destabilizing effect.

If the torque of the motor is insufficient for the full rotation of the orthogonal mechanism, the vibrating table performs librational damped oscillations or an aperiodic damped process. The study of the effect of friction on the librational oscillations of the orthogonal mechanism of the vibrating table can be the subject of further research

4 Conclusions

Based on the analysis and discussion of the results of the research the following conclusions can be drawn:

- Dynamics of the lever mechanism with orthogonal motion in view of friction in the presence of fixed load in the vibrating table has been studied
- It has been found that the average value of the oscillating angular speed of the driving link (motor shaft) depends on the coefficient of sliding friction
- The dependency diagram of the amplitude of deviation of this angular speed from its mean value on the coefficient of rotational friction is presented
- The dependence of the coefficient of non-uniformity on the rotation of the driving member on the coefficients of friction has been established, according to which, by reducing these parameters sufficiently, it is possible to obtain an acceptable value of the coefficient of non-uniformity necessary for optimal design of vibrating table orthogonal mechanism with non-ideal energy source.

Acknowledgments. The research has been financed by the Ministry of Education and Science of the Republic of Kazakhstan based on the Grant 2263/GF4.

References

1. Balthazar JM, Mook DT, Brasil RMLRF, Fenili A, Belato D, Felix JLP, Weber HI (2002) Recent results on vibrating problems with limited power supply. Meccanica 330:1–9
2. Balthazar JM, Mook DT, Weber HI, Brasil RMLRF, Fenili A, Belato D, Felix JLP (2003) An overview on non-ideal vibrations. Meccanica 38:613–621
3. Bissembayev K, Iskakov ZH (2015) Mathematical modeling of the mechanical system of the vibrating table with orthogonal movement and hydraulic turbine with inclined blades. In: Proceedings of ICoEV 2015, Slovenia, Ljubljana, pp 38–47
4. Bisembayev K, Iskakov Zh (2012) Mathematic model of the orthogonal mechanism of the press machinevibrating table. Bull Kazakh National Teachers Training Univ Named after Abai 3(39):32–38
5. Bissembayev K, Iskakov Zh (2014) Nonlinear vibrations of orthogonal mechanism of shaking table. Int J Appl Mech Eng 19(3):487–501
6. Bissembayev K, Iskakov Zh (2015) Oscillations of the orthogonal mechanism with a non-ideal source of energy in the presence of a load on the operating link. Mech Mach Theor 92:153–170

7. Cvetićanin L (2010) Dynamics of the non-ideal mechanical systems: a review. J Serb Soc Comput Mech 4(2):75–86
8. Jamalov, NK, Kamal AN (2015) Complex computer-aided synthesis and analysis of parallel linkages ASYAN. In: Proceedings of 6th international conference – issues of mechanics of the modern machines, vol 2, Ulan-Ude, pp 68–72
9. Tuleshov EA (2010) Dynamic analysis and design of mechanisms of automatic molding machines based on the crank machines. Dissertation of Candidate of Science {Engineering}: 05.02.18., Almaty

Design of Neural Network Predictor for Vibration Analysis of a Drill Column Machine During Drilling Plastic Work-Pieces

Şahin Yıldırım[(✉)] and Emir Esim

Erciyes University, Kayseri, Turkey
{sahiny,emiresim}@erciyes.edu.tr

Abstract. In spite of advantaged machining technology, there are still some vibrations and heating problems of drill machines during drilling work pieces. This experimental and simulation investigation is focused on drilling condition of drill column machines system's performance using neural network based approach for plastic material and different feed rates under different working speeds. Firstly, the system is tested with plastic material for different drilling speeds and feed rates. Moreover, different regions of the system were measured with vibration measuring system. Secondly, the experimentally measured vibration and acceleration parameters were predicted with two types of neural network predictors. The result were improved that neural networks can be used as predictor such systems in real time applications during drilling process.

Keywords: Drill vibration · Neural network predictor · Radial basis neural network · Drill machine

1 Introduction

Drilling process is very important process for manufacturing real time application. This process is depended and based on the use of energy delivered by the machine tool, which is an activity, is formed workpiece in the desired shape, dimensions and surface quality by taking a certain amount of cutting tool material. One of the most important parameters affecting production precision in machining is vibration. These vibrations can be harmonic resonance, chatter and outsourced. These critical vibrations of drilling machines; It not only affects poor surface finish, increase of tool wear but also affects tool life. Many researchers have studied vibrations cause loss of production and affect working performance of the machine tools. Some of these studies are given below.

Chatter has been distinctly researched in order to prevent vibration. Quintana and Ciurana presented a review on the chatter [1]. Most of these investigations on predictive approaches have been carried out. Stability Lobe Diagram (SLD) showing the stable conditions as a function of depth of cut and spindle speed. Altıntaş and Budak used this diagram to select optimal operating conditions (i.e., depth of cut and spindle speed) [2].

Schmitz et al. [3], performed analysis using the Receptance Coupling Substructure Analysis (RCSA) method for machine dynamics in their experimental work. Zhang et al. [4] presented a new method to predict the dynamic response of the machine-spindle-holder-tool assembly using the receptance coupling substructure analysis technique. In this study, Timoshenko beam model is used to perform the impact test for measuring the vibration of the spindle. And finite element method (FEM) is used in the dynamic analysis of tool-holder. It has been shown that the experimental and FEM results obtained confirm each other.

Ertürk et al. [5] analyzed the spindle, tool holder and tool combination for the frequency response of the tool tip using the analytical and FEM using the Timoshenko beam model. Pedranmehr et al. [6] performed the modal analysis of milling machine experimentally and compared the results with the FEM. Zhijun Wu and colleagues performed modal analysis and harmonic response analyzes using the FEM to analyze machine tool characteristics. It was seen that they supported one of the results obtained with the experimental work and the final elements they had done and they were in great agreement with each other [7]. Eski has studied vibration analysis with neural network on 3-axis CNC vertical milling machine [8]. Esim and Yıldırım have studied drilling performance analysis of a drilling machine with neural networks [9].

In this study, it is aimed to investigate the effects of vibrations occurring at two different spindle speeds and feed rates and to predict the effect of vibrations of these parameters using artificial neural networks.

2 Dynamics of Column Machine

The dynamic performance of the machine tool plays an important role in the quality of the workpiece. Determining the dynamic analysis performance of machine tools under the influence of real dynamic loads is very useful and contributes to the determination of the characteristics of the machine tools in the design phase of the machine tools. Modal analysis, harmonic response, random response and transient response analysis are used to determine the dynamic characteristics.

Dynamic analyzes are used under varying loads with the characterization of a structure when considering the system's stiffness and damping. Here, as a general approach, the system of bringing about a solution to describe the dynamic characterization of the system can be used for dynamic equations in which the system is divided into elements. Here, assuming that there is a linear relationship between the elements and the node model, the system is transformed into a multi-degree-of-freedom structure because of the many elements involved [7].

The multi-degree of freedom vibration model is expressed as:

$$[m]\ddot{x}(t) + [c]\dot{x}(t) + [k]x(t) = f(t) \qquad (1)$$

The expression of the system in the form of Laplace is given as follows:

$$([m]s^2 + [c]s + [k])x(s) = f(s) \qquad (2)$$

Where; $[m]$ Mass matrix, $[c]$ damping matrix, and $[k]$ stiffness matrix. Mass matrix and stiffness matrix are the symmetric matrix; damping matrix is defined as non-symmetric due to nonlinear construction [10].

The mechanical impedance of this multi-degree-of-freedom system is expressed as:

$$z(s) = [m]s^2 + [c]s + [k] \qquad (3)$$

If the mechanical impedance matrix is written in the dynamic equation:

$$z(s) = \frac{f(s)}{x(s)} \qquad (4)$$

The transfer function of the system can be defined by the inverse of the impedance matrix:

$$H(s) = z^{-1}(s) \qquad (5)$$

The response of the system to frequency under a dynamic load in a sinusoidal drive is described as a harmonic response. This is called some kind of forced vibration, so that you can learn how the effect is affected. Under harmonic loading, the dynamic equation is defined as:

$$x(s) = z^{-1}(s).f(s) = H(s).f(s) \qquad (6)$$

where x(s) can be calculated if the transfer function and the applied force are known.

3 Descriptions of Experimental Apparatus

This section focuses on main description on the proposed test machine. Nowadays, drill column machines have been used for manufacturing process. Drill machine which is used in this investigation is working automatically. It can be give automatic feed rate and can be arrange spindle speed like a CNC machine. The control panel of drill machine where can be arranged feed rates, spindle speed, automatic drill and threading settings. In experimental studies plastic material is used as a work piece. Also in experimental test a drill bit with high speed steel properties at 135° tool angles of 10 mm diameter is preferred as a drilling tool experimental test system is shown in Fig. 1. As illustrated in figure, the three-axial accelerometer was mounted on the working table.

In this paper, Brüel&Kjaer (B&K) portable and multi-channel pulse 3560 B type IDA (Intelligent Data Acquisition), accelerometer (B&K 4524 B 001 type) and computer were used for experimental vibration analysis. Then the vibration data obtained are used for estimating the vibration using artificial neural networks. Flow chart of measuring and simulation approaches shown in Fig. 2.

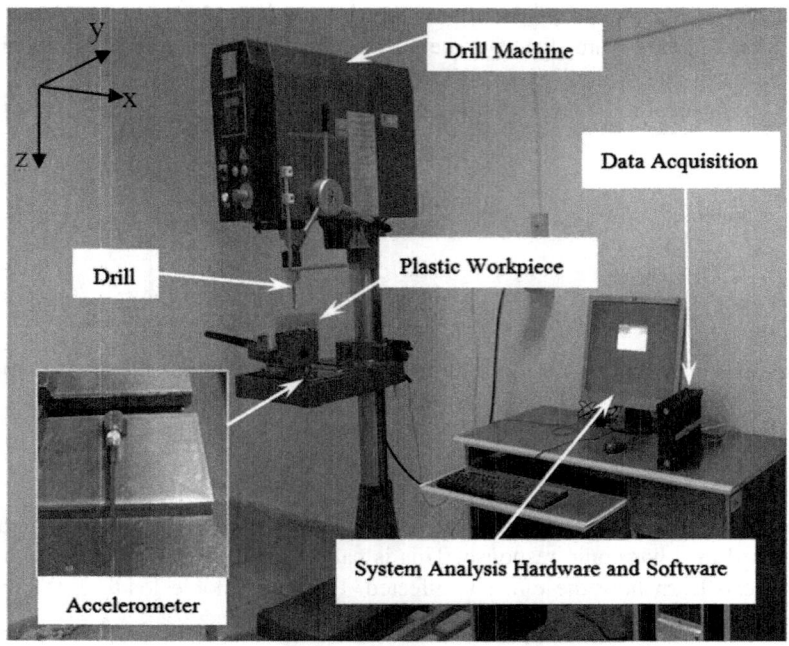

Fig. 1. Experimental test rig

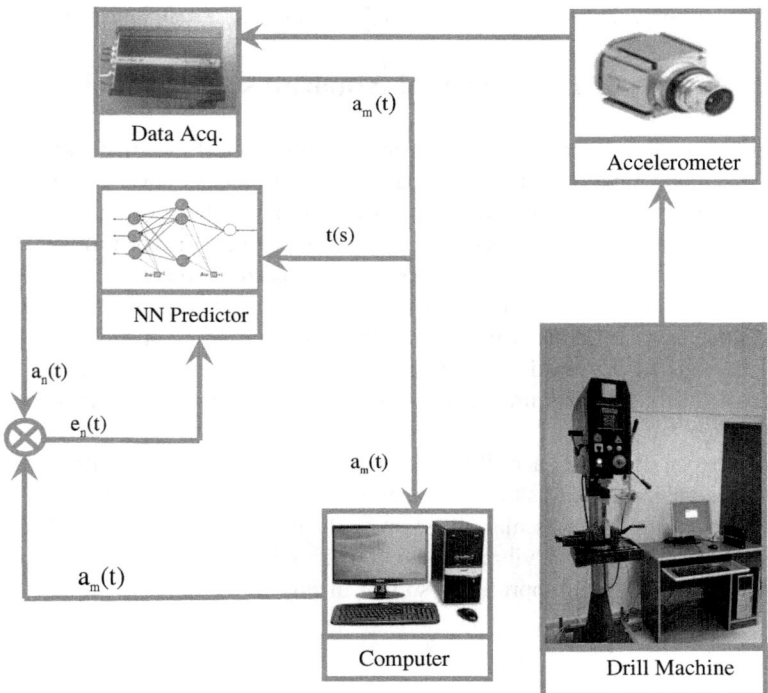

Fig. 2. Flow chart of experimental and neural network simulation

4 Feed Forward Neural Networks (FNNs)

In this experimental research work, a FNN type of predictor was employed to predict vibration parameters of proposed and data measured experimental test system see Fig. 1. A feed forward neural network is consisted of one input, three output and hidden layers (with ten non-linear neurons). See Fig. 3.

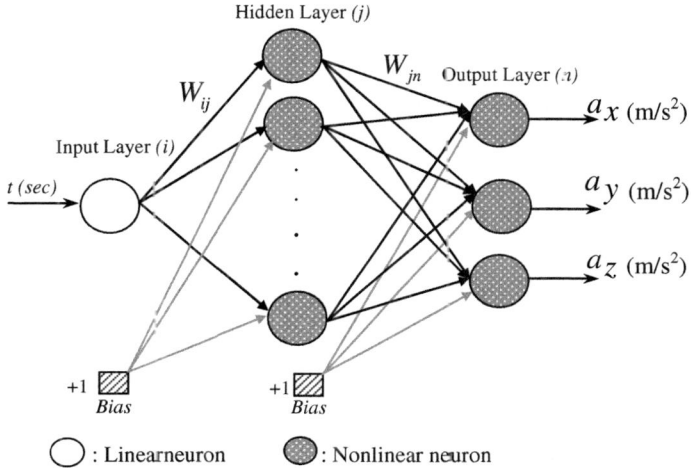

Fig. 3. Schematic presentation of the NN predictor

The main advantages of the FNN are fast learning and simple structure.

$$z_j = g\left(\sum_{i=1}^{1}\sum_{j=1}^{10} W_{ij} t_i + b_j\right) \qquad (7)$$

Where z_j are the output of the j_{th} neuron in the hidden layer, W_{ij} are the weight of the connection between the input and hidden layer neurons, b_j are the bias of the j_{th} neuron in the hidden layer. b_j can be regarded as the weight of the connection between a fixed input of unit value and neuron j in the hidden layer. The function g (.) is called the activation function of the hidden layer. The output signal of the neural network can be expressed in the following form;

$$a_n = g\left(\sum_{j=1}^{10}\sum_{n=1}^{3} W_{jn} z_j + b_n\right) \qquad (8)$$

Where W_{jn} are the weights between j_{th} neurons of the hidden layer and nth neurons of the output layer and b_n are the bias of the nth neurons in the output layer. Two types neural networks are used to analyze the drill column machine vibration. Descriptions of these types are outlined in the following subsections.

4.1 Back Propagation Neural Network (BPNN)

BPNN is most commonly used to update weights of NNs. The weights between input layer and the hidden layer are updated as follows [9];

$$\Delta W_{ij}(t) = -\eta \frac{\partial E_2(t)}{\partial W_{ij}} + \alpha \Delta W_{ij}(t-1) \tag{9}$$

The weights between the hidden layer and the output layer are updated in the following equation;

$$\Delta W_{jn}(t) = -\eta \frac{\partial E_1(t)}{\partial W_{jn}} + \alpha \Delta W_{jn}(t-1) \tag{10}$$

Where α is the momentum term, and η the learning is rate. $E_1(t)$ is the error between experimental and neural network output signals. $E_2(t)$ is the propagation error between hidden and input layers.

4.2 Radial Basis Neural Network (RBNN)

Traditionally, RBNN which model functions y(x) mapping x $\in$ Rn to y $\in$ R have hidden layer so that the model [11],

$$f_j(x) = \sum_{i=1}^{1} \sum_{j=1}^{10} W_{ij} h_{ij}(x) \tag{11}$$

The characteristic feature of RBNN is the radial nature of the hidden unit transfer functions, h_{ij} which depend only on the distance between the input x and the center h_{ij} of each hidden unit, scaled by a metric R_j,

$$h_{ij}(x) = \varphi \left[(x - c_{ij})^T R_j^{-1} (x - c_{ij}) \right] \tag{12}$$

where φ is some function which is monotonic for non-negative numbers. Gaussian basis transfer function is given as;

$$T.F = \exp(-\sum_{n=1}^{3} \frac{(x_n - c_{jn})^2}{r_{jn}^2}) \tag{13}$$

A direct approach to the model complexity issue is to select a subset of centers from a larger set which, if used in its entirety, would over fit the data.

5 Experimental and Simulation Results

Experimental and simulation results from a drill column machine are presented for plastic materials using FNNs. Plastics material were selected to investigate the vibrational effect of material selection during drilling process under 0.30 mm/rev and 0.15 mm/rev feed rate. Experimental measurements are performed for 14 s by using the two different drill speeds. Working conditions are given in Table 1. Then, BPNN and RBNN are used for predicting the vibration of drill column. Training and testing of the neural network are employed with neural network toolbox of MATLAB. 350 and 150 data are used in training and testing stages, respectively. After training Neural network structures are saved and used for estimating vibration of drill machine. Experimental and artificial neural network test results are presented in the following graphs.

Table 1. Working conditions, RMS error values of the proposed networks.

n (rpm)	S (mm/rev)	NN type	RMSEs
1000	0,15	BPNN	0.0744
		RBNN	2.18×10^{-15}
1500	0,15	BPNN	0.0980
		RBNN	2.53×10^{-15}
1000	0,30	BPNN	0.2222
		RBNN	5.03×10^{-15}
1500	0,30	BPNN	0.1057
		RBNN	2.27×10^{-15}

Furthermore, results of experimental approaches were indicated with BPNN and RBNN for the speed of 1000 and 1500 rpm with 0.15 mm/rev feed rate in Figs. 4 and 5, respectively. As seen in the figures, The RBNN approach follows experimental result. Besides, amplitudes of the vibrations naturally increase because drill speed differences.

For the case of 1000 rpm and 1500 rpm drill speeds with 0.30 mm/rev feed rate, neural network and experimental results are outlined in Figs. 6 and 7, respectively. The result of BPNN flow the experimental values. But, there is not full matching for experimental results. The outputs of RBNN fully fallow the experimental results.

As can be outlined from Table 1; RMSE's for all cases are very different between BPNN and RBNN. BPNN presents poor convergence ability for modeling and predicting the drill column machines' vibration characteristics, the proposed RBNN activation function is radial basis function that used for neural network has a small convergence error and can be considered as a function generator.

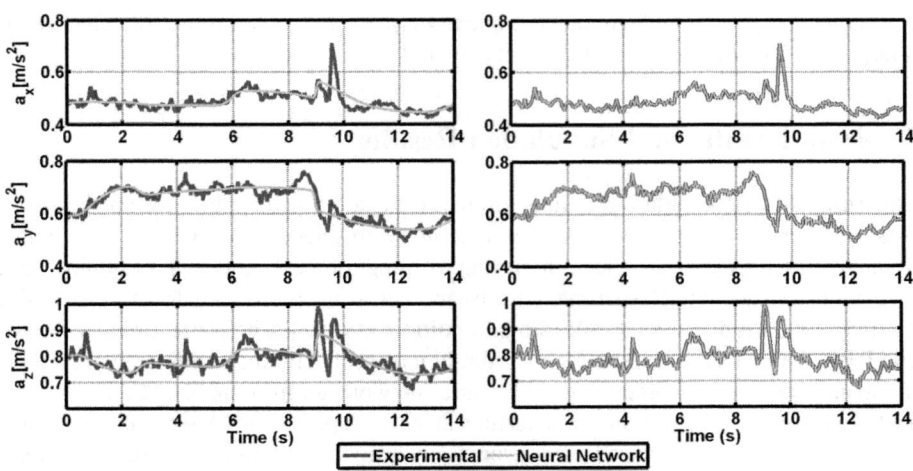

Fig. 4. Accelerations variation (**a**) BPNN (**b**) RBNN with 1000 rpm drill speed and 0.15 mm/rev feed rate

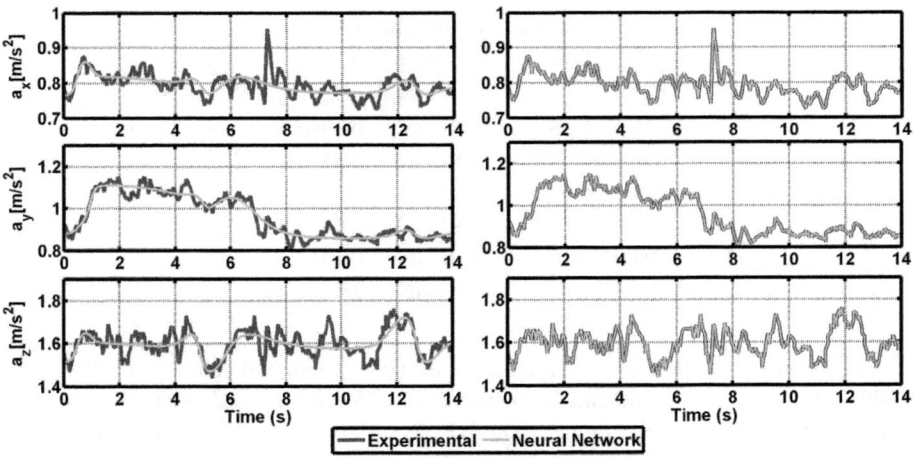

Fig. 5. Accelerations variation (**a**) BPNN (**b**) RBNN with 1500 rpm drill speed and 0.15 mm/rev feed rate

The capability of the proposed neural model shows that the vibration characteristics for the plastic materials, drill speed and feed rates can be determined correctly without experimental analyzing. The proposed artificial neural network can be used for vibration modeling of the machine tool with fast convergence and accuracy.

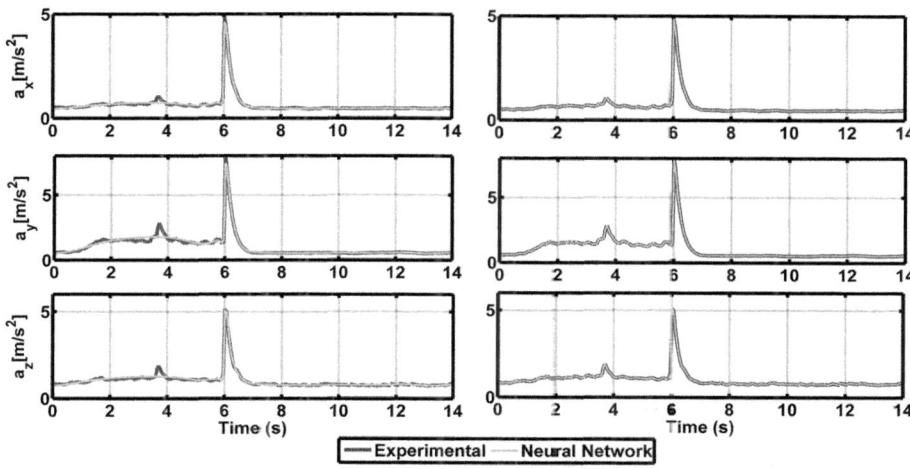

Fig. 6. Accelerations variation (a) BPNN (b) RBNN with 1000 rpm drill speed and 0.30 mm/rev feed rate

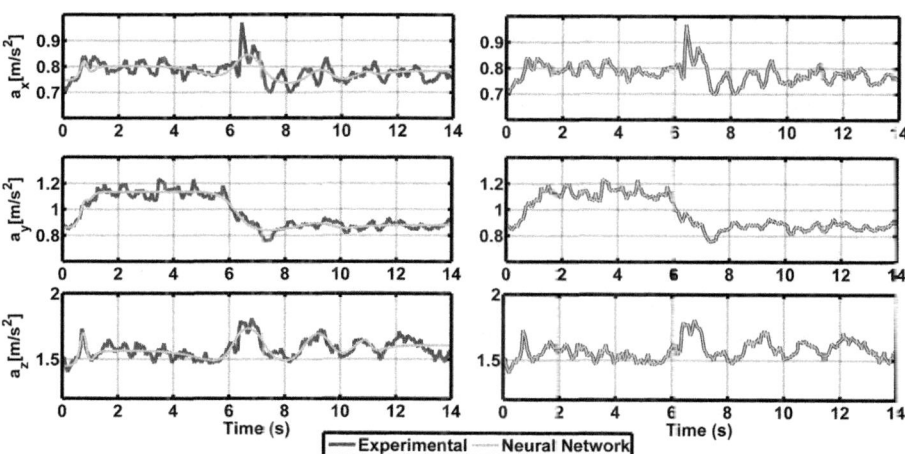

Fig. 7. Accelerations variation (a) BPNN (b) RBNN with 1500 rpm drill speed and 0.30 mm/rev feed rate

6 Conclusions

A BPNN and a RBNN are used to vibration analyses of drill column machines with different working speed. The main goal is to evaluate vibration of drilling system for 0.15 mm/rev and 0.30 mm/rev feed rates and plastic work. Experimental results show that the vibration amplitudes are changed with drill speed and feed rates.

According to the obtained results, while the RBF neural network presents good convergence ability for modeling and estimating the vibrations of drill column for the case of all working conditions. By using RBNN with radial-basis function has a robust characteristic in order to model and predict vibration of the machine tools.

References

1. Quintana G, Ciurana J (2011) Chatter in machining processes: a review. Int J Mach Tools Manuf 51:363–376
2. Altintas Y, Budak E (1995) Analytical prediction of stability lobes in milling. CIRP Ann Manuf Technol 44:357–362
3. Schmitz T, Davies M, Kennedy M (2001) Tool point frequency response prediction for high speed machining by RCSA. J Manuf Sci Eng 123(4):700–707
4. Zhang J, Schmitz T, Zhoa W, Lu B (2011) Receptance coupling for tool point dynamics prediction on machine tools. Chin J Mech Eng 24:1–6
5. Erturk A, Özgüven HN, Budak E (2006) Analytical modeling of spindle–tool dynamics on machine tools using Timoshenko beam model and receptance coupling for the prediction of tool point FRF. Int J Mach Tools Manuf 46:1901–1912
6. Pedrammehr S, Farrokhi H, Khani Sheykh Rajab A, Pakzad S, Mahboubkhah M, Ettefagh MM, Sadeghi MH (2012) Modal analysis of the milling machine structure through FEM and experimental test. Adv Mater Res 383–390:6717–6721
7. Wu Z, Xu C, Zhang J, Yu D, Feng P (2010) Modal and harmonic response analysis and evaluation of machine tools. In: International conference on digital manufacturing & automation, pp 928–935
8. Eski İ (2012) Vibration analysis of drilling machine using proposed artificial neural network predictors. J Mech Sci Technol 26(10):3037–3046
9. Esim E, Yıldırım Ş (2016) Drilling performance analysis of drill column machine using proposed neural networks. Neural Comput Appl 1–12. doi:10.1007/s00521-016-2322-8
10. Rao SS (2004) Mechanical vibrations, 4th edn. Pearson Education Inc., Upper Saddle River
11. Eski İ, Erkaya S, Savaş S, Yildirim Ş (2011) Fault detection on robot manipulators using artificial neural networks. Robot Comput Integr Manuf 27:115–123

Railway Vehicle Model Developed by ASELSAN

Mustafa Nicem Tanyeri and H. Murat Gültekin(✉)

ASELSAN, UGES-MTM, Ankara, Turkey
{mntanyeri,mgultek}@aselsan.com.tr

Abstract. In recent years, there has been an increasing amount of projects on underground public transport due to environmental and efficiency reasons. One of these project has been carried out by ASELSAN to refurbish metro vehicle of Ankara. A traction control algorithm is required to manage the electric traction motor of the vehicle to use maximum adhesive effort and to improve traction and braking performance. In order to design such a traction control algorithm, an accurate system plant is essential. The objective of this paper is to present a railway vehicle (metro) model that is developed in Simpack-Rail which is used as a plant for development of the traction control algorithms. First, the representation of the vehicle and its model is shown. Next, the model tuning according to data collected from real vehicle is presented. At the end, comparison of the vehicle model with the real vehicle by conducting dynamics tests is explained. This plant model can also be used in order to run simulations of the vehicle against derailment scenarios according to EN14363 running safety standard of railway vehicles.

Keywords: Multibody dynamics modelling of railway vehicles · Simpack · EN14363 · Traction control · Simulink

1 Introduction

There has been an increasing trend of multibody dynamics modelling of railway vehicles. Thanks to these models, todays engineers are able to observe dynamic behavior of railway vehicles without data acquisition from real vehicles. In addition, they are able to predict the consequences of the design changes in terms of vehicle dynamics and vehicle performance. Also, test scenarios described in many EN standards, such as EN14363 safety against derailment, may be simulated on these multibody models [4].

Multibody dynamics models help reduce the product development time and cost which are the main goals of engineering. It is also the safest method to perform tests with unpredictable results such as derailment and roll.

Such a multibody dynamics model was developed within the scope of refurbishment project of Ankara Metro vehicles. The main goal of this model was to construct a plant for the development of traction control algorithms and to perform simulations on derailment scenarios according to EN14363 [1, 2].

2 Vehicle

The modeled vehicle is one set of Bombardier H6 of Municipality of Ankara which consists of two "car A" and one "car B". The first and the third cars of the set are "car A" and the second car is "car B" which is driven by both of "car A"s. It is a bi-directional vehicle that can be driven in either direction, forwards or backwards. Under the normal operating conditions, it operates as a 6-car set consisting of two 3-car sets.

"Car A" is the car that has a traction system and the "car B" is the car without a traction system. Each car has two bogies and each bogie has two wheelsets. Wheelsets are connected to the bogie frame with primary suspensions (chevron springs) and the bogie frame is connected to the body with secondary suspension (air springs). Since car A has traction system, bogies of car A consist of two traction motors and two reduction gears.

3 Modelling

The vehicle is modeled in Simpack Rail® which is a multibody dynamics software for railway system dynamics.

3.1 Assumptions

All part of the vehicle, such as bodies, bogie frames and wheelsets, are assumed as rigid bodies. Weights of the devices connected to each body are added in the weight of corresponding body. Inertias of the bodies and bogies are calculated from CAD models.

3.2 Parameters

The most important parameters for the dynamics of the vehicle are mass of each body and its position on the vehicle. Masses of major parts such as, wheelset, bogie frame, motor and gearbox were weighed [3] (Fig. 1).

Masses of the CAD models of the bodies and the bogies were corrected accordingly and inertias were calculated from these CAD models.

Weight of the vehicle is measured with "TrainWeigh AX-300 bogie measurement scale" and it is validated with "TrainWeigh AX-300 wheelset measurement scale". As seen in Fig. 2, the yellow one is the bogie measurement scale and the grey one is the wheelset measurement scale.

Weight results were defined in the dynamic model. There are discrepancies between the left and right wheels. The reason of these discrepancies is cables and pneumatic lines. The exact position of the center of the gravity of the cables and pneumatic lines are unknown because it is not practical to detect.

Fig. 1. Weighing of wheelset (left), motor (middle) and bogie frame (right).

Fig. 2. Weighing of the modernized Bombardier H6. (Color figure online)

Another important parameter that effects the vehicle dynamics is the wheel-rail contact. The rail profile of Ankara Metro is UIC-54 with standard track gauge of 1435 mm. The wheel profile of the vehicle is S1002. Since the wheels are worn their diameter changes. Diameter of the wheels are measured and they are defined in the dynamic model accordingly.

In addition, technical specifications like stiffness and damping coefficients in three axle of primary (chevron springs) and secondary (air springs) suspensions are defined in the model according to the data from their supplier (Fig. 3).

Fig. 3. Bogie of the modernized Bombardier H6 and bogie of the dynamic model.

4 Validation Test and Simulation

After construction of the dynamic model of the vehicle, a validation test was performed in order to compare the model and the real vehicle. Test is performed between Ivedik and Akköprü stations. The track between Ivedik and Akköprü stations was chosen because it is possible to run the vehicle at higher speed in that part of the track and it has two curved sections with radius of 302.4 m and 350 m. Namely, it is suitable for vehicle dynamics testing. The track was also modeled in Simpack with its all specifications (super elevation, curve radius and elevation).

Data were acquired from body, bogie frame and axle box with accelerometers and an IMU (Inertial Measurement Unit) is placed on the floor of the vehicle. Data of IMU were used for capturing the pitch, roll, and yaw motions. To validate dynamic properties of primary and secondary suspensions accelerometers were used (Fig. 4).

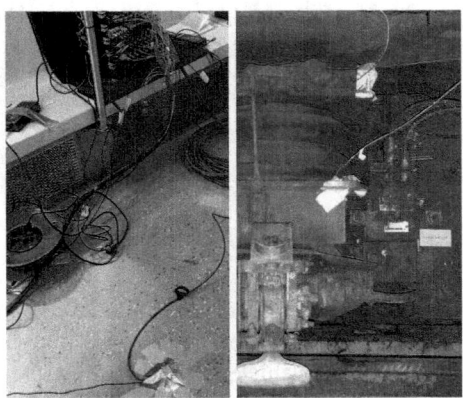

Fig. 4. Position of the IMU and position of the accelerometers on the bogie frame and on the bolster.

Model is tuned to achieve less than 5% deviation between the model and the real vehicle with respect to test and simulation data. Speed and roll data of test and simulation runs are compared in Figs. 5 and 6, respectively.

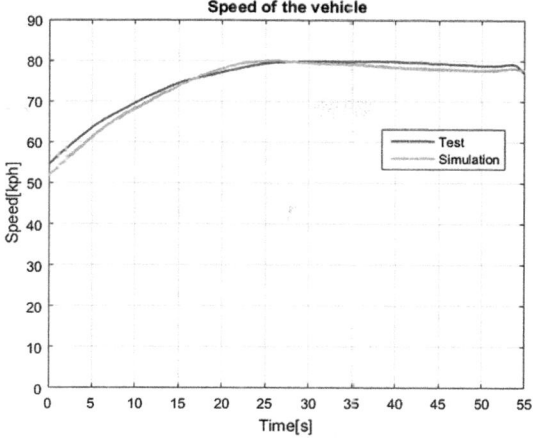

Fig. 5. Speed data from test versus speed data from simulation.

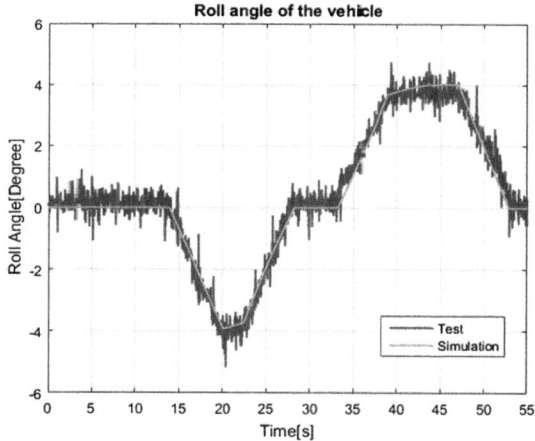

Fig. 6. Speed data from test versus speed data from simulation.

5 EN14363-Algorithm

According to EN14363, after the refurbishment, refurbished vehicle must be tested for the acceptance of running characteristics. Tests defined in EN14363 were simulated in Simpack. The main parameter in the standard is the Y/Q ratio. The ratio between horizontal guiding force Y and vertical wheel force Q. The limiting value for Y/Q was defined as 1.2 for a flange angle of 70° (this corresponds to $\mu = 0.36$). Results of the EN14363 derailment simulation is shown in Fig. 7 [1, 2].

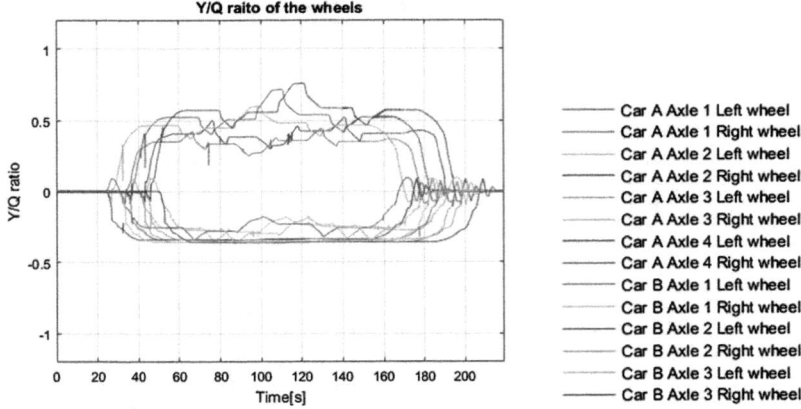

Fig. 7. Y/Q ratio of all wheels.

Moreover, an anti-slip/slide traction control algorithm was developed in Simulink by using SIMAT and MATSIM, which are Simpack interfaces to MATLAB & Simulink. The algorithm controls motor's torque to use a maximum possible adhesive effort in order to improve traction and braking performance. In Fig. 8 Simulink co-simulation block of the vehicle generated by Simpack is shown.

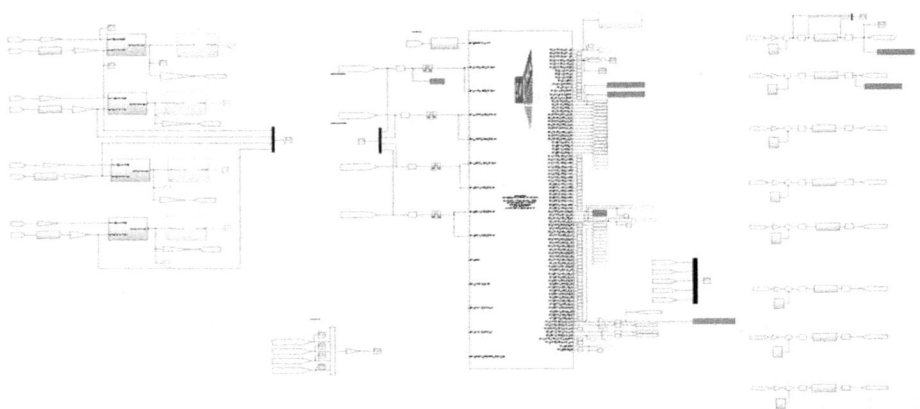

Fig. 8. Simulink co-simulation block of the vehicle generated by Simpack.

6 Conclusions

In conclusion, the multibody dynamics model of the refurbished Bombardier H6 vehicle set was developed. By vehicle dynamics tests the dynamic model was validated. This model was used as a plant in order to develop the anti-slid/slip traction control algorithms and to simulate derailment test scenarios defined in EN14363.

Multibody dynamics model of Bombardier H6 helped to reduce the product development time and cost of the project. Derailment tests were performed safely and economically on simulation environment. Until vehicle ride tests, anti-slip/slide control algorithm was developed and validated on the model resulting a significant cost and time reduction. Because test requirements on the vehicle is minimized.

All traction system of Bombardier H6 were replaced with the units developed by Aselsan. Development process of the units includes designs of hardware, software and mechanical systems. Integration of the units was also done by Aselsan. This study is first of its kind in Turkey, especially the development of the traction control system by using a multibody dynamics vehicle model was a challenge.

Acknowledgments. We take this opportunity to express gratitude to EGO, Aselsan and to our colleagues for their help and support.

References

1. EN 14363:2005 Standard - Railway applications — Testing for the acceptance of running characteristics of railway vehicles – Testing of running behavior and stationary tests (2005)
2. EN 14363:2016 Standard - Railway applications—Testing for the acceptance of running characteristics of railway vehicles – Testing of running behavior and stationary tests (2016)
3. Spiryagin M, Cole C, Sun YQ (2014) Design and Simulation of Rail Vehicles. Taylor & Francis Group, LLC, Boca Raton
4. EN 15827:2011 Railway applications—Requirements for bogies and running gears (2011)

Author Index

A
Abdoulaye Ben-Aziz, Kirakoya, 191
Akçura, Nail, 240
Alasli, Abdulkareem, 240
Asama, Junichi, 141
Ayit, Orhan, 181

B
Banica, Elisabeta, 56
Berger, Maik, 21
Bissembayev, Kuatbay, 261

C
Catera, Piervincenzo, 232
Ceccarelli, Marco, 121
Celik, Baris, 141
Çelik, Onur, 161
Çetin, Levent, 240
Chen, Chuan, 203
Chen, Yong, 251
Ciupe, Valentin, 111
Comanescu, Adriana, 56
Comanescu, Dinu, 56
Corves, Burkhard, 79, 171

D
Dede, Mehmet İsmet Can, 161, 181
Dong, Huimin, 48, 224

E
Esim, Emir, 269

F
Ferraresi, Carlo, 11
Franco, Walter, 11

G
Gruescu, Corina Mihaela, 111
Gültekin, H. Murat, 279

H
Hejnová, Monika, 89
Hernández, Alfonso, 131
Huesing, Mathias, 171
Hüsing, Mathias, 79

I
Işıtman, Ogulcan, 181
Iskakov, Zharilkassin, 261
Ivanov, Konstantin S., 212

J
Jamalov, Nutpulla, 261

K
Kiper, Gökhan, 31
Kurtenbach, Stefan, 79

L
Li, Xiaopeng, 48
Liu, Zhaohui, 251
Lovasz, Erwin-Christian, 111

M
Macho, Erik, 131
Margineanu, Dan, 111
Mohan, Santhakumar, 171
Mohanta, Jayant Kumar, 171
Müglitz, Jörg, 21
Mundo, Domenico, 232

N
Nguyen, Thi Thanh Nga, 79

O
Oiwa, Takaaki, 141
Ondrášek, Jiri, 99

P
Petuya, Victor, 131

Q
Quaglia, Giuseppe, 11

R
Russo, Matteo, 121

S
Şahin, Osman Nuri, 161
Selvi, Özgün, 67
Shweiki, Shadi, 232

T
Tanyeri, Mustafa Nicem, 279
Tatar, Santra, 111
Teichgräber, Carsten, 21
Terabayashi, Kenji, 141

U
Ulu, Burak, 153
Urízar, Mónica, 131

W
Wang, Delun, 48, 224
Wang, Zhi, 48

X
Xie, Jin, 251

Y
Yaman, Yavuz, 3
Yang, Yuhu, 203
Yaşır, Abdullah, 31
Yavuz, Samet, 67
Yıldırım, Şahin, 153, 269
Yildirim, Şahin, 191
Yu, Shudong, 48
Yu, Yue-Qing, 39

Z
Zhang, Chu, 224
Zhang, Jili, 224
Zhu, Shun-Kun, 39

Printed by Books on Demand, Germany